AF885427

SOTHIS PRESS

Sothis Press aims to completely reimagine the public domain book with an emphasis on universal access to captivatingly designed print editions. We believe that technology will enable every single book ever published to be perpetually accessible in beautiful print form. Sothis is beauty in ubiquity.

www.sothis.press
1 Waterhouse Square, London, ECIN 2ST

A Text Book of the Physics of Agriculture

F. H. King

SOTHIS PRESS

ISBN 978-1-0365-1439-6

A Text Book of the Physics of Agriculture
is in the public domain.

This edition is published by
Sothis Press © 2023. The company
is registered in England and Wales.
The Company Number is 14900389.

A TEXT BOOK

OF THE

PHYSICS OF AGRICULTURE

BY

F. H. KING

Formerly Professor of Agricultural Physics in the University of Wisconsin

Author of "The Soil;" "Irrigation and Drainage;" "Principles and Movements of Ground Water"

FIFTH EDITION

MADISON, WIS.
PUBLISHED BY THE AUTHOR

1910

All rights reserved

S
591
K5
1910

Copyright, 1899
By F. H. King

PREFACE.

The great need of agricultural practices at the present time is a keener appreciation and a more thorough comprehension of the principles which underlie them. The facts of agriculture are spread through so many and widely different fields, and are so numerous, that no one can hope to grasp them all or needs to do so. But the laws and principles which control *his* practice each farmer *must* know before he can secure his results with the greatest certainty and at the least cost.

In these pages the aim has been to present to the student who expects to be a farmer, some of the fundamental principles he must understand to become successful. They are presented from the standpoint of physics rather than of chemistry or of biology, and in dealing with the physical side of the problems the burden of effort has been to lead the student to see WHY he should practice more than WHAT, and it is hoped the student will pursue the various subjects treated in this spirit, not only in his study, but above all on the farm and in the field.

The book has been written from the standpoint of the general student and farmer rather than that of more technical scientific agriculture and only so much of laboratory methods and specific data of observation are given as may serve to demonstrate the fundamental principles treated.

F. H. KING.

University of Wisconsin,
Madison, Wis., May, 1901.

CONTENTS.

INTRODUCTION.

PHYSICS OF THE SOIL.

CHAPTER I.

NATURE, ORIGIN AND WASTE OF SOIL.

CHAPTER II.

Chemical and Mineral Nature of Soils.

CHAPTER III.

Soluble Salts in Field Soils.

CHAPTER IV.

Physical Nature of Soil.

CHAPTER V.

Soil Moisture.

CHAPTER VI.

Physics of Plant Breathing and Root Action.

CHAPTER VII.

MOVEMENTS OF SOIL MOISTURE.

CHAPTER VIII.

CONSERVATION OF SOIL MOISTURE.

CHAPTER IX.

RELATION OF AIR TO SOIL.

CHAPTER X.

Soil Temperature.

CHAPTER XI.

Objects, Methods and Implements of Tillage.

GROUND WATER, WELLS AND FARM DRAINAGE.

CHAPTER XII.

Movements of Ground Water.

CHAPTER XIII.

Farm Wells.

CHAPTER XIV.

Principles of Farm Drainage.

CHAPTER XVIII.

PRINCIPLES OF CONSTRUCTION.

CHAPTER XIX.

CONSTRUCTION OF SILOS.

FARM MECHANICS.

CHAPTER XX.

PRINCIPLES OF DRAFT.

CHAPTER XXI.

CONSTRUCTION AND MAINTENANCE OF COUNTRY ROADS.

CHAPTER XXII.

Farm Motors.

Contents.

CHAPTER XXIII.

Farm Machinery.

PRINCIPLES OF WEATHER FORECASTING.

CHAPTER XXIV.

The Atmosphere.

CHAPTER XXV.

Movements of the Atmosphere.

CHAPTER XXVI.

Weather Changes.

INTRODUCTION.

1. Physics.—Briefly defined, physics is the science of Matter and Energy. It aims to measure and investigate the movements of or within any body, whether living or dead, endeavoring to show how the forces of nature operate upon or within the body to produce the phenomena associated with it.

If we were endeavoring to ascertain how much the sun weighs, how much energy in the form of heat and light is being sent out from it daily, or how this energy is produced, our study would be one of *Solar Physics.* If we were measuring the diameter of the earth, or the volume of water in the oceans; if we were endeavoring to ascertain how the forces have operated to uplift mountain ranges or to cut out deep canons or broad valleys, then our problem would be one of *Terrestrial* or *Earth Physics.* If we were measuring the strength of a horse; how many pounds of feed he must use to plow 10 acres of ground; or endeavoring to show how the oxygen he breathes and the food he eats give rise to the energy of his muscles, our problem would be one of *Animal Physics.* If we were studying how the root forces its way through the soil; how water is forced into and through the roots to the leaves on the tree or how the sunshine breaks down the carbon dioxide in the green chlorophyll, our problem would become one of *Plant Physics.* If we are endeavoring to determine the dimensions of beams to use in a barn; how heavy a rod to use in a truss or how to brace a building so that it may safely withstand the pressure of the wind, then we are dealing with the *Physics* of *Architecture.* And so we might go on enumerating every science and every art to show that there is a physics of each or a necessary treatment of them from the standpoint of mechanical principles of matter and energy. Physics, therefore, a broad science, is one of wide application and fundamentally important to the understanding of almost any concrete subject when treated from the standpoint of cause and effect.

2. Matter and Force.—So far as we are at present able to comprehend, the various phenomena of nature are manifestations of two classes of agencies, *matter* and *force*. The river flowing steadily toward the sea is a mass of matter urged continually onward by the force of *gravitation*. Coal and oxygen burning in the firebox of the locomotive are two forms of matter urged into motion by the force *chemical affinity*. The time-keeping watch is a mechanism of brass and steel kept in uniform motion by the force *cohesion* uncoiling the wound-up spring; and the capillary rise of oil in the lamp wick and of water through the soil are other movements of matter actuated by the same force.

3. Constitution of Bodies.—All bodies or masses of matter with which we are acquainted possess such properties as to make it appear that there is room in them not occupied by the essential material which makes up the body. They are made out of definite units which have been named *molecules* much as a bank of sand is composed of grains or as a sack of shot is filled with spheres of lead.

The openness of structure of all bodies is a very important conception to have clearly in mind. It is this openness of structure which makes it possible for foul odors to be absorbed by milk or drinking water; for moisture to enter sprouting seeds; for the food we eat to pass through the walls of the alimentary canal to enter the blood vessels and out of these again to the muscles and nerve centers. It is the openness of structure of the lung lining which permits the oxygen of the air to enter the system and the carbonic oxide to escape from it; and were it not for this structure we could neither smell nor taste, for substances must penetrate these sense organs before the sensations are awakened.

That there is unoccupied space in bodies which appear to have a close structure may be demonstrated with the apparatus represented in Fig. 1. The bottle is filled with water and into this is dropped a large crystal of some salt, as potassium nitrate or sulfate, or 4 teaspoonfuls of granulated sugar. When this is done the rubber cork carrying the graduated glass tube is inserted and crowded down until the water rises in the tube and stands at one of the graduation marks. If any change in volume occurs with the solution of the salt it will be shown by a rise or fall of the water in the tube where the amount of change can be read. The bottle is placed in a large vessel of water for the purpose of maintaining a constant temperature during the experiment.

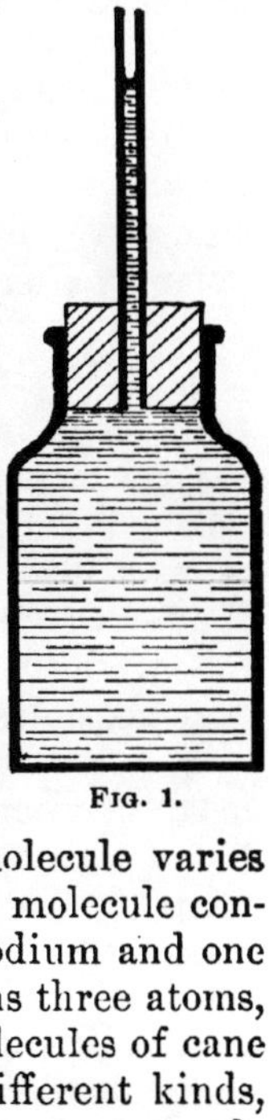

FIG. 1.

The molecules themselves are made up of smaller units which have received the name of *atoms* and the number of these atoms which enter into the construction of the molecule varies with the substance. In some substances the molecule consists of two atoms, as common salt, one of sodium and one of chlorine, while the water molecule contains three atoms, two of hydrogen and one of oxygen. In molecules of cane sugar there are forty-five atoms of three different kinds, carbon, hydrogen and oxygen, and there are many substances having molecules more complex than those of sugar.

4. Distances Between Molecules Change With the Temperature of the Body.—A bar of iron lengthens and shortens as its temperature rises and falls, and the wheelwright takes advantage of the fact to set the tires of the wagon. This change of volume with temperature is due to the fact that the mean distance between the molecules becomes greater the higher and less the lower the temperature is. From this it follows that ordinarily molecules are not in contact and that there is room in the interior of bodies, however compact they appear to be, not occupied by them. Observations with the ordinary mercurial thermometer prove the same general fact. As the temperature rises a portion of the mercury is forced out of the bulb into the stem showing that there is not room enough there for all of the mercury although the bulb has actually become larger. So, too, when the temperature falls the mercury again returns to the bulb although the bulb has itself become smaller than before.

5. Molecules of Bodies Always in Motion.—It follows from what has been said in the last section that with every change of temperature in bodies their molecules move. The general fact is that the molecules of all bodies whose temperature is not absolute zero are in rapid motion no matter whether the body be a solid, a liquid or a gas. The higher the temperature of the body the more rapidly do the molecules in it vibrate, the greater is their rebound after each collision and so the greater is the mean distan between them; this is why most bodies expand with increase of temperature and contract when cooling.

It is the fact of movement among molecules which causes the diffusion of sugar or salt through water after solution takes place, which causes the perfume of flowers to be constantly moving away from them, which gives solid camphor its odor and which causes snow and ice to evaporate at temperatures even below freezing.

The elastic power of air in the bicycle tire is due to the rapid movement of the molecules and their frequent and hard collision against the walls. It is the same fact which gives the steam its power to drive the engine. The larger the amount of air which is pumped into the tire of the bicycle the greater is the number of collisions per square inch of surface per second and so the harder the tire becomes. Then, again, when the wheel is left in the hot sun the greater tension which is developed is due to the fact that the absorption of heat causes all the molecules to travel faster, and, traveling faster, they must exert a greater pressure whenever collision occurs and their motion is arrested.

It has been computed that the mean rate at which the molecules of hydrogen gas travel at ordinary temperature and atmospheric pressure is some 6,000 feet per second. Under the same conditions molecules of oxygen gas which are 16 times as heavy travel only one-fourth as rapidly.

If it is difficult to think of a body like a horse-shoe or a hammer maintaining its form against great strains when the molecules composing it are neither at rest nor in contact it may be helpful to recall the conditions which exist in the solar system. Here we have the sun with all its planets and their satellites, together with asteroids, comets and meteors, each in rapid motion but separated by immense distances, and yet the whole system constitutes one gigantic body maintaining persistently its form as it moves through space.

6. The Size of Molecules.—Molecules are so very small that it is extremely difficult to form any just conception of them, yet there are many experiments and observations which prove them very minute. Nobert, for example, ruled parallel lines on glass at the rate of 101,600 per linear inch, proving that the point of the diamond which plowed the furrows must have been far less than $\frac{1}{100000}$ of an inch in diameter.

Lord Kelvin has computed that the number of molecules in a cubic inch of any perfect gas under a temperature of 32° F. and a pressure of 30 inches of mercury must be as great as 10^{23} or ten sextillions.

This is an enormous number, but that there is a probability of truth in it may be demonstrated by a simple experiment.

Dissolve .05 of a gram of aniline violet in alcohol and distribute it through 500 cu. in. of water in a large glass flask. Pour out half the colored water and fill to 500 cu. in. again. Repeat this operation as long as the eye can with certainty detect the color in the water. As many as nine divisions may be made and the eye detect the color when looking down through 12 inches of the water poured into a long glass tube held over white paper, using a similar tube with clear water as a standard for comparison.

If the division of the aniline is carried to this extent there will be in the last 500 cubic inches of water only

$$\frac{1}{512} \text{ of } \frac{5}{100} = \frac{1}{10,240} \text{ of a gram of aniline.}$$

It is reasonable to suppose that in the last 500 cubic inches of water there was at least one molecule of aniline in each cube of water .01 of an inch on a side, and if this is true there must have been at least

$$100 \times 100 \times 100 \times 500 = 500,000,000$$

molecules of aniline in the last vessel of water. Since at least this number of molecules is found in $\frac{1}{10240}$ of a gram of aniline one gram would contain not less than

$$10,240 \times 500,000,000 = 5,120,000,000,000 \text{ molecules.}$$

It is plain, therefore, from this straightforward line of observation and simple calculation that molecules of aniline at least must be very small and that a pound of the material would contain an enormous number.

From another line of observation Maxwell has computed that the molecules of hydrogen, oxygen and carbon dioxide are so small that the numbers in the table below are required to weigh one gram.

Number of molecules in one gram of

Hydrogen	Oxygen	Carbon dioxide
$2,174(10)^{23}$	$1,359(10)^{22}$	$9,881(10)^{21}$

That is to say, the number of molecules is so large in one gram of these three substances that 2,174, 1,359 and 9,881 must be multiplied by 10 used as a factor 23, 22 and 21 times respectively in order to express them.

7. Molecules and Commercial Fertilizers.—It is a very strange fact that 100 to 500 pounds of commercial fertilizers applied to a poor soil will produce such marked effects upon the growth of plants when these small amounts are spread over an acre of ground and then dissolved in and distributed through the soil water of perhaps the entire surface four feet. To know, however, that the molecules of these fertilizers are so extremely small and that there are such immense numbers of them in a single pound enables the mind to better comprehend how such marked effects are possible.

The surface four feet of good field soil when well supplied with moisture may contain the equivalent of 10 inches of water on the level. This amount of water expressed in cubic feet per acre is 36,300. The experiment with aniline indicates that a single gram has been divided into not less than 5,120,000,000,000 parts. Let us compute how many parts this number would give to each cubic inch of the 36,300 cubic feet of soil-water in the upper four feet of an acre.

$$\frac{5,120,000,000,000}{36,300 \times 1,728} = 81,624$$

We see, then, that a single gram of aniline may be divided enough to place 81,624 parts in every cubic inch of moisture of an entire acre of good soil to a depth of four feet.

But one gram of sodium nitrate would contain, according to Maxwell's results,

$$NaNO_3 : 2\,O :: \text{No. of O molecules} : \text{No. of } NaNO_3 \text{ molecules}$$

$$\text{or} \quad 85 : 32 :: 1,359(10)^{22} : x$$

$$\text{whence} \quad x = 51(10)^{23} = 5,100,000,000,000,000,000,000,000$$

Treating this result as we did that of the aniline we shall have

$$\frac{5,100,000,000,000,000,000,000,000}{36,300 \times 1,728} = 81,310,000,000,000,000$$

as the number of molecules of sodium nitrate in each cubic inch of water from which the plants may draw their supply of nitrogen. It will be seen that this number is so large that even a cube of water .01 inch on a side will contain 81,310,000,000, a number far too large for comprehension, and yet if 200 pounds of sodium nitrate were applied to the acre this number would have to be multiplied by the number of grams in 200 pounds to express the number of molecules there would be for each cube of soil-water one-hundredth of an inch on a side.

8. Molecular Structure in Relation to Poisons.—It is the extremely large number of molecules which may exist in a small space, coupled with the energy which these molecules may carry with them in their movements, which makes possible the violent disturbances in the life processes of animals and plants associated with the introduction into the system of such small quantities of substances known as poisons. It will be easily understood from what has been said regarding the vast number of fertilizer molecules per cubic inch of soil moisture, when only a single gram has been disseminated throughout the surface four feet of a full acre, that extremely small quantities of any poison, like strychnine, will contain molecules enough to charge the body of the largest animal with great numbers of the poisonous units.

The important practical lesson to be remembered in this connection is that, since such extremely small quantities of matter, when introduced into the plant or animal body, are sometimes capable of producing such profound disturbances as to cause death, extremely small quantities of other substances may have very important beneficial effects; and it is quite possible that it may be along such lines as these we must search for an explanation of some of the little understood points associated with the nourishment of both plants and animals.

9. Ability to Recognize Small Quantities of Matter.—We often marvel at the delicacy of the chemical balance and many other instruments of measurement, but the delicacy of the sense organs of men and animals, and particularly the sense of smell, is so extreme that it is difficult to form a just conception of the minuteness of the quantity of matter or of energy to which they will respond with the degree of intensity which permits accurate judgments to be formed.

The sensations of odors result from the disturbances produced on the organs of smell by molecules of different substances moving through the air when brought to the nose. But when the blind lady took the glove of a stranger and, walking up and down the aisles of a large audience room filled with people, handed the glove to the owner, made known to her only by the likeness of the odor from the glove to that escaping from the stranger, who will say what fraction of a gram of that volatile principle it was which produced so marked a sensation? The weight of volatile substance rising into the air from a man's track, made by a shoe rather than his bare foot, must be very small indeed, and yet the sense of smell in the dog is so keen that he will follow his master at a rapid run even when the tracks are two hours old and where many other people may have passed along the same course more recently than did his master.

The readiness with which flowers, fruits and vegetables may be identified by their odors alone, often at considerable distances, and with which animals scent their enemies or their food, are all of them concrete demonstrations at once of the extreme minuteness and vast numbers of molecules, while at the same time they prove how sensitive is the animal organization to such minute quantities of material.

10. Foul Odors and Flavors in Dairy Products.—Since the commercial value of dairy products is determined in a high degree by their flavors and odors and since these qualities are judged through the sense of smell, which we have seen is so extremely delicate and keen, and since such minute quantities of the odor or flavor producing substances are certain to awaken the undesirable impressions, it is clear that the greatest of care must be exercised in producing, handling and caring for them through all the steps preceding the delivery to the consumer. Since we have seen that so little fertilizer may be disseminated through so much soil moisture and since so little may be detected by the organs of smell, it is plain that too great care cannot be taken in keeping the milk clean and that only those who do this can hope to secure the custom of people willing to pay a high price for the milk, cream, butter or cheese which just suits them.

11. How Odors and Flavors Find Their Way Into Milk.—The substances producing these qualities in milk make their entrance there in three different ways: (1) from the blood at the time the milk is secreted; (2) from the outside after the milk is drawn; and (3) they are produced within the milk after it has been secreted before or after it is drawn.

12. Odors Entering Milk During Secretion.—Any volatile principle which may chance to be present in the blood of the animal at the time the milk is being drawn will find its way into the milk and will impart a quality to it, the intensity of the flavor or odor depending upon the amount of the volatile principle present and the readiness with which it evaporates.

Nearly all food stuffs contain substances which produce odors and if these substances are not destroyed during the processes of digestion they will again escape from the body of the animal through the channels of excretion; that is, through the skin, kidneys, lungs, rectum or udder. and if any of these principles still remain in the blood at the time the milk is being drawn they will appear in it. It follows, therefore, that the longer the interval of time between the taking of food into the body and the drawing of the milk the less danger there will be of the milk being tainted by it. The reason for this is found in the fact that the milk is excreted during the time of milking while the blood is coursing through the udder, carrying whatever odor producing substances may then be present.

13. Time to Feed Odor Producing Foods.—It is clear from what has been said that if it is desired not to have the milk charged with the undigestible odor-principles of food while it is being drawn these foods should be fed as soon as possible after milking and never just before in order that time enough may have elapsed to permit the odor principles to have been eliminated from the blood by the other organs. On the other hand, if the food contains a principle whose odor is desired in the milk, then the reverse rule as regards time of feeding should be practiced, namely, to feed these just before milking.

14. Introduction of Odors Into Milk From the Air.—It is the fact that the molecules of substances are not in contact and that they are in motion which makes it possible for milk when in an atmosphere containing odors to become charged with them. If the odors of manure, of urine, of ammonia, or any of those associated with the decay of organic matter are in the air above the milk the rapid motion of these molecules will cause some of them to plunge into the milk and accumulate there until they become so numerous that just as many tend to escape per minute as tend to enter. The milk is then saturated with the odor in question.

The warmer the air surrounding the milk and the warmer the milk the more quickly will the condition of saturation be reached, simply because the rapidity of molecular motion increases with the temperature, for when the molecules of foul odor are once inside the warm milk they travel or diffuse downward more rapidly because it is warm.

15. Odors and Flavors Resulting From the Introduction of Solids Into Milk.—It must be clear from what was demonstrated in (6) that when great care is not taken both in keeping the stable and cows clean and free from dust the fine particles of dirt falling into the milk, even though the amount is small, may readily dissolve and impart a strong flavor to it, and one careless milker may easily greatly injure the quality of that from the whole herd where all of the milk is pooled. The fundamental point to be kept ever in mind is that a very little dirt is capable of being divided to an extreme degree and that through the senses of taste and smell extremely small amounts may readily be detected.

16. Odors and Flavors Developed in Milk After It is Drawn.—Milk is a very nutritive fluid and for this reason great care must be exercised not only to keep dirt out but also to prevent those germs from entering it which have the power of developing rapidly there, producing undesirable odors and flavors and thus injuring the quality of the milk. These objectionable germs are liable to be introduced into the milk through the dust from the stable and the cow as well as from the lack of proper cleanliness of the vessels in which the milk is handled.

17. Deodorizing Milk.—The removal of odors from milk may be accomplished by greatly increasing its surface in a space containing none of the odors which the milk contains. The method known as the "Aeration of Milk" has for its purpose this rather than the exposure of the milk to the air, as the presence of the air hinders the escape of the odors rather than favors it and if the milk could be exposed in a vacuum their escape would be more complete and more rapid.

The escape of the odors from the milk depends upon the rapid motion of the odor molecules in it which forces them to escape whenever they approach the surface with sufficient velocity to overcome the surface attraction, and the division of the milk into a large number of small streams increases the chances for the odors to escape in proportion to the increase of the surface. The finer the milk streams, the farther they are apart and the longer the stream is in falling the more complete will the removal of the odors be. Where there can be a movement of air over the milk surface or among the streams of milk this will favor the removal by carrying the odor molecules away and thus preventing them from re-entering the streams.

Since the molecular movement is greater the higher the temperature it follows that the deodorizing process should be applied as soon after the drawing of the milk as possible before it has had time to cool and the molecular motion to slow down.

18. Place For Using the Deodorizer.—If the aerator or deodorizer is used in the barn or where there are many objectionable odors it must be remembered that exactly the same conditions which favor the escape of the odors which the milk contains when drawn are the best conditions to permit it to become charged with odors from outside, and hence the deodorizer or aerator should be placed where it is surrounded by a current of pure air.

19. Cooling Milk.—The cooling of milk immediately after it is drawn has a powerful influence in preventing odors from developing in it as a result of the growth of any germs which may have found their way into the milk because the low temperature makes their growth much slower. Cooling, then, is not a deodorizing process but one which prevents the formation of new odors. If, then, it is desired to remove the animal odors this if possible should be done first and then the milk cooled to prevent the formation of other odors.

20. Work.—Whenever any body is moved under the action of a force work is done and the amount of this work is measured by the intensity of the force and the distance through which it has acted. When a body weighing one pound is lifted one foot against the attraction of the earth the amount of work done is one foot-pound. The same weight lifted 10 feet represents 10 foot-pounds and 10 pounds raised one foot has the same value.

A team hauling a load over a road under a mean pull of 200 pounds is doing 200 foot-pounds of effective work for every foot traveled and in traveling 10 miles the total work done is

$$10 \times 5,280 \times 200 = 10,560,000 \text{ foot-pounds.}$$

When a larger unit than the foot-pound is desired that of the foot-ton may be employed and its value is 2,000 pounds lifted one foot high or 2,000 foot-pounds. If a wagon with its load weighing 4,000 pounds is moved along the road the work done will not be measured by the product of the load into the distance traveled but by the intensity of the pull necessary to pull the load into the distance traveled. On a good level macadam road 60 pounds will move a ton and 120 pounds two tons. To draw four tons over 10 miles of such level road means the doing of

$$\frac{4 \times 60 \times 10 \times 5,280}{2,000} = 6,336 \text{ foot-tons.}$$

So, too, if the pressure of steam on the head of the piston in a steam engine is 80 pounds per square inch and the area of the piston is 100 square inches the amount of work it can do per foot of stroke is

$$80 \times 100 = 8,000 \text{ foot-pounds.}$$

If this engine makes 200 strokes per minute, then the work it does per minute will be

$$200 \times 8,000 = 1,600,000 \text{ foot-pounds.}$$

21. Energy.—Energy is the ability of a moving body to do work and the amount of energy the moving body has is measured by the amount of work it can be made to do in coming to rest. If a weight suspended from a string be drawn to one side and then released it will begin falling and acquiring velocity, and on reaching the lowest level it will possess the ability of doing a certain amount of work. That amount will be enough to raise its own weight through the height from which it fell in the same time. If a bow is bent and the string is released against the arrow it will recover its form of rest but in doing so will impart to the arrow an amount of motion equal to that which the bow acquired in straightening out. When work is done in winding the clock the distorted spring has the power to develop an amount of energy equal to that expended in winding it up. In chopping wood the action of the woodsman's muscles increases the amount of motion in the ax until it falls upon the wood, when the energy which has been imparted to it does the work of cutting.

We cannot exert pressure enough with the hand alone to force the nail into the board, but by giving the muscles an opportunity to act gradually upon the hammer it is a simple matter to store in it enough energy to easily drive the nail into the wood. When coal or wood is burned in the fire-box of the engine and the heat developed converts water into steam under high pressure in the boiler we have still another case where energy is developed and accumulated in the rapidly moving molecules of steam which drive the piston whenever the valves are opened leading to it.

22. Conservation of Energy.—No discovery of modern science is more fundamental than the fact that neither matter nor energy can be destroyed or created. One form of energy may be transformed into another, and one kind of substance may be decomposed and others made from the components, but in these transformations there is neither annihilation nor creation. The small amount of ashes left from the winter's supply of coal or wood seems to point to a destruction of matter, but their weight added to that of the products which pass up the chimney is even greater than that of the original fuel by the amount of oxygen which was required to burn the fuel. So, too, the energy of 10 horses expended in threshing grain seems to be annihilated but it is only transformed. Heat of friction and concussion, sound and material raised into new positions, from which it may fall, when added together will make a sum equal to that developed by the horses. Again we appear to realize in the increase of our domestic animals or in milk produced much less weight than has been used by them in feed and drink but this is because such large quantities of the materials eaten, breathed and drank escape in an invisible form through the skin and lungs.

23. The Source of the Earth's Energy.—The real source of the earth's energy is the sun. All the rivers of the world flowing to the sea and the rush of the winds swaying the tree-tops and lashing the ocean into billows represent so much water and air lifted from a lower to a higher level by the sun's heat and now being pulled by gravity back to their original level to be raised again and to again return, just as a pendulum rises and falls while swinging through its arc.

The wood burned in the stove, the coal burned in the engine and the food consumed by the horse are all the product of sunshine which lifted the constituents of soil, moisture and air into such combinations as readily permits of their return to other forms, setting free the energy which was consumed in producing them.

24. Solar Energy.—When the sun rises the temperature increases, usually becoming higher and higher until past noon, then when the sun sets the temperature falls again, continuing to do so until once more the sun is above the horizon. So, too, as our days grow longer and longer with the approach of summer in the middle and higher latitudes, making more hours of sunshine in every twenty-four, the mean daily temperature increases but falls away again when the nights became longer than the days. Such and many other facts prove the sun to be a source of energy which in some manner is being transferred to our earth. More than this, since the earth travels entirely around the sun once each year and yet each day receives heat and light from it, it follows that solar energy is continually leaving the sun in all directions, so that the amount arrested by the earth forms a very small portion of the whole.

25. How Solar Energy Reaches the Earth.—To understand how the energy of the sun reaches us, coming across 93,000,000 of miles we must learn that the energy travels in the form of waves through some medium filling space, which has been named ether, but whose real nature is not yet understood.

Something similar to the process in question would be represented by a man at the center of a pond throwing its water into waves. These waves would spread in all directions and when reaching the beach a portion of the energy of the waves would be absorbed or transferred to whatever body they chanced to strike. The energy, therefore, generated in the muscles, is changed first into wave energy in the water and conveyed away from the man in all directions, but afterward when arrested at the beach, the waves may move the pebbles, making them grind upon one another, wearing themselves into sand, or their sliding may change a portion of the wave energy into heat and thus the person in a small degree may warm the pebbles lying on the distant margin of the lake, not directly by the heat of his body, but by the waves set up in the water, and much as the earth is warmed by waves sent out through the ether of space from the surface of the sun.

The rapid and intense molecular motion at the surface of the sun is transformed into wave motions in the surrounding ether of space, as the motions of the imaginary man were changed into waves in the water, and these ether waves travel away from the sun's surface in all directions at the rate of 186,680 miles per second. So many of these waves as the size of the earth permits it to stop are arrested and transformed into the various forms of motion which are manifested at its surface.

26. Amount of Energy Developed at the Sun's Surface.—Careful measurements and calculations have shown that the energy developed second by second at the sun's surface, amounts, according to Lord Kelvin, to not less than 133,000 horse power on each square meter or 1.09 square yards of its surface.

27. Rate at Which Solar Energy Reaches the Earth's Surface.—As the intense energy developed at the surface of the sun spreads away from it, it becomes weaker and weaker in the ratio that the square of the distance of the waves from the sun increases, as represented in Fig. 2, and

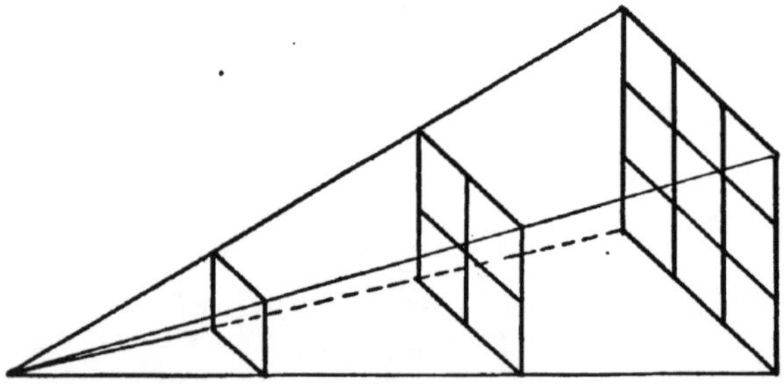

FIG. 2.

so at the earth's surface the amount of energy has become so much reduced that Lord Kelvin places it at only a little more than 1.3 horse power per each square yard of surface.

But small as this amount of energy is when compared with that leaving a like area at the sun's surface it is nevertheless very large.

It may seem strange that so much energy falling upon the earth does not keep its surface at a higher temperature than is observed, but when it is stated that the temperature of the space which surrounds the earth outside its atmosphere is —273° C. and that only the thin atmosphere shields the surface from this intense cold, it is plain that a large amount of heat must be required to hold the mean temperature even as high as 45° F. which is

$$273° + 7° = 280° \text{ C above absolute zero.}$$

If we add to the necessity of holding the earth's surface at a temperature 280° C. to 300° C. above the space in which it moves, the demand for energy needed to maintain the movements of water and of winds, together with that embodied in activities of animal and plant life, then 1.3 horse power per square yard of surface does not appear so much too large.

28. Kinds of Ether Waves.—The energy reaching the earth from the sun in the form of wave motion is not all alike in that the waves have different lengths, or, what is the same thing, greater numbers of one kind reach the earth in a unit of time. Waves which are so frequent that from 392 to 757 billions reach us per second produce the sensation of light when falling upon the eye; the slower ones producing red light and the more rapid ones the extreme violet colors of the rainbow. Associated with these color waves there are many other dark waves to which the human eye is not sensitive. Some of these are much shorter than the color waves and are especially powerful in breaking down the molecular structure of different substances; that is, in producing chemical changes such as occur on the photographer's plate when the negative is made and such as take place in the green parts of plants when carbon dioxide is broken down and the chemical changes are produced which result in building the sugars, starches and cellulose of plants. Others of these waves are much longer than the light waves and these have a wonderful power in producing heating effects when they fall upon certain substances, one of which is water.

When bright sunshine is allowed to pass through a large lens the glass is but little warmed by the passage, but if paper is held at the light focus it is quickly set on fire by the dark or invisible rays. That it is the dark rays may be proved by allowing the light to pass first through a solution of iodine in bisulphide of carbon which permits the dark waves to easily pass while it cuts down or stops the light waves. When these dark waves are brought to a focus in water it is made to boil quickly under their influence.

On the other hand if sunlight is first passed through a solution of alum in water, which stops the dark waves but allows the light waves to pass, then when they are focused upon water but little heating effect is noted.

29. How Water is Evaporated.—It is the fact that water does not allow the long dark waves from the sun to pass readily through it which causes it to evaporate so rapidly from ocean, lakes and streams, and from the soil and the leaves of vegetation. When these waves fall upon water they set its molecules in such rapid vibration that the surface tension, or force of cohesion, is overcome and many of the water molecules are thrown out into the air in the form of invisible vapor. Were the water not so opaque to the dark waves, neither snow nor ice would be as rapidly melted in the spring nor would there be so much evaporation from the ocean as we now have, hence rains would be less frequent and the land less productive.

30. How Chemical Changes Are Produced by Ether Waves.—When the light waves and especially the shorter dark waves fall upon many substances they appear to set up vibrations within the molecules themselves, which in time may become so intense as to overcome the force by which the components are bound together and the molecule is thrown into parts, setting them free so that when their motion slows down they may join in new combinations. It is much as if some giant power were to seize upon a steel chain, throwing it into such intense vibrations that its several links are broken.

31. Nature of Heat and Cold.—When a body becomes warm the rate of vibration of the molecules which compose it is increased and the path through which they move becomes longer. If the body becomes cold the rate of movement of the molecules becomes less rapid and the distance through which they move less. The higher the rate of molecular motion within a given body the warmer that body is and vice versa. If the molecular motion of a body could be completely brought to rest its temperature would be absolute zero. Under this condition it is supposed that any body would have its smallest volume; and all liquids and gases would become solid.

32. Temperature.—When the temperature of a body is given it is intended to state the degree of molecular vibration within it. The temperature at which a Fahrenheit thermometer marks zero is not that of no molecular motion but simply 32 degrees of that scale slower than the rate at which pure water becomes a solid; while zero indicated by a Centigrade thermometer is the rate of molecular motion which permits water to become solid and is a temperature 273 degrees above what is assumed to be absolute zero or the condition of absolute rest.

33. How Temperature is Measured.—It is a general law that those substances which may exist as solids, as liquids or as gases, as is the case with water, which we know as ice, water and steam, or invisible vapor, change from the solid to the liquid form and from the liquid to the gaseous form when the rate of molecular motion has reached a certain degree, and this being true the freezing and boiling points of various substances may be taken as standards of temperature.

Water being a common substance which changes its state at convenient and common rates of molecular motion has been selected to fix two degrees of temperature called the freezing and boiling points of water. When a thermometer scale is to be graduated its bulb is placed under the influence of melting or freezing water, and the place at which the moving point comes to rest marked; then it is placed under the conditions of boiling water and the new point also marked. The space between these two points on the scale is then divided into 80, 100 or 180 divisions, according to the system which it is designed to follow. Since this range in molecular vibration is divided into 180 degrees on the Fahrenheit scale its degrees are the shortest, while those of the Reaumer scale are the longest because the same range is divided into but 80 divisions.

The Centigrade and the Fahrenheit scales are the two commonly used in this country, the degree of the former being equal to $\frac{9}{5}$ of the latter.

34. Accuracy of Thermometers.—The bulbs of most thermometers shrink after they are blown and if they have not been permitted to stand for a number of years to season before fixing the zero and boiling points of the scale, these points will change and the thermometer will give incorrect readings in time and the cheaper grades of thermometers are liable to be subject to this error.

The accuracy of the freezing point may be approximately tested by surrounding the bulb with snow or crushed ice out of which the melted water may drain, allowing the thermometer to remain until the temperature becomes stationary.

The accuracy of the boiling point may also be approximately determined by holding the bulb in rapidly boiling soft water.

A thermometer may be correct at the freezing and boiling points and inaccurate at most intervening degrees, growing out of the unequal diameter of the tube in different portions and the fact that all degree marks may be made of the same length. Errors of this sort can be detected only by comparing the thermometer with a standard.

35. Units of Work and Energy.—It has been found necessary in dealing with the numerical relations of work and energy to adopt standards of measurement just as has been done for lengths, volumes, surfaces and mass, and various units are in use.

36. Foot-pound and Foot-ton.—A common unit of work is the foot-pound, which is a mass or weight of one pound lifted vertically against or in opposition to the force of gravity.

If a body is moved one foot in any other direction than against the force of gravity and the intensity of the pull or push necessary to do this is equal to that required to lift one pound, then in this case the work done is one foot-pound. If 2,000 pounds is lifted one foot high then 2,000 foot-pounds of work have been done, and this is sometimes designated a foot-ton. The same intensity of pull in any other direction may be expressed in the same terms.

Time is not a factor taken into account in simply expressing the amount of work done for the reason that a very small force when permitted to act for a very long time may raise the same weight through one foot, which would require a very intense force if permitted to act but a very short time.

37. Horse-power.—When the rate at which work is done and the intensity of the force required to do the work at the stated rate are to be expressed quantitively, then a unit involving time must be chosen and the *horse-power* is one of these. The horse-power used by English and American engineers is the amount of energy which can do 550 foot-pounds of work per second or 33,000 foot-pounds per minute, equal to 16.5 foot-tons in the same time. To raise grain in an elevator to a hight of 20 feet at the rate of 16.5 tons per minute would require 20 horse-power.

If a horse is walking 2.5 miles per hour and exerting a steady pull on his traces of 100 pounds then the effective energy he is developing is

$$\frac{100 \times 5,280 \times 2.5}{60 \times 60 \times 550} = \tfrac{2}{3} \text{ H. P.}$$

and this for a well fed horse weighing 1,000 pounds, working 10 hours per day at the rate of 2.5 miles per hour, is called a fair day's work. If a 1,500-pound horse could do work in proportion to his weight then his ability to develop energy would be equal to the standard English horse-power of 550 foot-pounds per second. Gen. Morin, however, has placed the ability of the average horse to do work at the rate of 435.8 foot-pounds per second.

38. Heat Unit.—In the steam engine the energy of heat is converted into work, and since heat is a form of molecular motion its quantity must have a fixed relation to the temperature of a given amount of material. The English and American heat unit is the amount of heat energy which is required to raise the temperature of one pound of pure water from 32° F. to 33° F., and since one form of energy may be converted into another the value of a heat unit may be expressed in foot-pounds. The English scientist, Joule, was the first to measure the number of foot-pounds of work which one heat unit could do and found it to be 772, which when corrected for the mercurial thermometer became at 15° C. 775 foot-pounds. Rowland obtained the value 778.3 foot-pounds. This means that the source of heat which is able to raise the temperature of one pound of water one degree every second would also be able to raise 778.3 pounds one foot high in the same time.

39. Determination of the Mechanical Equivalent of Heat.—In order to ascertain the value of the heat unit in foot-pounds, Joule arranged a vessel containing water in such a way that by means of nicely adjusted weights he could cause them to drive a set of paddles in the water and by the mechanical agitation warm it. By knowing the number of pounds in his weights, the distance they were allowed to fall and the rise in temperature which was observed in a given weight of water, he found the relation to be that stated in (**38**).

40. Specific Heat.—We have learned (**32**) that temperature is a measure of the rate of molecular motion within a given body; it is not, however, a measure of the amount of work which must be done upon that body to change its temperature through a given number of degrees; neither is it a measure of the amount of work which may be secured from that body when its temperature falls a given amount.

When the same number of heat units is imparted to like weights of different substances their temperatures are not raised through an equal number of degrees. The same amount of heat, for example, which will raise the temperature of one pound of water from 32° F. to 33° F. will raise a pound of sand from 32° F. to 37.23° F. For some reason more work must be done on water than on the sand to secure the same change of temperature, but, true to the law of the conservation of energy, when the water again cools down it gives out as much more heat in doing so as was required to produce the rise in temperature. It is this fact which causes large bodies of water to make the winters of adjacent lands warmer and the summers cooler. Soils change in temperature more rapidly than would be the case were their specific heats higher, and for this reason in part a wet soil is cooler than the same soil when dryer.

41. Latent Heat.—When ice at 32° F. has heat applied to it its temperature does not rise so long as there is still ice to melt, the whole of the energy given to it being consumed in changing the solid ice into liquid water, that is, in doing the work of melting. The amount of heat required to melt one pound of ice is 142 units when expressed in round numbers; or if the work done is expressed in foot-pounds it will be

$$142 \times 778.3 = 110,518.6 \text{ foot-pounds}$$

and the time required for one horse power to do the work would be

$$\frac{110,518.6}{33,000} = 3.35 \text{ minutes.}$$

When crushed ice and salt are mixed in the ice-cream freezer the changing of the two solids to a liquid requires so much energy, and it is used so rapidly, that the cream is quickly frozen, its molecular motion being used in doing the work.

When water has been brought to the boiling temperature it ceases to become warmer so long as boiling continues, all of the heat energy entering from the fire being required to do the work of changing liquid water into steam. The amount of heat required to change one pound of water at 212° F. into steam at the same temperature is 966.6 heat units. When expressed in foot-pounds it becomes

$$778.3 \times 966.6 = 752,305$$

and the time required for one horse-power to do this work is

$$\frac{752,305}{60 \times 550} = 22.8 \text{ minutes.}$$

When a pound of water at 32° F. becomes ice at 32° F. there reappears as heat the 142 heat units which were required to melt it, and so too when one pound of steam condenses into water there reappears 966.6 heat units. Before the nature of these changes were as well understood as they now are, it was supposed that the heat became hidden or *latent* but that it was heat still.

42. Measuring the Energy Required to Melt Ice.—This may be determined approximately by taking equal weights of water at 212° F and of ice at 32° F., putting the two together and noting the temperature at the moment the ice is all melted. When this has been done it will be found that the combined water has a temperature of about 51° F.

If, however, equal weights of water at 32° and 212° are mixed there will be found a temperature of

$$\frac{212 + 32}{2} = 122$$

one volume of water having lost as much as the other gained.

In the first case, however, the water lost

$$212 - 51 = 161$$

while the ice gained only

$$51 - 32 = 19.$$

There was therefore in this case an apparent loss of

$$161 - 19 = 142^\circ$$

If a pound of water and of ice had been taken for these experiments it is plain from (38) that the 142 would also represent 142 heat units.

43. Measuring the Energy Required to Evaporate Water.—If a pound of steam at 212° F. be condensed within 5.37 pounds of water at 32° F. there will result 6.37 pounds of water having a temperature very close to 212° F. The one pound of steam has therefore raised the temperature of 5.37 pounds of water through

$$212^\circ - 32^\circ = 180^\circ$$

$$212° - 32° = 180°$$

without having its temperature materially lowered. The molecular energy, therefore, which the one pound of steam contained was

$$180 \times 5.37 = 966.6 \text{ units.}$$

This large amount of energy in steam explains how it is able to do so much work when acting upon the engine piston and why a burn from steam may be so much more severe than that from boiling water.

44. Evaporation Cools the Soil.—We have seen that one pound of steam in condensing into water generates 966.6 heat units, and that the reverse statement is also true, namely, to convert a pound of water into the gaseous state, under the mean atmospheric pressure, requires the absorption by that pound of 966.6 heat units. When one pound of water disappears from a cubic foot of soil by evaporation, it carries with it heat enough to lower its temperature, if saturated sand, 32.8° F.; and if saturated clay loam, 28.8° F.

To dry saturated sandy soil until it contains one-half of its maximum amount of water requires the evaporation of about 9.5 pounds to the square foot of soil surface when this drying extends to a depth of one foot, while the similar drying of clay loam requires the evaporation of 11.5 pounds, and

$$11.5 - 9.5 = 2 \text{ lbs.}$$

or the amount of evaporation which must take place in the clay loam to bring it to the same degree of dryness as the sandy soil. But to evaporate two pounds of water requires

$$966.6 \times 2 = 1933.2 \text{ heat units.}$$

and this, if withdrawn directly from a cubic foot of saturated clay loam, would lower its temperature 57.6° F. Here is one of the chief reasons why a wet soil is cold.

That the evaporation of water from a body does lower its temperature may be easily proved by covering the bulb of a thermometer with a close fitting layer of dry muslin, noting the temperature. If the muslin be now wet, with water having the temperature noted, and the thermometer rapidly whirled in a drying atmosphere its temperature will quickly fall, owing to the withdrawal of heat from the bulb by the evaporation of water from the muslin.

45. Regulation of Animal Temperatures.—All of our domestic animals require the internal temperature of their bodies to be maintained constantly at a point varying only a little from 100° F., and this necessity requires provisions both for heating the body and cooling it. The cooling of the body is accomplished by the evaporation of perspiration from the skin, and the amount of perspiration is under the control of the nervous system. When the temperature becomes too high, because of increased action on the part of the animal, or in consequence of a high external temperature, the sweat glands are stimulated to greater action and water is poured out upon the evaporating surfaces and the surplus heat is rapidly carried away; each pound evaporated by heat from the animal withdrawing about 966.6 heat units.

46. Bad Effects of Cold Rains and Wet Snows on Domestic Animals.—When cattle, horses and sheep are left out in the cold rains of our climate the evaporation of the large amount of water which lodges upon the bodies, and especially in the long wool of sheep, creates a great demand upon the animals to evaporate this water. The theoretical fuel value of one pound of beef fat is 16,331 heat units, and that of average milk is 1,148 heat units. A pound of beef fat may therefore evaporate

$$\frac{16,331}{966.6} = 16.8 \text{ lbs. of water}$$

and a pound of average cow's milk

$$\frac{1148}{966.6} = 1.18 \text{ lbs. of water}$$

On this basis, if a cow evaporates from her body four pounds of rain she must expend the equivalent of the solids of 3.39 pounds of milk.

A wet snow-storm is often worse for animals to be out in than a rain-storm, because in this case the snow requires melting as well as evaporating, and the number of heat units per pound of snow is

$$142.65 + 966.6 = 1109.25 \text{ heat units,}$$

and the heat value of a pound of milk is barely sufficient to melt and evaporate a pound of snow.

47. Cooling Milk with Ice and Cold Water.—If it is desired to cool one hundred pounds of milk from 80° F. down to 40° F. it is practically impossible to do so with water in the summer season in Wisconsin. It is difficult even to cool it as low as 48° F. for most of the well and spring water has a temperature above 45° F. and much of it is above 50° F. If lower temperatures than 48° F. are desired during the warm season some other means must be resorted to. Since it requires 142 heat units to melt a pound of ice, one pound is capable of cooling from 80° to 40° F.

$$\frac{142+8}{40} = 3.75 \text{ lbs. of milk,}$$

supposing the specific heat of milk to be the same as that of water, which is not quite true. To cool 100 pounds of milk from 80° F. to 40° F. will require, therefore, about

$$\frac{100}{3.75} = 26\tfrac{2}{3} \text{ lbs. of ice,}$$

supposing it to be used wholly in cooling the milk.

If the water has a temperature above 40° F. before the milk and ice are placed in it, there will be required enough more ice to cool the water down to the temperature desired for the milk.

The greatest economy in the use of ice will be secured, therefore, when the creamer contains just as little water as will cover the cans and give the needed space for the ice, and when the walls of the creamer are made of so poor a conductor of heat as to admit as little as possible from without.

48. Washing with Snow or Ice.—When ice or snow are used in winter for washing purposes there is a large loss of heat incurred in simply melting the ice and raising the temperature of the water from 32° F. up to 45° F., the temperature it may have in any well protected cistern. To melt a pound of ice and raise its temperature to 45° F. will require

$$142 + 13 = 155 \text{ heat units,}$$

If 300 pounds of water are required for a washing then the lost heat will be

$$300 \times 155 = 46{,}500 \text{ heat units.}$$

The fuel value of one pound of water-free, non-resinous wood, such as oak or maple, has been found to be 15,873 heat units; that of ordinary dry wood, not sheltered, containing 20 per cent. of water, is 12,272 heat units. At this latter value it will require, supposing 50 per cent. of the fuel value to be utilized in melting the ice and heating the water,

$$\frac{2 \times 46{,}500}{12{,}272} = 7.58 \text{ lbs. of wood}$$

more than would be needed to do the same washing with water at 45° F., to say nothing of the expense of getting the snow or ice and the unhealthfulness of handling it.

48a. Burning Green or Wet Wood.—Whatever water wood or other fuel may contain when it is placed in the stove, so much of the fuel as is required to evaporate this water must be so expended and is prevented from doing work outside of the stove. We have seen (48) that when wood contains 20 per cent. of water there is required

$$15{,}873 - 12{,}272 = 3{,}601 \text{ heat units}$$

per pound of wood to evaporate the water contained, which is 22.7 per cent. of the total value. Wood, after being in a rain of several days, contains more water than this and green wood much more, sometimes as high as 50 per cent., while well seasoned sheltered wood may contain less than half that amount.

It is frequently urged that when some green or wet wood is burned with that which is dry there is a saving of fuel. There is some truth in this, especially in stoves having too strong a draft and too direct a connection with the chimney and if the radiating surface is small or poor. The evaporation of the water prevents so high a temperature from occurring in the stove, which makes the draft less strong, and this gives more time for the heat to escape from the stove before reaching the chimney, and hence less is lost in this way. Then as the fire burns more slowly there is not the overheating of the stove, at times, which may occur with lack of care when very dry wood is used, and a considerable saving occurs in this way. These statements apply more particularly to heating stoves than to cooking stoves. Dry wood is best for the kitchen stove under most circumstances, the slower fire being secured when needed by using larger sticks and by controlling the draft.

SURFACE TENSION, SOLUTION AND OSMOSIS.

49. Surface Tension.—The free surface of any liquid behaves much as though it were covered by an elastic membrane and it is this surface action which draws the raindrop into the form of a sphere as it falls through the air. It is surface tension that causes water to form into spheres on a dusty floor, on a hot stove or on cabbage leaves. The dewdrop owes its shape to surface tension, and it is this which is employed to mould the melted lead into perfect spheres as it falls from high towers, cooling into solid shot before reaching the bottom.

The cause of surface tension is the cohesive attraction of the molecules for one another. This attraction extends throughout the liquid and is very strong; but only the molecules at the surface show its influence because it is only these which are not pulled evenly in all directions by water molecules on every side, thus leaving the interior ones free to move in any direction, while those on the surface are only pulled toward the sides and toward the interior. In Fig. 3 is illustrated one method of showing the action of surface tension. Here the dry camel's hair brush at the left shows the individual hairs standing apart, and when the brush is placed in the water they still stand apart, but when it is removed, as shown on the right, the whole are closely compacted by the pull of the surface film.

Fig. 3.—Illustrating the action of surface tension.

50. Rise of Water in Capillary Tubes.—It is surface tension which causes the rise of water in capillary tubes, as represented in Fig. 4, above the level of the water in the open vessel in which they are placed, when the water wets the glass tube—that is when the attraction of the glass for the water is stronger than the attraction of the water, as explained in **(191)** and **(192)**. The rows of molecules of glass just above the water level attract and lift the water closest to them but, as these are moved upward, they draw after themselves more water also. This attraction is felt over a distance which Quinke estimates at not far from $\frac{1}{500000}$ of an inch. The first lifting of the water by the glass brings it near enough to other glass molecules next above and the water is drawn still higher; in this way

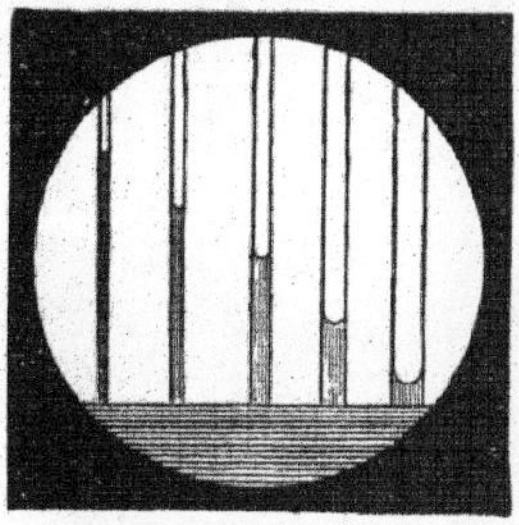

Fig. 4.—Rise of water in capillary tubes.

the column rises higher and higher until the strength of the pull is balanced by the weight of the water column so lifted.

51. Evaporation.—Very many substances, when they have a free surface exposed, change in weight by evaporation. The action of heat causes the molecules to vibrate rapidly and by colliding with one another near the free surface some are thrown out by the force and direction of the blows. It is in this way that clothes dry rapidly on a warm day, and that water evaporates from the surface of the soil, from the leaves of plants and from the bodies of animals. Even snow and ice evaporate, as camphor does, without first becoming liquid, but the process and the cause are the same as the evaporation of water, namely, rapid vibration and collision at the surface due to the absorption of heat energy from outside.

It is often said that the air takes up the moisture and the more rapidly the dryer it is or the stronger the wind blows. The air itself is not the cause of the more rapid evaporation observed, neither does evaporation stop when the air becomes saturated with moisture. Evaporation may be even more rapid in a vacuum and in rarefied air than where the air is dense and the pressure heavy; and when the air is saturated with water vapor and evaporation appears to stop, it may be going on just as rapidly, only condensation may be taking place at the same rate, that is, just as many molecules of water return from the space above the surface as leave it in a unit of time.

52. Solution of Solids.—The solution of solids, like sugar or salt in water, is not fundamentally different, either in cause or in manner, from the evaporation of water or of camphor referred to in (51), and is due to the absorption of heat from without which causes some of the surface molecules to be thrown into such rapid motion that the attractive force which draws them toward the center is no longer able to retain them in place and they are thrown out.

When a lump of salt is dropped into water its surface molecules are drawn outward by the surrounding water so that the *effective* pull upon them toward the center is made less by it. It is therefore easier for a given temperature to throw out into the water some of the molecules in the surface layer of the salt. Stated in another way the water surrounding the salt so weakens the surface tension of the solid lump of salt that solution takes place at a lower temperature.

53. Influence of Temperature on Solution.—Since it is the absorption of heat which causes solution it is clear that the higher the temperature the more rapid will the solution be. It is even true that any solid will evaporate or dissolve if only it is given a high enough temperature, provided its molecules are not decomposed at a lower temperature than that which is required to overcome the force of cohesion which makes them solid. It is a matter of common observation that substances dissolve more rapidly in warm than in cold water and it is equally true that the soluble salts in the soil will form more rapidly when the soil is warm than when it is cold and it is because of this fact, in part, that crops grow better under the higher temperature.

54. A Saturated Solution.—When conditions are favorable for the solution of a solid in water there comes a time when there is no increase in the concentration of the solution. A condition is reached which is analogous to the air saturated with moisture, when as many molecules pass from the solution and become fixed upon the face of the solid as are thrown by heat from the face of the solid into the solution. When a solution reaches this condition it is said to be saturated.

In the case of the soil water, where the roots of plants are brought in contact with it, if the roots are removing the materials which are dissolved, their action hastens the rate of solution, for they prevent it from becoming saturated and thus prevent the return to the soil grains of particles once removed.

55. Diffusion.—When water has evaporated into the air; when a salt has dissolved in water, there is a tendency for these separated molecules to travel in any and all directions until the whole body of the liquid in which the solution is taking place contains the same number of the dissolved molocules per cubic inch. A lump of sugar placed in the bottom of a cup of tea dissolves in time and becomes scattered uniformly through the whole mass, making all parts equally sweet. This scattering of molecules is called diffusion and the rate varies with the temperature and the individual velocities of the molecules dissolved.

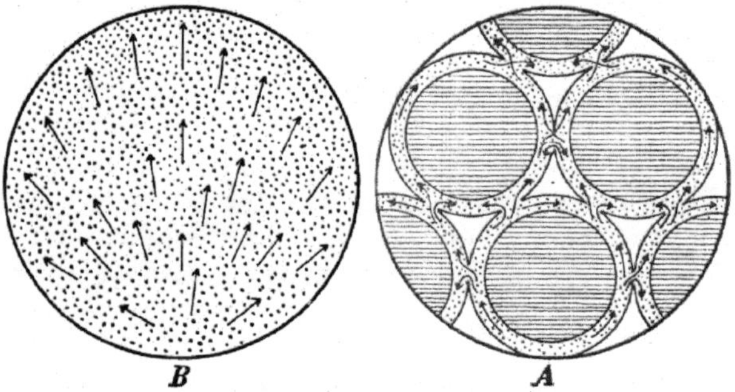

FIG. 5.—Illustrating the difference in the rate of diffusion in soil and in liquids

The rate of diffusion of salts in a vessel of water is much more rapid at the same temperature than it could be if the water were filled with sand. This will be understood from a study of Fig. 5, where A is supposed to be a place from which salts are diffusing through the water surrounding a set of soil grains, while B is a corresponding point from which diffusion is taking place in directions indicated by the arrows of that figure. Where the diffusion must take place through the films surrounding the soil moisture not only is there less water to travel in but the course of the molecules must be many times arrested by the soil grains themselves.

56. Gaseous Pressure.—The pressure which is exerted

by gaseous bodies like air or steam, upon the walls of confining chambers or vessels, is due to the combined energy of the blows of the molecules against these walls. The greater the number of molecules in a given space and the more rapidly they move the greater is the pressure they exert. If the temperature of a gas is increased, leaving the volume the same, the pressure is increased in the same ratio, because the velocity with which the molecules are moving is increased. So, too, if the number of molecules of a given gas in a given space is increased the pressure is increased in a like ratio, if the temperature remains the same, because then there are more molecules to strike a unit area in a given time.

To double the pressure on a gas will reduce its volume one-half and to double the volume of a gas will reduce its pressure one-half. So, too, will doubling the absolute temperature of a gas double its pressure if it is not allowed to expand. It is on these accounts that the higher the steam pressure in a boiler the hotter it is and the more work it is capable of doing.

57. Osmosis.—Abbé Nollet, who lived between 1700 and 1770, appears to have been the first to record that, if a glass vessel be filled with wine and covered with a bladder and then immersed in water, the contents of the vessel would increase and sometimes to such an extent as to rupture the membrane. Such a phenomenon has been named osmosis, and there are many familiar phenomena of every day experience which are of the same nature.

When dry beans, peas or grain of any kind are put into water they swell, increasing in volume as the wine in Nollet's covered dish did when placed in water. So, too, if dried raisins, prunes or apples are placed in water they increase in size, thus exhibiting the process of osmosis.

On the other hand if fresh fruit of almost any kind is placed in a strong solution of sugar it at once begins to shrivel and decrease in size; this again is due to osmosis, the juices being forced from the fruit by the pressure of the dissolved sugar as will be explained in the next section.

58. Osmotic Pressure.—The power which causes the swelling of the dried grains and fruits and that which causes the shrinkage of the fresh fruit, in the cases cited in (57), is known as osmotic pressure and is caused in fundamentally the same manner as that of gaseous or steam pressure, but in this case by molecules of substances in solution moving in the same manner as the molecules of gas move when developing gaseous pressure.

59. Conditions Under Which Osmotic Pressure Becomes Manifest.—In order that the molecules of a dissolved substance may exert pressure analogous to gases they must be dilute solutions, so that the individual molecules of the kind manifesting the pressure are too far from one another to be influenced by their individual attractions; besides this, in order that the pressure may become manifest, the dissolving substance and the substance dissolved must be separated by a membrane through which the molecules of one of the fluids pass more readily than the other, as represented in Fig. 6.

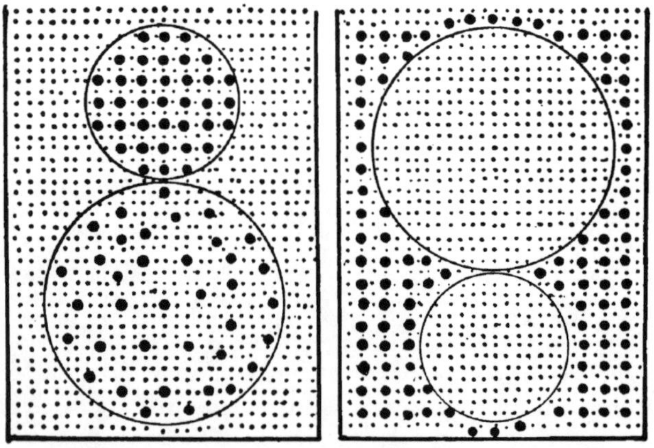

FIG. 6.—Illustrating the principle of osmotic pressure.

The diagram on the left may represent a raisin, dry bean or other dry seed placed in water, the upper circle showing the conditions just as placed in the water, and the larger one after the water has diffused through the wall, dissolving the substances on the inside, which produce the osmotic pressure. In this diagram it is supposed that the surrounding membrane is porous enough to let the molecules of water pass readily through by the ordinary laws of diffusion, but the molecules of the substance inside, which is dissolved by the entering water, are too large to be able to pass out into the water by diffusion, and the result is they simply strike against the membrane, distending it in exactly the same way that the molecules of air in a rubber ball or rubber bicycle tire distend that.

In the diagram on the right in Fig. 6, the reverse conditions are represented. A green or fresh fruit is supposed to be placed in a solution of sugar, whose molecules are too large to readily pass into the fruit through its wall, while the contained sap of the fruit readily diffuses outward into the sweetened water. Under these conditions the molecules of sugar, striking against the fruit on all sides at once, develop so much pressure that the juices of the berry are squeezed out of it as water might be forced out of a sponge, and its volume is reduced from the large size in the upper part of the diagram to the smaller one in the lower portion.

60. Measurement of Osmotic Pressure.—It was a long time after the discovery of osmotic pressure before satisfactory means for measuring its full intensity were devised. The parchment and animal or vegetable membranes which were first used in such studies were either not sufficiently impervious to the pressure producing molecules or else their strength was not great enough to allow the full measure of pressure to develop and the result was the early experiments failed to show how powerful osmotic pressure may become when the conditions are all favorable.

Traube discovered the possibility of producing membranes by chemical precipitation, at the plane of contact between two solutions, which could be used instead of organic membranes to study osmotic pressure, and later Pfeffer devised the apparatus represented in Fig. 7, with

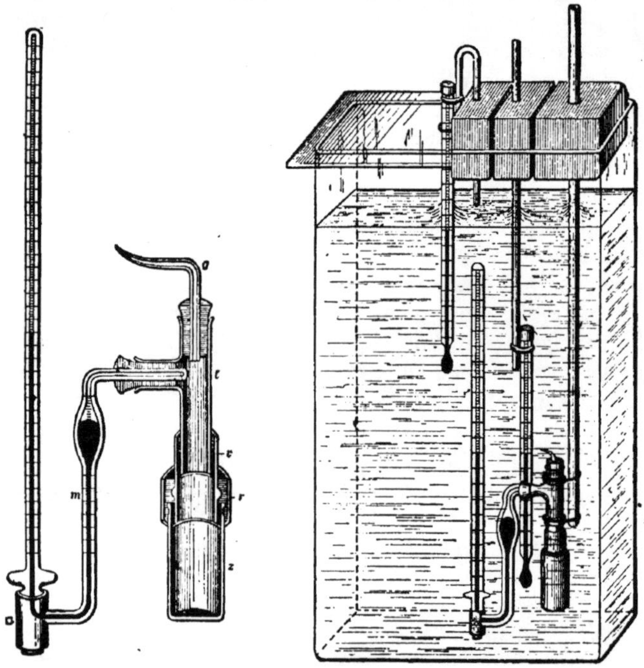

Fig. 7.—Pfeffer's apparatus for measuring osmotic pressure.

which he was able to measure osmotic pressures of very high intensity and with a fair degree of accuracy. He used a porous porcelain cell Z, with three glass pieces, t, v, r, put together with sealing wax in the manner represented in section in the left of the figure, where the method and arrangements for measuring the pressure are also shown at a, m. The complete apparatus, in working order, is represented at the right in the same figure.

To secure the manifestation of osmotic pressure with this apparatus there is developed, on the inner wall of the porous porcelain cell, a precipitation membrane which is impervious to the solution whose pressure is to be measured but which is readily permeable by water, placed in the outer vessel. The function of the porous cell is to act as a strong framework capable of permitting the precipitation membrane to withstand the pressure developed without a sensible increase in the volume of the cell taking place. When the pressure-producing fluid is placed on the inside and the apparatus is placed in water the case becomes analogous to the left diagram of Fig. 6, except that the wall is now incapable of expansion and the pressure becomes manifest through the rise of mercury in the pressure gage.*

61. Osmotic Pressure of Cane Sugar.—Pfeffer, working with his apparatus and different strengths of cane sugar in solution on the inside, was able to show that pressures were developed having the intensities indicated in the table below:

Table showing osmotic pressure of solutions of cane sugar of different degrees of concentration.

Strength of solution.	Pressure in m.m. of mercury.	Pressure per sq. inch.	Height of column of water sustained, in feet.
		lbs.	
1 per cent..................	535	10 36	23.9
2 per cent...	1,016	19.68	45.4
2.74 per cent...............	1,518	29.41	67.8
4 per cent...................	2,082	40.34	93.0
6 per cent......	3,075	59.57	137.4

From this table it appears that Pfeffer was able to secure pressures ranging from about 10 to 60 pounds per square inch, or enough to sustain a column of water from 24 to 137 feet high.

* Detailed descriptions of the method of forming the membrane and setting up the apparatus can be found in Gray's Botanical Text Book, 6th Ed., Vol. II, p. 227, and in Jones' The Modern Theory of Solutions, p. 3.

Using a 3.3 per cent. solution of potassium nitrate in his apparatus Pfeffer secured a pressure of nearly 85 pounds per square inch or enough to sustain a column of water 195 feet high. This force has been looked upon as the cause of the movement of sap in plants and it was a search for a cause for this movement which led Pfeffer to make the observations here referred to.

62. Influence of Temperature on Osmotic Pressure.—Pfeffer extended his observations so as to measure the influence of different temperatures on the osmotic pressure of the same solution in the same piece of apparatus and some of the results he obtained are given in the next table.

Table showing the influence of temperature on the intensity of osmotic pressure.

1	With temp.	14.2°C.	pressure=51 c. m.	but with temp.	32°C.	pressure=	54.4 c. m.
2	" "	6.8	" =50.5	" " "	13.7	"	= 52.5 "
3	" "	15.5	" =52.	" " "	36.0	"	= 56.7 "

In order to understand the relation of osmotic pressure to temperature it is necessary to state them in terms of degrees above absolute zero (32) rather than above the temperature at which water freezes. When the results are stated with reference to the absolute zero of temperature they stand as below:

1	With temp.	287.92°C	pressure=51 c. m.	but with temp.	305.72°C	pressure	=54.4 c.m.
2	" "	280.52	" =50.5	" " "	287.42	"	=52.5 "
3	" "	289.22	" =52.	" " "	309.92	"	=56.7 "

These observations are in harmony with others regarding plant growth which show that a low soil temperature may cause plants to wilt even in the night when evaporation from the leaf surface is small, while a high soil temperature may increase the root pressure to such an extent as to cause drops of water to form at the tips of leaves in a bright day.

62a. Osmosis and Diffusion in Plant Feeding.—If in a plant cell water is being used in the production of some substance such as starch, sugar or cellulose, the water molecules will be removed from solution and prevented from exercising pressure, thus causing a reduction of the osmotic water pressure in that cell; this will permit more water from the adjacent cells to be driven in to make good the loss.

So, too, if water is being lost by evaporation from the leaves, this loss will result in a reduced osmotic water pressure in the leaf cells which will permit the heavier pressure in the cells extending backward toward and to the root hair in the soil to force more water onward toward the leaves and thus maintain the flow of water by powerful osmotic pressure toward the leaves as long as evaporation continues.

When nourishment is being stored in the seed the substances in solution in the sap are being taken out and laid down in solid form, thus tending to maintain at that place a reduced osmotic pressure which permits the substance of that sort to be forced continually toward the place where the formation is going on. In this way the starch and other products are supposed to be brought from the leaves and stems to the seeds or places in stems, like the potato, where food products are being stored.

In the gathering of nitrogen from the nitrates in the soil water, too, the process would be the same. Wherever the nitrate is being transformed there its osmotic pressure would be falling and this permits more to be forced to the same point.

The so-called selective power of plants, whereby they obtain those substances dissolved in the soil water which they need, is thus explained. It should be understood that unless the molecules of a substance in solution are too large to pass from cell to cell through the walls these substances will do so until the solution inside the plant has a strength equal to that outside, but if this substance chances to be one which the plant does not use there will be no further concentration of that substance in the plant unless it be at places where evaporation is taking place.

If a poisonous principle exists in the soil water osmotic pressure will tend to force this substance into the plant tissues and the plant is helpless to prevent this entrance.

63. Dissociation of Salts in Solution.—There is a large class of substances which, when they go into solution in water, increase its electrical conductivity. It is also true that the osmotic pressure which they may develop is greater than can be explained on the basis of the number of molecules which were contained in the salt before its solution occurred. To account for both the greater electric conductivity and the higher osmotic pressure in such cases it has been assumed that, at the time of solution, more or less of the molecules dissolved separate into two groups, each of which may take part in developing osmotic pressure, making it greater than it could otherwise be.

When a very dilute solution of potassium nitrate, for example, is made, it is supposed that the molecules are broken into two groups, each of which may absorb heat energy and so strike a greater number of blows per unit of time against the confining membrane, and in this way produce a higher pressure. The two ions, as they are called, act like two hammers and each is able to absorb and deliver more energy when moving separately than when combined as a single but heavier hammer.

PHYSICS OF THE SOIL.

CHAPTER I.

NATURE, ORIGIN AND WASTE OF SOIL.

64. Nature of the Soil.—The great bulk of most soils is made up of small fragments of rock of various kinds, but nearly always there is associated with these varying amounts of organic matter derived from the breaking down of plant and animal tissue.

On the surface of the soil grains, too, there is always adhering more or less of substances which have been dissolved in the soil-water but which have been deposited again when the water was evaporated.

In most soils, but chiefly in the clayey types, there occurs some aluminium silicate having water combined with it, which is regarded as giving to them their sticky, plastic quality when wet. The amount of this material in a good soil is always small, seldom reaching more than 1.5 per cent., but the particles are so extremely minute that very little by weight has a marked effect upon its character.

65. Soils and Sub-soils.—In climates where the rainfall is sufficient for large crops it is common to speak of the surface few inches of rock fragments as the *soil* while that below is known as the *sub-soil*. The fundamental reason for making this distinction is found in the fact that the latter is less productive than the surface soil. So general is this difference in fertility that the term "dead-furrow" has been universally applied to the finishing of a land in plowing where the two furrows are thrown in opposite directions, leaving the sub-soil exposed, and where crops are always smaller. On the other hand, where two furrows are thrown together to form the "back-furrow" and the depth of soil increased crops are notably more vigorous.

We do not yet know just why a sub-soil when exposed to the surface is less productive than the true soil, but the difference seems in some way to be associated with the larger per cent. of the extremely minute particles which sub-soils contain.

In arid regions where the rainfall is not sufficient for crop production it seldom occurs that the deeper soil is markedly different in productiveness from that at the surface. Soil taken from the bottom of cellars and even from depths as great as 30 feet is found quite as productive when placed upon the surface as the top soil. So generally true is this that when it is desirable to level fields for purposes of irrigation in arid climates the soil from the higher places may be scraped to the lower levels without fear of lessening the productiveness of the fields.

66. Uses of Soil.—In the agricultural sense the most important use of soil is to act as a storehouse of moisture for the use of plants; and the productiveness of any soil is in a very large degree determined by the amount it can hold, by the manner in which it is held and by the readiness and completeness with which the plant growing in it is able to withdraw that water for its use as needed.

In the second place, the soil is a storehouse from which plants derive the ash ingredients of their food, the lime, the potash, phosphoric acid and other materials of this class, all of which are derived from the slow decay and solution of the soil grains.

Besides these the soil is a laboratory in which a great variety of microscopic forms of life are at work during the warm portions of the year, breaking down the dead organic matter of the soil, converting it into nitric acid and other forms available to higher plants, and the student must never forget that the magnitude of the crop taken from the field is always in proportion to the size of the crop developed by the micro-organisms in the soil.

Then again, the soil is a medium in which plants may place their roots in such a manner as to enable them to stand erect in the open sunshine and moving air currents above.

Finally, the soil is a means whereby the sunshine is changed into forms of energy available to the needs of soil organisms and the roots of plants and without which this life could not exist; for all of its movements must originate primarily from the sunshine altered in the soil or in the tissues of the plant above the soil.

67. Formation of Soil.—There are many agencies at work in the formation of soils and the processes of soil growth are in continuous operation day and night, winter and summer. Since all soil material originates from the breaking down of the various rock structures which make up the earth's surface all of the agencies which are operative in rock destruction may also contribute to soil formation.

68. Influence of Rock Texture on Soil Formation.—Nearly all kinds of rock are made up of fragments or crystals of various sizes and shapes and these are held together by interlocking, by some cementing material, or else by direct cohesion when extreme pressure has brought the grains close enough together to make this possible. It is seldom true, however, that the structure is so close or the cementing so complete as to make the rock impervious to water and the closest granite or the finest marble may absorb as much as .1 to .4 of a pound of water to 100 pounds of rock. If this water is changing it will dissolve away the cementing materials and the faces of the crystals themselves, making the rock still more open and the grains may even fall apart as is frequently observed in those cases known as "rotten stones."

The water may freeze in the stone and by its expansion cause it to crumble. Or again, when the sun shines on rocks made up of minerals of different kinds the crystals do not all expand at the same rate and this unequal expansion and contraction tends to loosen crystals and fragments, breaking the rock down, and thus form soil.

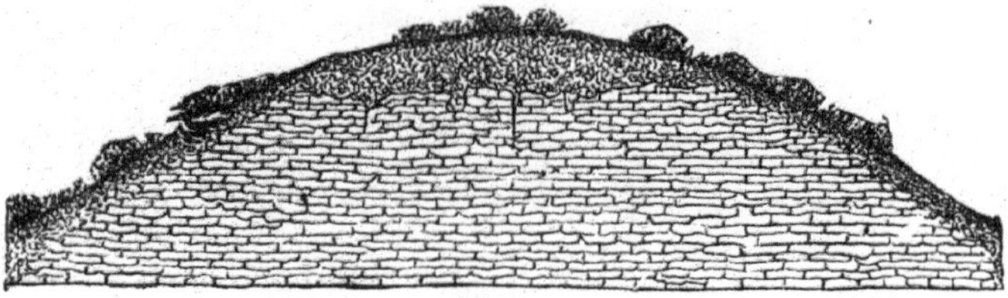

FIG. 8.—Section of limestone hill showing rock changing to soil. (After Chamberlin.)

69. Formation of Soil From Limestone.—If one will visit any limestone quarry where the soil and rock are exposed in section as represented in Figs. 8 and 9 it will be clearly seen how the rock is slowly converted into soil. In such cases as these, the water containing carbonic or other acids dissolves away the lime and magnesia, leaving the more insoluble portions of the lime rock to form the soil mantle which is left. These more insoluble portions are usually clay and very fine sand so that soils formed in this way are oftenest clayey soils, sometimes containing even less lime than other soils not derived from limestone.

FIG. 9.—Section of flat limestone surface showing rock changing to soil. (After Chamberlin.)

The mantle of soil seen above gravel beds in railroad cuts and where hills have been graded down on wagon roads has usually most of it originated from the decomposition of the gravel in place in the same manner as a soil from the limestone itself. So, too, in countries where granite and other crystalline rocks lie beneath the soil, these have been broken down and the over-lying soil derived from them.

70. Influence of Rock Fissures.—An examination of almost any quarry where considerable surfaces are exposed reveals the presence of systems of fissures which divide the stone layers into blocks of various sizes and at the same time provide easy avenues for the entrance of surface waters. These features are shown clearly in Figs. 10, 11, 12 and 13, and into them the roots of trees sometimes make their way where by expansion, due to growth, such strong pressures are devoloped as sometimes to throw down large blocks of stone. Then again, in cold climates these fissures may become filled with water which, when freezing, overturns and throws down many fragments, thus hastening their passage into soil.

FIG 10.—Fort Danger, Wis., showing rock fissures which lead to rock destruction. (After Chamberlin.)

FIG. 11.—Bee Bluff, Wis., showing rock fissures which lead to rock destruction. (After Chamberlin.)

71. Soil Removal.—It follows from what has been said that the same processes which result in soil formation must also contribute to its destruction in one place or removal to another. All are familiar with the creeping of soils from the brows of steep hill-sides toward their bases and out upon the more level plains which stretch away from them. These downward movements are caused by several agencies: (1) The beating of falling rain-drops and the carrying power of the streamlets which form as these gather together; (2) the expansion and contraction of the soil due to the alternate wetting and drying, there being less resistance to expansion downward than upward against gravity. These movements are analogous to those of the steel rails of the railroad which tend to creep down grade under the influence of changing temperature, which causes them to first lengthen and push down hill and then shorten and again draw downward because of less resistance in that direction. (3) Then, again, every disturbance of the soil produced by animals burrowing or walking up or down the hillside, tends usually to work the soil from higher to lower levels. Even the action of the wind is on the whole downward.

FIG. 12.—Giant's Castle, near Camp Douglas, Wis., showing cliffs of rock crumbling into soil. (After Chamberlin.)

72. Soils Produced by Running Water.—Rivers and streams are continually at work at this double process of soil building and soil removal. When one watches the bed of a stream as the water ripples over the uneven surface it is easy to note how rapidly soil and sand grains are being rolled and tumbled along the bottom. If it is desired to measure this rate of movement a shallow pan or box may be sunk in the bed of the stream, leaving its rim flush with the surface over which the water rolls. After a sufficient interval remove the box and dry and weigh the material collected.

Fig. 13.—Pillar Rock, Wis., showing rocky cliff in the last stages of decay. (After Chamberlin.)

At each bend in a stream soil is being taken from the concave side and carried onward toward the sea, while on the opposite side new soil is being formed from that dragged along the bottom. In this manner streams change their courses and wander from side to side across the valley, each time making a new soil on the side from which they are retreating and carrying away an older soil from the encroaching side. It is in this way that broad and flat river valleys are formed, with their terraces, such as are shown in Fig. 14. It is in this way, too, that the "ox-bows" of the Mississippi below Vicksburg were formed, some of which are represented in Fig. 15.

These abandoned river channels are at first long and narrow lakes but ultimately, with the repeated overflows of the stream, they become filled. Sometimes they remain for long intervals depressions in which swamp or humus soils develop.

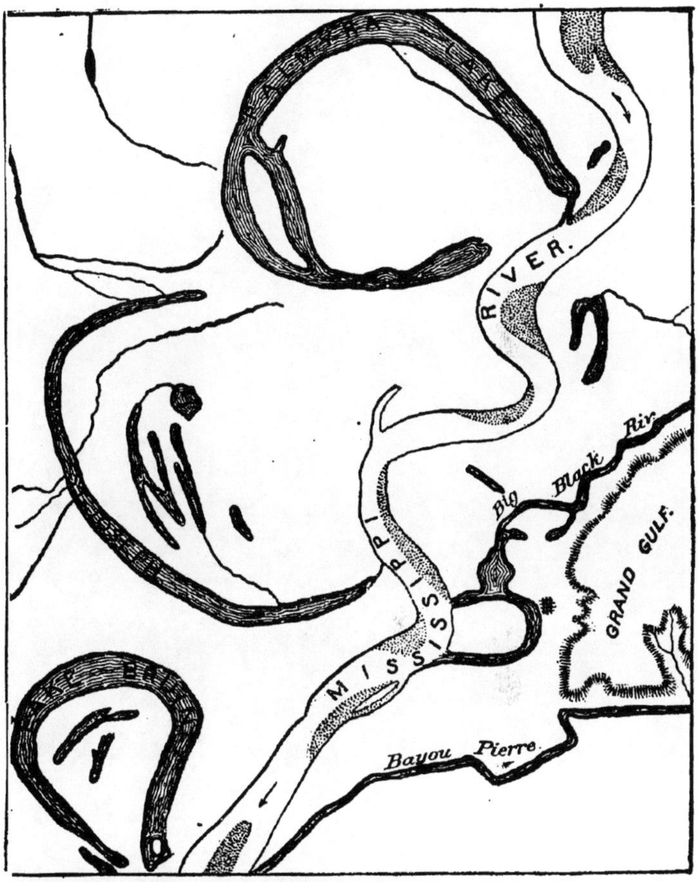

FIG. 15.—Showing the shifting of river channels, the formation of "ox-bows" and alluvial soils.

73. Glacial Soils.—In those portions of the world where the temperature is so low that most of the moisture falls as snow and where these snows do not all melt during the warm season there come to be such vast accumulations that the great weight compresses the snow into ice. So extensive and massive are these snow and ice fields in Green-

Fig. 16.— Showing a terminal moraine near Eagle, Wis.

FIG. 17.—Showing how glacial action has broken, ground up and accumulated rock fragments. Section of a terminal moraine near Whitewater, Wis.

land and in parts of Alaska today that they move over the face of the country much as a broad river would move, except at a much slower rate. The same type of phenomena occur, too, in the elevated mountain districts of Europe and in the Sierras of this country, the ice streams converging and flowing into the lower valleys in the form of glaciers. As these ice streams move over the uneven surface of their valleys and crowd against their sides, the rocks, gravel and sand taken up by the moving ice act with great effectiveness to abraid into soil the rigid rock surfaces over which they move.

FIG. 18.—Showing rock surface over which glaciers have passed, scratching and polishing it.

In a recent geological epoch the whole of the North American continent north of the Ohio and Missouri rivers and much of northern Europe and Siberia were under enormous moving ice sheets which resulted in the formation of the extensive glacial soils of these countries; consisting largely of a rock flour ground to varying degrees of fineness, and naturally very fertile where the materials have not been sorted by the waters from the melting ice in such a way as to form siliceous sandy plains. Figs. 16, 17, 18 and 19 are views illustrating different phases of soil formation by glacial action.

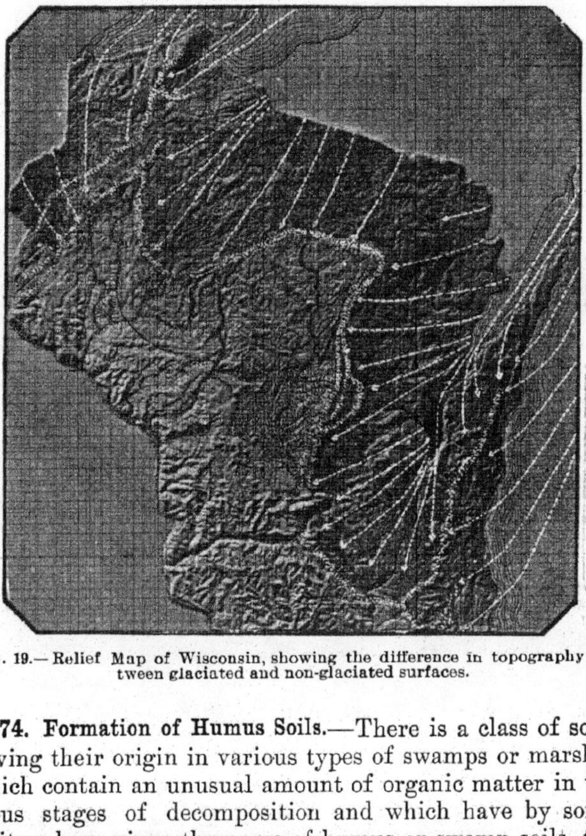

FIG. 19.—Relief Map of Wisconsin, showing the difference in topography between glaciated and non-glaciated surfaces.

74. Formation of Humus Soils.—There is a class of soils having their origin in various types of swamps or marshes which contain an unusual amount of organic matter in various stages of decomposition and which have by some writers been given the name of humus or swamp soils, the former name referring to the large amount of humus these soils contain and the latter to the physical conditions under which they have been formed.

In many places in the higher latitudes and at considerable elevations nearer the equator where the surface is too flat for ready drainage, and where the winter snows remain so long upon the ground that the summer is too short to permit the soil to become dry enough to allow the air to penetrate deeply and freely, the organic matter accumulates and soils are formed containing a large proportion of humus; even beds of peat may develop.

Fig. 20.—Showing a method of formation of peat marshes and swamp soils about lakes.

Under other conditions, where rivers approach their outlet across a very flat country and are no longer able to scour their channels and keep them clean, the moving sediment finally raises the banks and the bed until the water is flowing above the surrounding country. Under these conditions with a continual seepage and frequent overflows swamps are developed in which marsh vegetation grows luxuriantly and, falling under conditions where free oxidation cannot occur, the remains only partially decay, giving rise to beds of peat and rich humus soils.

In other cases, where a river often shifts its course and especially where the cut-offs or ox-bows illustrated in Fig. 15 are formed, these places, with the poor drainage which they must have and with the occasional overflows to keep the cut-offs filled with water, are maintained wet long and continuously enough to allow humus soils to form.

With the final withdrawal of the great ice sheet from the glaciated parts of America and Europe there were left large numbers of shallow lakes whose flat margins were wet enough to support marsh vegetation and very often this vegetation came to form a floating fringe steadily encroaching upon the lake in the manner represented in Fig. 20. As the vegetation continued to grow and die the fringe became heavier and sank more deeply in the water until finally the whole lake was overgrown and until the organic matter, together with the sediments brought down by the rains and the winds and washed in from the surrounding higher land, became so heavy and so thick as to rest upon the bottom of the lake, converting it into a marsh of peat or humus soil.

On the margins of larger lakes and especially along the seashore, sand bars or reefs are thrown up behind which bodies of water are shut off and in these organic matter may accumulate in the same manner as that just described, giving rise to the same type of soils.

In still other cases, on the margins of the sea bottom, there flourishes a peculiar type of vegetation known as eel grass, which lives always beneath the water at low tide in a position represented in Fig. 21. These grasses offer a natural obstruction to the incoming and outgoing tidal waters, causing them to throw down their sediments and thus build up the sea floor with silt containing large amounts of organic matter under conditions unfavorable to rapid decay. As the sea floor rises in this way above low tide level the eel grass dies and another type of swamp vegetation takes its place, as between a and b in the figure, and here again the formation of humus soil is continued under somewhat different conditions.

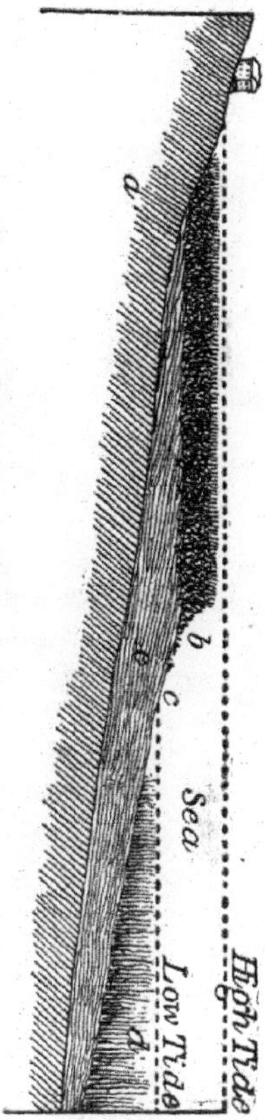

Fig. 21.—Showing a mode of formation of humus soils on the borders of the sea.

75. Wind-Formed Soils.—The wind moving continuously over the face of the land is now and long has been a potent factor in soil removal and soil building. Indeed, it is probable that nowhere can soils be found which do not contain many wind-borne particles. Every raindrop which falls and every snowflake, however white, brings to the field upon which it falls one or more particles of soil which has been drifting in the higher air from unknown distances.

The drifting of dust from roads during dry times and from fields in the spring are strong reminders of the potency of wind action at times, but it is the less evident but continuous action that counts most in the long run and, were it not for the steady wearing away and rearrangement of the soil surface, wind-formed soils would be much more evident and general than they are.

On the leeward margins of arid regions and on sandy coasts the building and eroding power of the wind becomes most evident, and the most extensive deposits which have been assigned to this cause are the loess beds of China which have great horizontal extent and in some places depths reaching even 1,200 and 2,000 feet. These deposits have been described by Richthofen as having been formed from dust accumulations drifted by the prevailing winds from the high desert plateaus of Central Asia.

In Europe, and in this country in the Mississippi valley, there are deposits of a similar character. They are distributed along the border of a former ice sheet of the glacial period and from thence they spread down the main streams, along the Mississippi from Minnesota to near the Gulf, along the Missouri from Dakota to its mouth, and along both the Illinois and the Wabash. These deposits are thickest, most typical and coarsest along the bluffs nearest to the streams and they thin out and become finer as the distance back increases. It is thought that the fine silts borne along by the waters of the glacial streams in times of high water were spread out over broad flats and as the waters withdrew they were left to dry in the sun and then picked up by the winds and drifted away. The loess soils are almost always extremely fertile and very enduring.

76. The Work of Animals as Soil Producers.—There are many animals which have contributed largely to the formation of soil through a grinding of pebbles and the coarser sand and soil grains into finer materials.

Darwin, through a long and careful study, reached the conclusion that in many parts of England earthworms pass more than 10 tons of dry earth per acre through their bodies annually and that the grains of sand and bits of flint in these earths are partly worn to fine silt by the muscular action of the gizzards of these animals. Their method of action in moving through the soil is this: They eat a narrow hole, swallowing the earth, when the point of the head is held fast in the excavation while an enlarged portion of the œsophagus or swallow is drawn forward, forcing the cheeks outward in all directions, thus crowding the soil aside and making the opening wider, when more dirt is eaten and the operation repeated, allowing the animal to advance through the soil.

Domestic fowls and all seed-eating birds, in picking up pebbles for service in grinding their food, do the same sort of work as the earthworms in producing fine soil, as every housewife can testify from the worn condition of bits of glass and pottery taken from the gizzard of the chicken.

77. Soil Convection.—There is another very important line of work done by earthworms, ants and all burrowing animals, in bringing the sub-soil to the surface and carrying the surface soil into the ground, thus maintaining a sort of soil-convection which, in effect, amounts to the

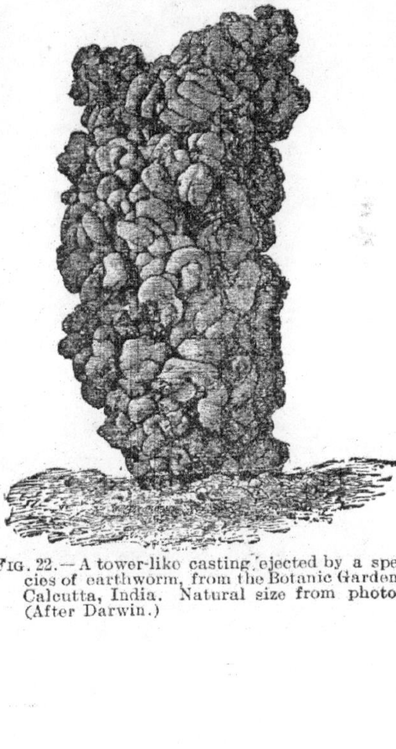

FIG. 22.—A tower-like casting, ejected by a species of earthworm, from the Botanic Garden, Calcutta, India. Natural size from photo. (After Darwin.)

FIG. 23.—Showing the work of the common earth worm during a single night after a heavy rain.

same thing as plowing except that its influence extends much deeper.

Both earthworms and ants often burrow in the ground to a depth of four feet, and in some cases more than nine, bringing the material to the surface and forming passageways down which the rains may wash the finer surface soil. Fig. 22 shows a single pile of earth cast up by an earthworm in the Botanic Gardens of Calcutta, and Fig. 23 shows the work of our common earthworm during a single night in bringing up soil after a rain.

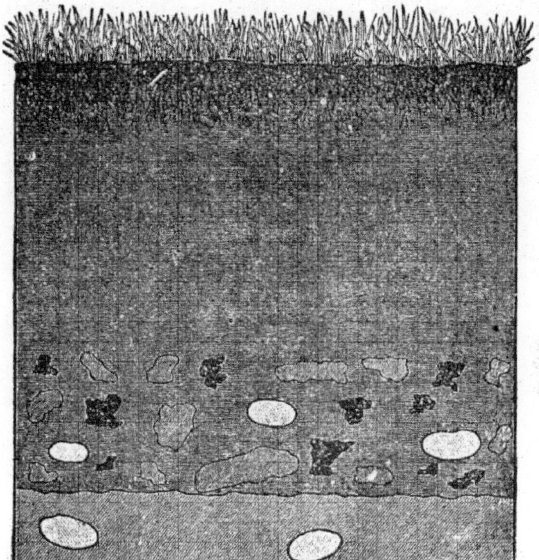

FIG. 24.—Section of vegetable mould in a field drained and reclaimed 15 years before; showing turf, vegetable moulds without stones, mould with fragments of burnt marl, coal cinders and quartz pebbles buried under the influence of earthworms. One-third natural size. (After Darwin.)

This frequent bringing of earth to the surface tends to bury objects and gradually to lower them into the ground, and Fig. 24 represents the results of one of Darwin's studies, showing the amount of soil which has accumulated above bits of burnt marl, cinders and pebbles during 15 years, largely through this action of earthworms and ants in bringing to the surface portions of the subsoil. It will be seen that the amount accumulated is more than three inches, or at the rate of an inch in 5 years.

CHAPTER II.

CHEMICAL AND MINERAL NATURE OF SOILS.

73. Unsatisfactory State of Present Knowledge.—It is now pretty generally conceded that the capacity of a soil to feed crops of a given kind cannot be foretold with much certainty from the results of chemical analyses as it has been the custom to make and present them. It has been found, for example, in the arid west, that soils notably deficient in humic nitrogen and which for this reason should be comparatively unproductive, have, nevertheless, been found capable of giving large yields when irrigated. Then again, in moist climates there are types of soil exceptionally rich in both humic and nitric nitrogen which are comparatively unproductive until they are given dressings of coarse farmyard manure. The analyst would place them among the richest of soils and yet they are among the poorest until given farmyard manure; and, what appears stranger still, straw and coarse litter may be much more beneficial to them than liquids from the stable cistern.

79. Essential Constituents of a Fertile Soil.—While it is true that our chemical knowledge of soils is very unsatisfactory, it has nevertheless been thoroughly established that a fertile soil *must* contain certain substances in order to permit any crop to come to maturity upon it and these are potassium, calcium, magnesium, phosphorus, sulphur, iron, nitrogen and probably chlorine. Let any one of these elements be absent from a soil, or its moisture, and crops fail to develop upon it. It has not, however, been established yet in what form of combination these elements must or may exist nor in what proportions to give the best results. It is known that they do not exist in the soil in the elementary form and that they are combined in a great variety of ways.

Furthermore, from these combinations, under favorable conditions, plants are able to supply their needs.

80. Functions of the Essential Plant Foods.—From the standpoint of plant physiology it is again unfortunate that little has yet been positively demonstrated regarding the part played by each of the essential elements of plant food taken through the soil and soil moisture. It is known that nitrogen is an essential constituent of the protein compounds of living tissues, and that to most of the cultivated crops it becomes available in the form of nitric acid or of a nitrate of lime, magnesia, potash or some other base. Potassium does not appear as an essential ingredient of plant tissues or of its storage products like starch or gluten, but Nobbe, Schroeder and Erdmann have shown that when Japanese buckwheat, placed in nutritive solutions entirely free from potash salts, after a few weeks' growth came to a standstill and that all organs of the plant came to be nearly or quite free from starch; but when a potassium salt was added to the solution starch began to develop and growth became normal.

In regard to phosphorus the clearest indications go to suggest that it is usually taken into the plant in the form of phosphates and, because its compounds are often associated with the soluble albuminoids, that it assists in some way in the transfer of these from place to place in the plant.

Some compound of iron must exist in soil solutions and must enter the plant before the normal development of the green coloring matter, chlorophyll, can take place; so extremely small quantities, however, are needed that no soil is ever lacking in sufficient available forms.

Sulphur is apparently largely if not wholly taken into the plant in the form of sulphates, and these are thought to be decomposed by the oxalic acid, setting the sulphuric acid free, which is then broken down and the sulphur appropriated to enter as an essential constituent of the albuminoid compounds.

But little is known of the part played in plant life by

the salts of magnesium except that they must be present in the seed.

The action of lime is held to be medicinal, its function being to neutralize the poisonous oxalic acid liberated as an intermediate product in the oxidation of carbohydrates.

Large amounts of silica and alumina and smaller amounts of many other substances are found in the ash of plants but their presence there is regarded as accidental, growing out of the simple fact that they chanced to be dissolved in the soil-water and passed into the tissues with it during growth.

81. Chemical Composition of Soils.—From what has been said regarding the origin of soils and the manner in which their particles have been moved from place to place, it is evident that there must necessarily be a strong similarity among them, of both chemical and mineral composition, wherever found. It has been customary in analyzing soils to digest a certain weight of dry soil for a stated time in a certain strength of hot hydrochloric acid and to examine the solution for the compounds it might contain, calling the part not dissolved the *insoluble residue.* The tables on pages 74–75 show the results of some of these analyses, taken from the papers of Hilgard in the Tenth Census of the United States.

82. Chemical Difference Between Clayey and Sandy Soils. —Studying the table of clayey and sandy soils it will be noted that out of every 100 pounds of the clayey soil there were, as an average, 31.791 pounds which dissolved in hot hydrochloric acid, while only 6.79 pounds were soluble in like weight of the sandy soil. In other words, a quarter of the weight of the clayey soils more than of the sandy soils is soluble in a unit of time in hot hydrochloric acid. There is about 2.5 times as much potash and organic matter, nearly twice as much phosphoric acid, 7 times as much lime, 9 times as much magnesia and 1.4 times as much sulphuric acid in the clayey as in the sandy soil, which may be dissolved out in equal times by the solvent used.

These ratios, however, are sometimes a long ways from

true when single cases are compared, and this is shown in a striking manner in the single case of clay soil given below the line of averages in the table of sandy and clayey soils. This is described by Hilgard as a fair upland soil yielding 700 to 800 pounds of cotton per acre, gray in color, not heavy, 6 to 8 inches deep, and underlaid by a subsoil quite heavy in tillage and dark orange in color; and yet its insoluble residue is about 91 per cent. and there are two of the sandy soils where the per cents. are 90 and 92 respectively, showing that the two are more nearly alike chemically than they are physically.

83. Observed Chemical Differences, Partly Due to Differences in Amount of Soil Surface.—It is a common experience that the more finely a substance is subdivided the more rapidly will it dissolve. Fine salt and powdered sugar, for example, dissolve much more rapidly in water than the coarser grained varieties do. In the clay soils the particles have a much smaller diameter than they do in the sandy soils and hence the number of grains in a given weight of soil will be much larger, but the number of grains cannot be increased without also increasing the surface upon which the solvent may act, and hence with the same strength and amount of acid, for equal weights of the coarse and fine grained soil, having exactly the same chemical composition, there should be dissolved in equal times a larger per cent. of the soil having the largest amount of surface. The sandy soils therefore are not likely to be as different from the clayey ones as the table of analyses indicate.

84. The Chemical Differences Between Soils and Their Subsoils.—In humid climates there is usually a marked difference in the producing capacity of the soils and their subsoils as was pointed out in (**65**), and a study of the table of subsoils, pp. 74, 75, will show that there is a chemical difference also. It will be seen that the surface soils contain more lime, phosphoric acid and organic matter, less soluble silica, alumina and iron and about the same amounts of potash, magnesia and sulphuric acid.

85. Comparison Between Clay Soils and Swamp Soils.—If a comparison is made between the clayey soils, which are generally productive naturally, and the humus soils it will be seen that the latter contain about twice as much potash, magnesia, sulphuric acid and organic matter, six times as much lime and a little more phosphoric acid, and yet for some reason the humus soils, when well drained, may not naturally be as productive as the clay soils are and here is where the present methods of soil analysis fail to tell the whole truth.

86. Comparison Between Clayey Soils and Loess Soils.—The loess soils do not show a much larger percentage amount of the essential ingredients of plant food than do the clayey ones. Indeed there is less of organic matter and only a little more of potash, phosphoric and sulphuric acids. The chief and great difference lies in the large amount of lime and magnesia which they contain, the first being more than 9, and the latter more than 8 times as large. If it is true that these soils are largely wind-formed it is to be expected that these two substances would appear at the surface to be taken up by the winds more than any other of the essential ingredients, first, because they are comparatively soluble and hence likely to be brought up by the capillary waters and left after evaporation where the wind has free access to them; and second, because they are not so soluble as to be completely dissolved by the heavy rains and carried back into the ground again.

87. Difference Between Arid and Humid Soils.—The soils which have accumulated in the arid climates of the world are quite markedly different from those of the more humid portions, both in physical and chemical properties. The per cents. given in the table of arid and humid soils are those of Hilgard and are averages of 466 analyses from humid climates and 313 from arid.

It will be seen that the arid soils contain more than 3 times as much potash, nearly 13 times as much lime and 6

Chemical composition of soils.

Essential ingredients in per cent. of dry soil.

POTASH.		LIME.		MAGNESIA.		PHOSPHORIC ACID.		SULPHURIC ACID.		WATER AND ORGANIC MATTER.	
Sand.	Clay.	Sand.	Clay.	Sand.	Clay.	Sand.	Clay.	Sand.	Clay.	Sand.	Clay.
.100	.416	.120	.080	.040	.691	.051	.103	.028	.061	2.055	1.906
.156	.176	.081	.090	.069	.112	.101	.071	.037	.055	2.642	8.891
.045	.186	.064	.071	.005	.063	.066	.204	.091	.285	2.422	8.953
.117	.134	.058	.219	.012	.289	.092	.069	.058	.035	1.807	8.309
.110	.242	.090	.387	.025	.508	.191	.071	.105	.055	3.477	6.843
.067	.032	.119	.036	.090	.070	.111	.082	.054	.054	2.881	6.167
.275	.431	.055	.540	.048	.838	.105	.187	.024	.009	3.682	6.922
.095	1.104	.076	1.319	.083	1.665	.049	.304	.015	.024	2.354	7.369
.209	.150	.141	3.054	.031	.029	.103	.212	.046	.089	3.113	4.962
.034	.253	.045	.310	.013	.296	.014	.079	.035	.079	1.636	4.962
.121	.319	.085	.617	.018	.456	.087	.141	.055	.075	2.607	6.528
	.137		.173		.203		.038				3.394

SWAMP AND LOESS SOILS.

Humus	Loess	Humus	Loess	Humus	Loess	Humus.	Loess	Humus.	Loess.	Humus.	Loess.
.639	.435	3.780	5.82[illegible]	.886	3.60[illegible]	.150	.200	.148	.090	13.943	1.205

SOILS COMPARED WITH THEIR SUB-SOILS.

	Potash Sand	Potash Clay.	Lime Sand	Lime Clay.	Magnesia Sand	Magnesia Clay.	Phosphoric Acid Sand	Phosphoric Acid Clay	Sulphuric Acid Sand.	Sulphuric Acid Clay.	Water and Organic Matter Sand.	Water and Organic Matter Clay.
SOILS.	.157	.214	.115	1.761	.076	.182	.128	.207	052	.000	2.853	6.014
SUB-SOILS.	.143	.344	.096	1.481	.073	.240	.124	.150	.060	.071	1.943	4.780
	+.014	−.130	+.019	+.280	+.003	−.058	+.004	+.018	−.008	+.019	+.910	+1.234

ARID AND HUMID SOILS COMPARED.

Potash Humid.	Potash Arid.	Lime Humid.	Lime Arid.	Magnesia Humid	Magnesia Arid.	Phosphoric Acid Humid.	Phosphoric Acid Arid.	Sulphuric Acid Humid.	Sulphuric Acid Arid.	Water and Organic Matter Humid.	Water and Organic Matter Arid.
.216	.729	.108	1.362	.225	1.411	.113	.117	.052	.041	3.644	4.945

Chemical composition of soils.

Inert ingredients in per cent. of dry soil.

Insoluble Residue.		Soluble Silica.		Soda.		Brown Oxide of Manganese.		Peroxide of Iron.		Alumina.	
Sand.	Clay.	Sand	Clay.	Sand	Clay.	Sand	Clay.	Sand	Clay.	Sand.	Clay
93.630	72.746	1.682	8.926	.060	.112	.102	.106	.761	12.406	1.532	2.473
94.770	73.690	.486	3.370	.069	.004	.156	.146	.706	5.989	.7[illegible]3	7.305
93.362	60.3[illegible]0	1.721	2.000	.018	.119	.220	.196	.941	9.709	1.339	18.066
95.630	73.422	.879	2.703	.064	trace	.049	.164	.224	4.054	.473	10.598
92.090	63.444	1.220	11.325	.035	.079	.126	.052	.963	3.894	1.959	13.454
90.230	77.860	1.940	1.790	.009	.011	.313	.056	1.927	5.646	2.141	7.538
90.6[illegible]1	54.565	1.885	13.219	.130	.277	.172	.079	1.837	7.089	1.436	16.071
92.460	51.063	1.550	20.704	.036	.325	.040	.119	.843	5.818	2.649	10.539
94.428	79.580	.529	3.628	.069	.065	.101	.195	.661	8.420	1.195	4.988
94.822	75.350	1.037	7.310	.022	.258	.020	.038	.930	5.784	1.576	5.567
93.210	68.203	1.293	7.498	.051	.128	.130	.115	.979	6.381	1.503	9.660
	91.498		1.722		.054		.066		1.372		1.522

SWAMP AND LOESS SOILS.

Humus.	Loess.	Humus.	Loess.	Humus.	Loess	Humus.	Loess	Humus.	Loess	Humus.	Loess
85.886	68.853	20.825	4.918	.109	.165	.098	.164	7.040	3.569	14.476	2.812

SOILS COMPARED WITH THEIR SUB-SOILS.

SOILS.

Sand.	Clay.	Sand	Clay.	Sand	Clay.	Sand	Clay.	Sand	Clay.	Sand.	Clay.
93.222	73.978	1.019	5.034	.072	.085	.124	.133	1.162	5.205	1.145	6.908

SUB-SOILS.

90.714	66.290	2.212	7.446	.064	.085	.080	.125	1.739	6.947	2.276	12.086
+2.508	+7.688	-1.193	−2.412	+.008	.000	+.044	+.008	−.577	-1.742	−1.131	-5.088

ARID AND HUMID SOILS COMPARED.

Humid.	Arid.	Humid.	Arid.	Humid.	Arid.	Humid.	Arid.	Humid.	Arid.	Humid.	Arid.
[illegible]4.031	70.565	4.212	7.266	.091	.264	.133	.059	3.131	5.752	4.296	7.888

times as much magnesia as do the humid soils with which they have been compared. They also contain some more of each of the other essential plant foods except sulphur, the sulphuric acid being less.

If, however, a comparison is made between the arid soils and the mean of the 10 clay soils given in the first table, it will be seen that, excepting potash, lime and magnesia, these contain more of the essential ingredients of plant food than do the arid soils, and so, too, there is more soluble silica.

88. Humus.—It is this product in the soil which gives to it usually its dark color, but so far as its chemical composition is concerned its nature is not yet well understood. It is a very important ingredient of fertile soils and is the product of decaying organic matter.

In torrid climates where the soil is warm the whole year and in arid regions where the soil is more open on account of deficient moisture as well as on sandy soils wherever found, the rate of complete decay is so rapid that the amount of humus is generally relatively small; but in temperate climates, where the soil is damp, its texture close and rains frequent, the organic matter decays more slowly and the amount of humus in the soil is relatively greater.

The great importance of humus in agricultural soils is found in the fact that it is relatively insoluble under good field conditions and does not leach away and in this form becomes the food of niter-forming germs which convert it by degrees into nitric acid, as one of their waste products, but the essential form of nitrogen for the food of most higher plants. A soil entirely devoid of humus must necessarily be manured or given nitrogen in some other form in order to make it fertile.

89. Difference Between the Humus of Arid and Humid Climates.—Hilgard and Jaffa have made the important discovery that the humus of arid soils is relatively richer in nitrogen than is that of humid soils and hence that smaller amounts of it will meet the needs of niter-forming germs and thus allow large crops to be produced where, with a poor form of humus, this would be impossible.

The results of their studies in this line are stated in the table below:

	No. of samples.	Humus in soil.	Nitrogen in humus	Humic nitrogen in soil.
		Per cent.	Per cent.	Per cent.
Arid soils	18	.75	15 87	.101
Semi-arid soils	8	.99	10.03	.102
Humid soils	8	3.04	5.24	.132

In speaking of these results they say, "It thus appears that, on the average, the humus of the arid soils contains three times as much nitrogen as that of the humid, that in the extreme cases the nitrogen percentages in the arid humus actually exceeds that of the albuminoid group, the flesh-forming substances."

"It thus becomes intelligible that in the arid region a humus percentage, which, under humid conditions, would justly be considered entirely inadequate for the success of normal crops, may, nevertheless, suffice even for the more exacting crops. This is more clearly seen on inspection of the figures in the third column, which represent the product resulting from the multiplication of the humus percentages of the soil into the nitrogen of the humus."

90. Chemical Composition of Soils Compared With the Rock from Which They Are Derived.—When a soil accumulates in place from slow decomposition of the underlying rock there is sometimes a close resemblance in chemical composition between the rock and the derived soil, but in other cases there is little resemblance between them. If the rock is made up of a large percentage of relatively soluble materials, as is the case with most limestones, then the solvent power of water, combined with the effects of leaching, tend to cause a concentration of the relatively insoluble ingredients, thus giving rise to a soil very different in chemical composition from the parent rock.

If, on the other hand, the rock is made up of minerals of nearly equal solubilities, or if in any way the soil results from a mechanical breaking up of the rock, then the soil may have much the same relative amounts of ingredients as the parent rock shows. In the table which follows are given the composition of some rocks and of soils derived directly from them:

Composition of rocks and residual soils.[1]

	Trenton Limestone		Bermuda Limestone		Gneiss.		Granite.		Diorite.	
	Rock	Soil.	Rock	Soil.	Rock	Soil.	Rock	Soil.	Rock	Soil.
	Pr ct.	Pr ct.	Pr ct.	Pr ct.	Pr ct.	Pr ct.	Pr ct.	Pr ct.	Pr ct.	Pr ct.
Silica (SiO_2)	.44	43.07	.032	45 16	60.69	45.31	69.33	65.69	46.75	42.44
Alumina (Al_2O_3)	.042	25.07	0 54	15 473	16.89	26 55	14.33	15.23	17.61	25.51
Ferric oxide.....	......	15.16	......	13.898	9.16	12.18	3.60	4.39	16.79	19.20
Lime (CaO)	34.77	0 63	54 496	3.948	4 44	tr.	3.21	2 63	9.46	0.37
Magnesia (MgO)	tr.	0.03	1.751	0.539	1.06	0 40	2.44	2.64	5.12	0 21
Potash (K_2O) ...	not d.	2.50	0 068	0.133	4.25	1.10	2.67	2.00	0.55	0.49
Soda (Na_2O)....	not d.	1.20	0.252	0.007	2.42	0.22	2.70	2.12	2.56	0 56
Carbon dioxide..	42.72	tr.	44.251	2.533	......	0.00	......	......	0.00	0.00
Phos. acid (P_2O_5)	......	......	......	......	......	0.47	0.10	0.06	0.25	0.29
Water and volatile products ..	1.03	12.98	.328	18.265	.62	13 75	11.22	4.70	0.92	10.93

The two limestones, it will be seen, have given rise to a soil containing almost as much silica, alumina and iron oxide combined as is contained in the three soils from the other three kinds of rock, the per cents. standing, in round numbers, 83, 75, 84, 85 and 87. In other words there is a strong tendency to bring all soils approximately to one composition. Indeed it may be said that in any soil the essential ingredients of plant food make up but from 3 to 8 per cent. of the total dry weight. It will be observed that in the case of the soil derived from the Bermuda limestone, not less than 98 pounds of every 100 pounds of rock

[1] Rocks, Rock Weathering and Soils. Merrill.

are dissolved and carried away by the water for each 2 pounds of soil formed, the chief materials carried away being the lime, magnesia and carbon dioxide.

91. Amount of Essential Plant Food Removed from the Soil by Crops.—It is very important, in the management of soils, to know something of the draught upon them which crops of different kinds make, and in the table which follows is given the amount of materials removed from the soil in 1,000 pounds of fresh or air-dried product.

Table showing lbs. of plant food in 1000 lbs. of air-dried product.

(WOLFF.)

	MAIZE.		OATS.		WINT'R WHEAT		SPRING WHEAT		WINT'R RYE.		BARLEY		RED CLOVER	
	Straw.	Grain.	Straw.	Grain.	Straw.	Grain.	Straw.	Grain.	Straw.	Grain.	Straw.	Grain.	Straw.	Grain.
Total ash........	45.3	12.4	61.6	26.7	46.0	16.8	38.1	18.3	38.2	17.9	45.9	22.3	57.6	38.3
Potash (K_2O) ...	16.4	3.7	16.3	4.8	6.3	5.2	11.6	5.6	8.6	5.8	10.7	4.7	18.6	13.5
Soda (Na_2O) ...	.5	0.1	2.0	1.0	0.6	0.3	1.0	0.3	0.7	0.3	1.6	0.5	1.1	0.4
Magnesia (MgO)	2.6	1.9	2.3	1.9	1.1	2.0	0.9	2.2	1.2	2.0	1.2	2.0	6.3	4.9
Lime (CaO)	4.9	0.3	4.3	1.0	2.7	0.5	2.6	0.5	3.1	0.5	3.3	0.6	20.1	2.5
Phos acid (P_2O_5)	3.8	5.7	2.8	6.8	2.2	7.9	2.0	9.0	2.5	8.5	1.9	7.8	5.6	14.5
Sul. acid (SO_3)..	2.4	0.1	2.0	0.5	1.1	0.1	1.2	0.2	1.6	0.2	1.8	0.4	1.9	0.9
Sulphur	3.9	1.2	1.7	1.7	1.6	1.5	....	....	0.9	1.7	1.3	1.4	2.1	....
Nitrogen..........	4.8	16.0	5.6	17.6	4.8	20.8	5.6	20.5	4.0	17.6	6.4	16.0	19.7	30.5

From this table it appears that each ton of clover hay withdraws from the soil 39.4 lbs. of nitrogen; 37.2 lbs. of potash; 12.6 lbs. of magnesia; 40.2 lbs. of lime; 11.2 lbs. of phosphoric acid; and 14.2 lbs of sulphuric acid, making an aggregate of ash ingredients alone of 154.8 lbs.

92. Amount of Plant Food in an Acre-foot of Soil.—If we take 4,000,000 pounds as the dry weight of an acre-foot of all soils, except the humus and that at 2,000,000 (**149**), and the percentages of essential plant food given in the tables on pages 74 and 75, the amount of plant food per acre-foot may then be computed, giving the results in the table below:

Table giving the tons of essential plant food per acre-foot of different types of soil.

	Sandy soil.	Clay soil.	Loess soil.	Humus soil.
	Tons.	Tons.	Tons.	Tons.
Potash (K_2O)	2.42	6.38	8.70	6.39
Lime (CaO)	1.70	12.34	116.40	37.86
Magnesia (MgO)	.96	9.12	73.84	8.68
Phosphoric acid (P_2O_5)	1.74	2.82	4.00	1.60
Sulphuric acid (SO_3)	1.10	1.50	1.80	1.48

From this table it appears that the amount of plant food per acre-foot of field soils, not including nitrogen, ranges from about 2 to 8 tons of potash, 2 to 116 tons of lime, 1 to 73 tons of magnesia, 2 to 4 tons of phosphoric acid, and 1 to 2 tons of sulphuric acid.

93. Number of Crops Required to Remove the Plant Food of an Acre-foot of Soil.—The ratio of dry weight of the kernels to that of the straw and chaff in a crop of wheat has been found to be as 1 to 1.1 in a dry season, but to be as high as 1 to 1.5 when there has not been an undesirable stimulation to the growth of straw. Taking this ratio of 1 to 1.5, a yield of 40 bushels of wheat per acre would mean a crop of 2,400 lbs. of grain and 3,600 lbs. of straw. From these two figures, the data in the table of (**91**) and that of (**92**), it is possible to compute the number of crops of wheat yielding 40 bushels per acre which would remove the amount of plant food in an acre-foot of one of the several types of soil represented in the table of (**92**). Solving the problem for the potash in the clay soil the case would be

$$\frac{6.38 \times 2,000}{(2.4 \times 5.2) + (3.6 \times 6.3)} = 362.9$$

where 6.38 is the tons of potash per acre-foot,
2,000 is the number of lbs. in one ton,
2.4 is the number of 1,000 lbs. of grain in 40 bush. of wheat,
5.2 is the number of lbs. of potash per 1,000 lbs. of grain,
3.6 is the number of 1,000 lbs. of straw with 40 bush. of wheat
6.3 is the number of pounds of potash per 1,000 lbs. of straw,
362.9 is the number of crops of wheat.

When the problem is solved for each of the essential plant foods used by the wheat crop, the results will stand for the clay soil as given below:

Potash enough for 363 crops of wheat of 40 bush. per acre.
Magnesia enough for 2,082 crops of wheat of 40 bush. per acre.
Lime enough for 2,260 crops of wheat of 40 bush. per acre.
Phosphoric acid enough for 210 crops of wheat of 40 bush. per acre.
Sulphuric acid enough for 108 crops of wheat of 40 bush. per acre.
Nitrogen enough for 78.5 crops of wheat of 40 bush. per acre.

In computing the nitrogen in the soil for this table .132 per cent., from the table in (89), was taken and the same weight of soil, 4,000,000 pounds per acre-foot as used for the other plant foods.

It has been assumed that 40 bushels of grain and 3,600 pounds of straw per acre are taken from the ground each crop and that nothing is returned to the soil, and yet chemical analyses would indicate that there is enough of everything but nitrogen for more than a century of cropping, and this is saying nothing regarding the plant food which is known to exist in the second, third and fourth feet of soil in which the roots of plants regularly feed. Plainly we have very important knowledge yet to discover regarding the feeding of plants from the soil.

94. Experiments at Rothamstead.—The classic experiments which have been made by Sir J. B. Lawes and his associates regarding the conditions which determine the fertility of the soil, have thrown much needed light upon this problem. By growing the same crop year after year on the same ground to which no nitrogen-bearing manures were applied, they learned that when fertilizers containing the essential ash ingredients of the plant were added to the soil larger yields and more nitrogen could be taken from the ground.

They found that when wheat grown continuously for 32 years on the same soil without manure of any sort could obtain but 20.7 lbs. of nitrogen per acre, the same crop on adjacent and similar land given fertilizers without nitrogen could gather 22.1 lbs. or 6.76 per cent. more. Barley, which, with no fertilizers, during 24 years could gather but 18.3 lbs. per acre per annum, did, when aided with other ash ingredients, remove from the soil 22.4 lbs. of nitrogen per acre. Beans, which gathered from untreated land 31.3 lbs. of nitrogen per acre during 24 years, took off from the land under the other treatment 45.5 lbs. per acre. So, too, in a rotation of crops, 7 courses in 28 years, no fertilizers gave 36.8 lbs. of nitrogen, while with superphosphate of lime the yield was 45.2 lbs. per acre. Again in the mixed herbage of grass land 20 years without fertilizers gave 33 lbs. of nitrogen per acre, but where mixed mineral fertilizers containing potash were given the yield was 55.6 lbs. of nitrogen per acre.

95. Store of Nitrogen in the Soil.—The mean amount of nitrogen in eleven arable and grass soils at Rothamstead is placed by Lawes and Gilbert at .149 per cent. and for eight other Great Britain soils at .166 per cent. Voelcker found in four Illinois prairie soils .308 per cent., and C. Schmidt gives for seven rich Russian soils .341 per cent. The mean of these 30 analyses is .219 per cent. and yet a soil containing but .1 per cent. will carry 4,000 lbs. or enough for nearly 60 40-bushel crops.

96. Amount of Nitrogen in Four Manitoba Soils.—As an example of soils exceptionally rich in nitrogen the table below gives the distribution and amount per acre in each of the upper four feet of four Manitoba soils:

	Niverville.	Brandon.	Selkirk.	Winnipeg.
	Lbs.	Lbs.	Lbs.	Lbs.
First foot	7,308	5,236	17,304	11,984
Second foot	5,408	3,488	8,448	10,464
Third foot	2,484	2,592	2,736	5,688
Fourth foot	1,520	870	1,457	4,015
Total	16,720	12,186	29,975	32,181
Tons	8.36	6.093	14.987	16.09

Thus it is seen that in the upper four feet of these rich soils there was found from 6 to 16 tons per acre of nitrogen.

97. Forms in Which Nitrogen Occurs in the Soil.—Nitrogen occurs in the soil in several distinct forms:

1. In humus, described in (**88**), which is by far the most important form and the substance which carries the largest proportion of that which the soil contains.

2. In organic matter in the form of roots, stubble and farmyard manure, which by slow degrees is converted into humus to make good that which has been used.

3. As free nitrogen in soil-air which is seized upon by some forms of microscopic life described in (**101**) and converted into organic form for their use.

4. As nitrates of lime, magnesia, potash and soda, and this is the form from which most of the higher plants get their supply.

5. As ammonia, nitrous acid and nitric acid, which are transition stages to one of the nitrates named above and which are formed either from the humus or organic matter or are brought down with the rain.

98. Distribution of Nitrogen in the Soil.—In humid climates the largest amount of nitrogen is found in the surface 6 to 12 inches, but as already shown in (**96**) large quantities are found as deep as four feet below the surface.

Warington determined the distribution of nitrogen in some of the Rothamstead soils to a depth of 9 feet in 9-inch sections. The results he found are given in the table below:

Nitrogen in soils at various depths.

	Arable soils.	Old pasture.
	Lbs. per acre	Lbs. per acre
First 9 inches contained	3,015	5,351
Second 9 inches contained	1,629	2,313
Third 9 inches contained	1,461	1,580
Fourth 9 inches contained	1,228	1,412
Fifth 9 inches contained	1,090	1,301
Sixth 9 inches contained	1,131	1,186
Surface 3 feet contained	7,333	10,656
Second 3 feet contained	4,365	....
Third 3 feet contained	4,559	....
Total	16,257	....

In these two cases the nitrogen decreases downward until about four feet and below this depth to nine feet the amount remains nearly constant. It will be seen that the amount is very large in the aggregate. Enough for more than 240 crops of wheat, 40 bushels per acre, could it all be used.

99. Amount of Nitric Acid in Soils.—The amount of the available nitrogen in soils, or nitric acid, is seldom a large quantity and while crops are growing the quantity is still smaller.

Warington states that the nitric nitrogen in the soil seldom reaches 5 per cent. of the total amount present, and in the surface three feet of the arable soil referred to in (**98**) this would represent 366.6 lbs. of nitric nitrogen and 1,650 lbs. of nitric acid per acre; enough, if it could all be used, to give a yield of 211.4 bushels of spring wheat per acre.

100. Nitric Acid in Fallow Ground.—The amount of nitric acid in fallow ground was determined to a depth of 4 feet in one-foot sections on May 24 and again on Aug. 22, and the results are given in the table below:

Nitric acid in fallow ground in pounds per acre.

	1st foot.	2nd foot.	3rd foot.	4th foot.
May 24	78.03	21.43	8.13	4.76
August 22	293.72	116.17	23.50	16.72
Gain	215.69	94.74	15.37	11.96

These figures are a mean of the amounts found in nine different sub-plots, the soil being a clay loam changing into sand in the third foot. It will be seen that the total amount of nitric acid at the close of May was 112.35 lbs., containing 24.97 lbs. of nitrogen, enough for only about 14.3 bushels of wheat. On the 22nd of August, however, there had been an increase to 450.11 lbs. per acre, containing 100.02 lbs. of nitrogen, enough for nearly 60 bushels of wheat per acre.

101. Source of Soil Nitrogen.—Until recently it was maintained that the nitrogen for the growth of all plants was derived from the humus of the soil and from the small amount of ammonia and nitrous and nitric acids brought down by the rains. It is now known that the free nitrogen of the atmosphere is the ultimate source of soil-nitrogen, and that the soil-nitrogen is being continually returned to the air again just as was long ago recognized to be the case with the carbon of living forms.

1. The immediate source of humic nitrogen is the slow decay of organic matter, whether this be the roots, stems or leaves of plants or the tissues and waste products of animals, and a large part of the life processes of the world take place between the conversion of humus into living tissues and dead tissues back into humus again.

2. The formation of nitrous and nitric acids through an oxidation of the nitrogen of the air by electrical discharges

such as occur during thunder storms is generally conceded. It is also thought that a part of these combinations may be brought about through the action of ozone upon ammonia. Warington is also of the opinion that the peroxide of hydrogen in the air causes the conversion of some atmospheric ammonia into nitric acid, and hence that not all the nitric acid brought down by the rains was formed as new materials in the atmosphere from direct union of oxygen and nitrogen gases.

The amount of nitrogen brought to the soil with the rains seldom equals 5 lbs. per acre per annum in the open country, as shown by the following table:

Nitrogen as ammonia and nitric acid, in pounds per acre per annum, in rain.

	Rothamsted. 8 years.	Lincoln, New Zealand. 3 years.	Barbadoes. 3 years.
	Lbs.	Lbs.	Lbs.
Nitrogen as ammonia	2.53	0.74	0.93
Nitrogen as nitric acid.......... ..	0.84	1 00	2.84
	3.37	1.74	3.77

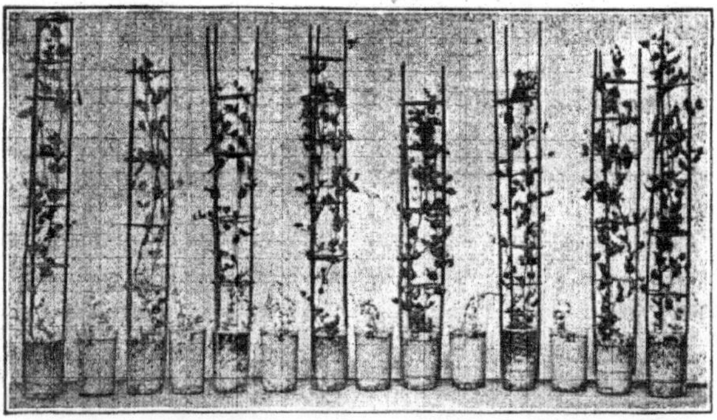

FIG. 25.—Showing the influence of free-nitrogen-fixing germs on the growth of peas. The large plants all grew in sand containing the nitrogen fixing bacteria, while the small plants grew in soils identically the same except that all bacteria were excluded from them. After Hellriegel

These amounts, it will be seen, are far too small to be of great importance to plant life.

3. The process of *symbiosis* is a third method by which the nitrogen supply of the soil is maintained and next to the decay of organic matter is the most important of any yet well understood. It was in 1888 that Hellriegel published the results of his studies, which thoroughly established the fact that great numbers of microscopic forms of life inhabit the roots of leguminous plants, forming upon

FIG 26.—Showing the growth of rye, oats, peas, wheat, flax and buckwheat in soils fertile in all elements of plant food except nitrogen, and illustrating the power of the pea, through its root tubercles, to procure nitrogen from the air. After P. Wagner.

them tubercles in which these organisms live and withdraw free nitrogen from the soil-air for their needs. It had long been known to farmers that in some way clover in rotation with other crops left the soil richer in nitrogen, and it is now known that the bacterium which lives on the clover roots, deriving a part of its food from the clover plant, at the same time increases the nitrogen supply available to the clover crop and so we have two forms of life living together in what has been named symbiotic relations. There are other forms of bacteria which live upon the bean, pea, lupine and other members of this family, also having the power of fixing free nitrogen from the soil-air in forms available to higher plants.

It is known that other forms of bacteria live in symbiotic relation with soil algae and in this way increase the supply of soil nitrogen as shown by Frank, Schlösing, Jr., and Laurent in 1891, followed by Kosswitsch in 1894; and the great demands for the fixing of free nitrogen to make good the rapid return of it to the air and loss in drainage waters appears to call for other agencies than those named.

FIG. 27.—Showing oats growing under conditions identical with those of Fig. 26, except that the several pots received Chile saltpetre, 1, 2 and 3 grams respectively, thus enforcing the immense importance to such plants of nitric nitrogen. After P. Wagner.

4. Winogradsky has shown that there is a form of bacillus in the soil which, when supplied with sugar and isolated from the influence of oxygen, is capable of thriving and fixing free nitrogen from the air, and this discovery may lead to a knowledge of still a fourth mode of increasing the world's supply of nitrogen.

Some of Berthelot's experiments are thought by him to show that soils destitute of all visible vegetation may gain large quantities of nitrogen when simply exposed to the air, and he thinks he has realized gains as large as 70 to 130 lbs. of nitrogen per acre in 11 weeks. Such conclusions, however, require careful verification as they are at least apparently contradicted by field practice.

102. Nitrification.—The formation of nitrates in the soil involves at least four distinct phases or stages: (1) the ammonia stage, (2) the nitrous acid stage, (3) the nitric acid stage and (4) the nitrate forming stage. When humus or dead organic matter is placed under the right conditions of temperature, moisture and air in the presence of ammonia-forming germs, these organisms feed upon portions of it and throw off ammonia as a waste product. Ammonia is extremely soluble in water and is retained by it in large volumes. Even dry soil has the power of condensing and retaining it. In a fertile soil where ammonia has been formed there are also present nitrous acid germs which are able to use ammonia in their life processes but throwing off nitrous acid as a waste product. The niter germs or "mother of petre" utilize the nitrous acid in their work and throw off as a by-product nitric acid. This nitric acid readily attacks any of the bases in the soil which are held by carbonic and other weak acids, displacing them and forming nitrate of lime, magnesia, potash or soda, as the case may be.

In the old days of "niter farming," when nitrate of potash for gunpowder was obtained from the soil, great pains were taken to form a soil rich in organic matter and to keep it warm, well supplied with moisture and thoroughly aerated. These, too, are the points to be secured in the best management of soil for farm and garden crops.

103. Denitrification.—Pitted against the processes of fixing free nitrogen from the air, which have been described, there are other processes which reverse these operations and set free again the nitrogen of organic compounds and of nitrates so that it is again returned to the atmosphere as free nitrogen gas.

(1) Dr. Angus Smith showed in 1867 that nitrates in sewage waters are decomposed and the nitrogen set free as a gas. (2) Schlösing showed that when moist humus-bearing soils are placed in an atmosphere free from oxygen they quickly lose all traces of nitrates. (3) Warington demonstrated that sodium nitrate in a water-logged soil is decomposed and the nitrogen liberated as a gas. (4) So great is the demand for oxygen in rich water-logged soils that according to the experiments of Müntz even such compounds as chlorates, iodates and bromates are deprived of their oxygen, leaving iodides, chlorides and bromides in their place. (5) When black marsh soils are stirred up with water and allowed to stand Prof. J. A. Jeffery and the writer have shown that the nitrates rapidly disappear and nitrogen gas is set free.

In all of these cases there are microscopic organisms in the soil and water whose needs for oxygen are so great that when that which is free in the soil-air or water-air is not sufficient they have the power of decomposing nitrates and even some organic compounds for the oxygen they contain and in this way liberate free nitrogen.

(6) There is still another condition under which denitrification takes place in which the loss is large, rapid and nearly complete. It is when human excrements are covered with pulverized dry soil, as is done in the dry-earth closets. The late Colonel Waring kept two tons of dry earth for a number of years, having it used over and over again in order to see how long it might be used without losing its efficiency. The closets were filled with the dry earth and excrement about 6 times each year, and when they were emptied the material was thrown in a heap on a floor of a well ventilated cellar to dry. After the same soil had been used over not less than 10 times it was analyzed for the amount of nitrogen it contained, and in 4,000 lbs. of the soil was found

no more than 11 lbs. of nitrogen and yet not less than 230 lbs. had been added to it and the soil at the start contained at least 3 lbs. There had been set free therefore

230 — 8 = 222 lbs. of nitrogen.

Nor was this all, for so completely had all the carbonaceous materials been oxidized that even the paper used had entirely disappeared.

How far these processes take place under field conditions when farmyard manure is applied we have yet to learn.

CHAPTER III.

SOLUBLE SALTS IN FIELD SOILS.

All the food of plants is taken by them in the form of liquids or of gases, and hence the fertility of a soil must be determined by the rate at which plant food may be dissolved in the soil water and carried to them at the time the crops are growing. If the ash ingredients and the nitrogen used by plants while growing are supplied in the soil water as rapidly as the crop can use them, then maximum yields will be certain if the temperature and sunshine are also right.

104. Amount of Soluble Salts in Field Soils.—There is a very wide difference in the amount of salts dissolved in soil water under different conditions. In arid regions, where there is little soil leaching, the salts become in places so abundant that plants are unable to grow and alkali lands are the result. In humid climates, especially where the soils are sandy, the salts may be so small in amount that plants starve. In the table below these differences are shown for the surface foot.

	Water soluble salts in soils of arid climates.		Water soluble salts in soils of humid climates.	
	Where barley will not grow.	Where barley grows 4 ft. high.	Fertile clay loam.	Poor sandy soil.
Lbs. per million of dry soil	8,585	4,877	272	21
Lbs. per acre of 4,000,000 lbs	34,340	15,508	1,088	84

These figures show a range of total salts soluble in water from 17 tons per acre foot to less than .05 tons.

105. Maximum Amount of Water Soluble Salts Which Limit Plant Growth.—Hilgard concludes from his studies that the maximum amount of soluble alkali salts which are consistent with a full crop of barley hay is 25,000 to 32,000 lbs. per acre in the surface four feet of soil, provided this is not more than one-half its weight sodium carbonate.

Whitney places the limit of possible plant production in the soils of the Yellowstone Park at 15,000 lbs. per acre in the surface foot, where the black alkali or sodium carbonate is absent.

Grapes grow in Algeria in alkali soils containing 600 lbs. per million of dry soil but die when it reaches 1,700 lbs. per million in the surface soil and 3,700 in the subsoil; but grain crops grow normally when the soil contains 2,000 lbs. per million.

106. Why too Much Soluble Salt in Soil Kills Plants.—De Vries found, as represented in Fig. 28, that when the liv-

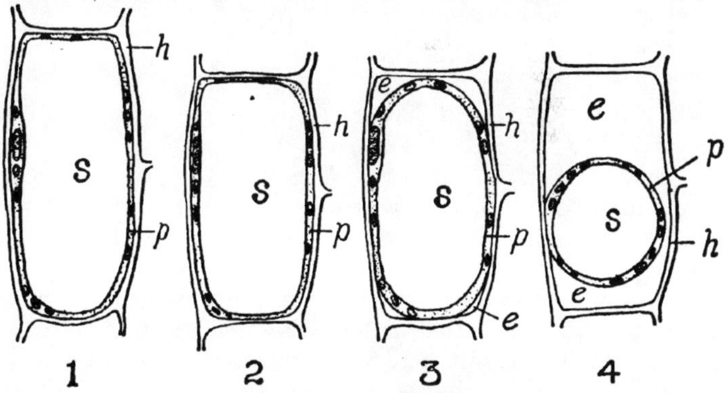

FIG. 28.—Showing the effect of too strong solution of salts on the protoplasm of plant cells.

ing cells of a plant were immersed in a 4 per cent. solution of potassium nitrate, there was first a shrinkage in volume through a loss of water, as shown between 1 and 2. When the solution was given a strength of 6 per cent. the protoplasmic lining, p, began to shrink away from the cell wall h, as shown at 3, and when the strength of the solution was made 10 per cent., the conditions shown in 4 are produced. When the cells of plants are affected in this way they wilt and growth ceases.

A soil containing 20 per cent. of water and also 2,000 lbs. of water soluble salts per million of dry soil would contain 2,000 lbs. in 200,000 lbs. of water, or 1 part in 100, which is 1 per cent. If the soluble salts constitute 2 per cent. of the dry weight of the soil then with 20 per cent. of moisture present the strength of the soil solution would be equal to that which De Vries found fatal to plants, or 10 per cent.

The salts in the surface three inches of soil upon which Hilgard found barley to grow four feet high were 1.2 per cent., while they were 2.44 per cent. in the same level where the barley died. With 20 per cent. of moisture in the soil, and all the salts dissolved, the soil solution in the first case would represent a strength of 6 per cent. and in the second case 12.2 per cent., which is larger than the amount De Vries found fatal.

107. Concentration of Salts in Zones.—Where long continued drought has occurred in soils rich in soluble salts the tendency is for the salts to collect in the surface two or three inches and in this way become injurious to plants when they would not be so with an abundance of water in the soil.

When heavy rains follow such a concentration of salts at the surface, or if the land is irrigated so as to produce percolation, the result is to wash the salts down in a body to the depth reached by percolation, and hence it may happen that a layer of soil very rich in salts may occur at the surface at one time and later at a distance of 12, 18, 24 or 30 or more inches below, determined by the depth of percolation.

108. Origin of Soluble Salts.—The excessive amounts of salts found in alkali lands are usually the result of long continued rock decay under conditions where little or no leaching has taken place. Rains enough fall to produce decay, but not enough to carry the salts formed into the drainage channels and out of the country. This is why alkali lands are largely peculiar to desert or semi-arid climates.

109. Leaching Necessary to Fertile Soils.—It is clear from **106** and **108** that if there was not some leaching to take up and carry away the extremely soluble salts not available as plant food all soils would in time become "alkali lands;" so that while excessive leaching is undesirable, a sufficient amount is indispensable.

The prevention of the accumulation of undesirable soluble salts in the soil of irrigated lands in dry climates is one of the most serious of practical problems.

110. Soluble Salts in Marsh Soils.—The black marsh soils of humid climates often contain unusually large amounts of soluble salts, sometimes reaching 2,366 parts per million of the dry soil in the surface 6 inches after maturing a crop. This would make the water contain 1.18 per cent. of salts if the water content of the soil was 20 lbs. per 100 of dry soil. Many of these soils behave much like alkali lands, being unproductive, the crops often dying when there is no evident reason for it.

111. Correction for Alkali Lands.—It has been found that when a soil is unproductive from too high a per cent. of sodium carbonate or black alkali and there is not enough of other soluble salts to be injurious, this may be corrected in part by the use of gypsum, or land plaster, which has the effect of converting the carbonate into the sulphate or "white alkali," like amounts of which are less harmful.

It often happens that waters which must be used in irrigation contain black alkali, and where this is the case it is well to correct the water by using land plaster in the reservoirs or distributing canals, for the water to run over or through, before reaching the field.

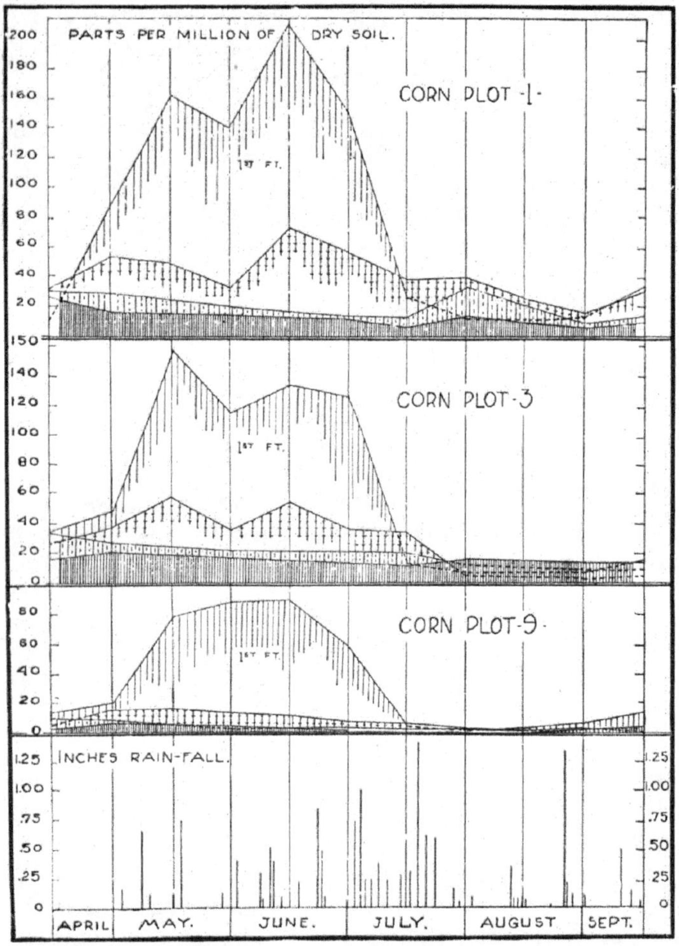

Fig. 29.—Showing the seasonal changes in the amounts of nitrates in each of the surface four feet of soil under growing corn.

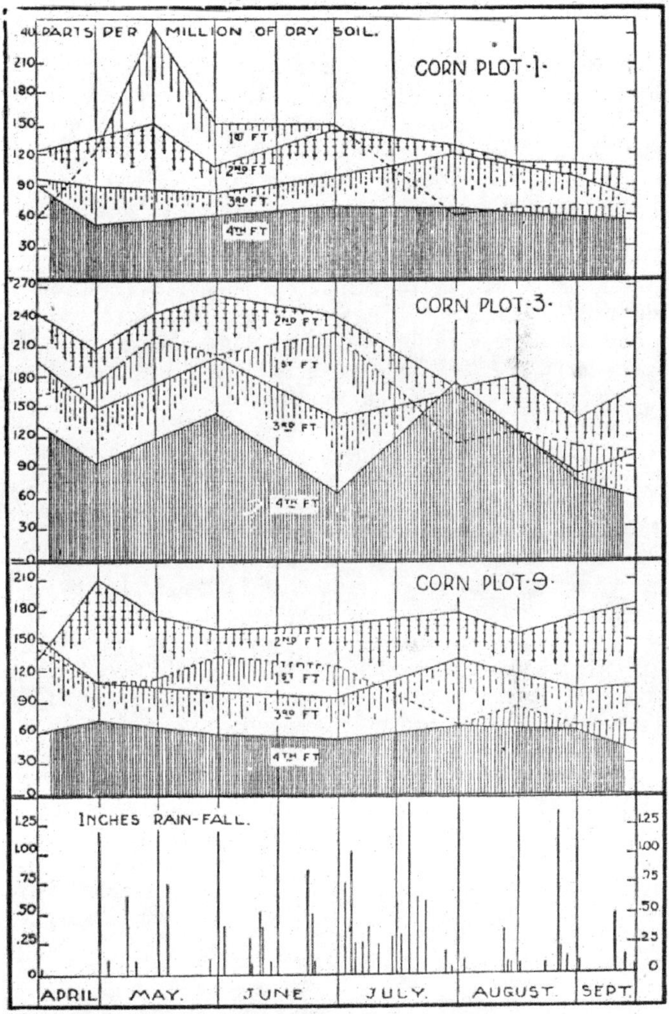

FIG. 30.—Showing the seasonal changes in the amounts of soluble salts in the soil under growing corn.

112. Drainage the Ultimate Remedy.—Drainage must be the ultimate remedy for any alkali land, as it can be only a matter of time when any fertile soil will develop enough undesirable soluble salts to render it sterile or less productive, unless the soluble salts not needed are removed, and only drainage can do this.

113. Deep and Frequent Tillage Helpful.—It is clear that whatever means will prevent the excessive evaporation of water from the surface will in so far lessen the concentration of salts there, and hence frequent and deep cultivation, to form effective mulches, will lessen the rise of water, and therefore of salts, to the surface and in this way permit crops to be grown on soils which are critically near the limit of sterility on account of the high salt content.

114. Change in Soluble Salts with Season.—In Figs. 29 and 30 are represented the changes in the nitrates and total soluble salts in the surface four feet under three fields of corn, beginning with April and ending with Sept. Referring to the nitrate curves it will be seen that the nitrates start in April nearly equal in the four feet, but increase rapidly in the first foot until the middle of June, when the corn begins to draw on the supply. From this time they decrease rapidly until the middle of July, when they are less than in April and less than in the second foot. By the middle of August, when the crop has ceased to draw much but water from the soil, there is a slow increase again and then one more rapid after the corn is cut, Sept. 1.

The change in the total salts is much less marked, but evident, there being a general decrease. The mean amount of salts at the beginning and at the end of the season are:

	April 18.	Sept. 1.
Total salts....................	540	363
Nitrates.......................	86	32
Difference	454	331

From these figures it appears that the salts, other than nitrates, have decreased during the season 123 lbs. per million of the dry soil for the four feet, or 1,968 lbs.

115. Variation of Soluble Salts with Different Crops.— There is a marked difference in the amount of soluble salts, and especially in the amount of nitrates, in soils under crops like corn and potatoes, where inter-tillage is practiced, and under such crops as clover and oats, where the ground is not cultivated at any time of the season. This is very clearly shown in Fig. 32; the nitrates are plotted in the lower two sets of curves and the total soluble salts in the upper two sets.

The nitrates in the first foot under the corn and potatoes increased rapidly until July 1st, when they were five times as concentrated as in the fourth foot; but in 30 days more the nitrates had been reduced from over 400 lbs. to 40 lbs. per acre.

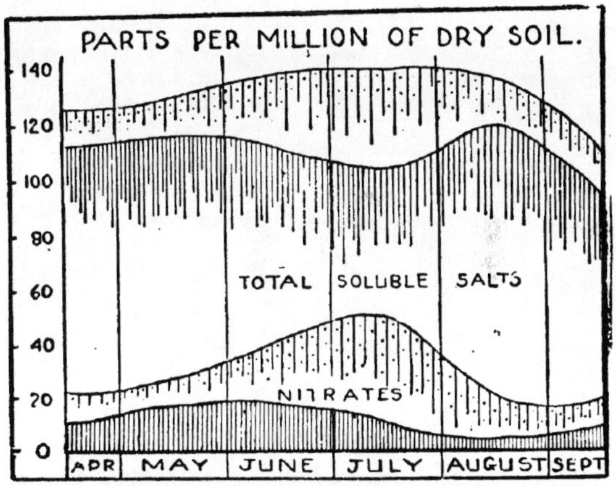

FIG. 31.—Shows the mean amount of nitrates and total soluble salts in the surface four feet of soil under cultivated and not cultivated crops. Heavy shading is uncultivated ground.

In the case of the uncultivated crops the fields started with about 40 lbs. per acre and increased to only 70, June 1st, when they were highest; from this date they fell to little more than 10 lbs. per acre in the surface foot, but rose again to 60 lbs. at the end of August.

With the total soluble salts there was at first a more

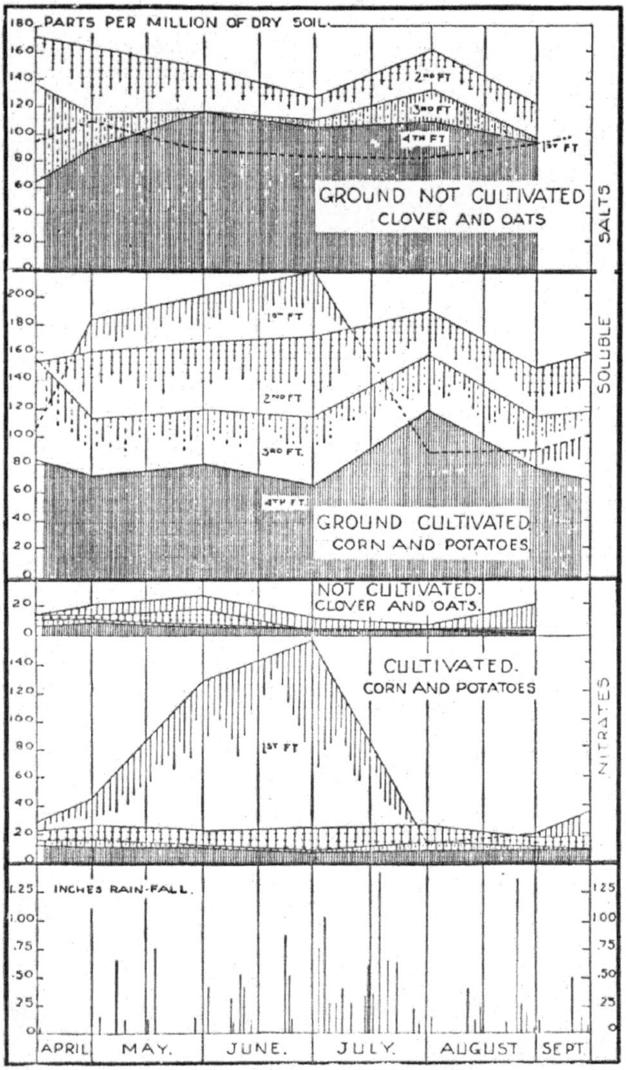

Fig. 32.—Showing the difference between the amounts of nitrates and of total soluble salts in the soil under cultivated and not cultivated crops.

rapid rise, from nearly 300 lbs. per acre in the surface foot on the cultivated ground April 18, to about 500 lbs. per acre, but falling again on August 1st to 250 lbs.

On the clover plots the start was at 250 lbs. per acre in the surface foot, rising to 290 lbs. in 12 days. From this date there was a slow decrease, falling to 220 lbs. on the date when the cultivated grounds were highest, at 600 lbs. per acre.

116. Relation Between Nitrates and Total Soluble Salts.— As a general rule when the nitric nitrogen in clay loams is very high the total soluble salts, as indicated by the electrical method, are very low. It will even happen that the electrical resistance will show but little more salts than are required to account for the nitrates, and this is perhaps what should be expected for, if nitric acid is being formed in the presence of carbonates, these would be decomposed to form nitrates, and if the rate of nitrification were sufficiently rapid, it might be that all the carbonates would be decomposed and little else but nitrates left.

The ratio of total soluble salts to nitrates in the surface foot of the five cultivated fields represented by the curves was a mean for the season of 2.14 to 1, while in the surface foot of the clover fields it was 4.8 to 1.

For the second, third and fourth feet the ratio is 7.29 to 1 for the corn and potatoes, and 9.97 to 1 for the clover, alfalfa and oats; and these ratios are what would be expected if the formation of nitric acid destroys the carbonates and bi-carbonates in the soil water.

117. Closeness of Plant Feeding.—It was pointed out in (7) what small amounts of a fertilizer can be widely distributed through an acre of soil, and we may now consider how extremely close plants do feed the nitrates of a soil. In the table which follows are given the amounts of nitrates which were found in each foot of nine field plots, represented by the curves, between July 18 and Sept. 1.

Table showing mean amounts of nitrates under different crops between July 18 and Sept. 1, in lbs. per acre of dry soil.

	Plot 1.	Plot 2.	Plot 3.	Plot 4.	Plot 5.	Plot 6.	Plot 7.	Plot 8.	Plot 9.
	Corn.	Clover.	Corn.	Oats and clover.	Potatoes.	Potatoes.	Clover.	Alfalfa	Corn.
	Lbs.	Lbs.	Lbs.	Lbs.	Lbs.	Lbs.	Lbs.	Lbs.	Lbs.
1st foot..	50.94	58.22	24.11	15.07	130.21	105.32	44.91	18.84	10.85
2nd foot.	127.35	23.74	48.81	14.42	155.95	172.62	15.63	10.65	8.85
3rd foot..	83.52	10.25	59.44	18.81	49.65	50.66	1.75	9.53	10.79
4th foot..	40.83	14.80	64.82	27.05	21.08	59.82	4.59	9.73	12.51

When these amounts are expressed as parts per million of the dry soil in the form of nitrogen, they stand 3.38, 1.61, 0.72 for corn; 3.87, 2.98, 1.00 for clover; 1.25 for alfalfa and 6.99 for potatoes, and yet with these small amounts of nitrogen in the soil during the time when the chief growth was being made, large yields were produced.

118. Limits of Nitric Nitrogen at Which Corn and Oats Turn Yellow.—Taking samples of soil from the surface foot upon which oats were turning yellow and under adjacent areas where the plants were normal green it was found that two sets of duplicate determinations gave

		Oats yellow	Oats green.
Parts of nitric nitrogen per million of dry soil....	June 10	.025	.213
	June 11	.027	.297

These amounts, when expressed in pounds per acre and as nitrates, are only .392 lbs. and 3.843 lbs., respectively, for the yellow and green oats.

Table showing the amounts of nitric nitrogen under corn rows where leaves are turning yellow and where they are yet normal green.

Depth.	Plot 9.		Marsh soil.		Randall field.	
	Yellow.	Green.	Yellow.	Green.	Yellow.	Green.
1st foot...............	0.61	0.92	0.95	3.62	0.10	0.95
2nd foot............	0.14	1.70	0.40	1.41	0.06	0.60
3rd foot.............	0.41	2.95	0.07	0.52	0.25	0.37
4th foot.............	0.42	1.82	0.00	0.00	0.30	0.30

Small as these amounts of nitric nitrogen are the yield of corn on plot 9 was a mean of 8,000 lbs. of water-free matter per acre. On another plot where the yield was 11,440 lbs. of water-free matter per acre the nitric nitrogen was reduced as low as 1.446 parts per million in the first foot and .726 parts in the second foot.

It must be understood that in these cases the demands for nitrogen were so urgent that the plants were taking it up almost as rapidly as it could be produced, leaving the amounts so low, as the figures show.

119. Nitrates of Fallow and Cropped Ground.—In the table which follows are given the amounts of nitrates found under different crops and, at the same time, under immediately adjacent fallow ground which had been cultivated and kept free from weeds.

	Oats.		Fallow.		Barley.	
	Nitrates.	Total salts.	Nitrates.	Total salts.	Nitrates	Total salts.
1st foot	5.94	70.94	246.40	199.3	2.62	61.72
2d foot	8.12	114.6	26.75	123.5	5.10	87.68
3d foot	4.73	124.7	6.50	108.0	4.01	112.6
4th foot	4.60	89.44	2.84	42.10	3.03	51.76
	Oats.		Fallow.		Peas.	
1st foot	3.25	80.35	143.05	206.1	8.38	77.00
2d foot	3.22	162.1	29.50	254.3	18.57	197.2
3d foot	2.95	102.7	8.87	115.0	6.59	135.8
4th foot	2.70	58.24	4.10	95.32	2.66	44.62
	Oats.		Fallow.		Spring rye.	
1st foot	2.47	78.56	129.15	211.3	1.24	77.34
2d foot	2.46	102.9	35.60	254.7	2.62	102.1
3d foot	3.83	72.98	9.11	117.8	2.07	94.82
4th foot	3.16	83.99	4.08	61.92	2.78	48.85

If the mean amount of nitrates in the surface foot of the fallow ground and under the crops are expressed in pounds per acre they stand 473.65 to 10.88. This difference is enough for 85 bushels of oats per acre, where the ratio of grain to straw stands as 3 to 5.

120. Loss of Nitrates from Fallow Ground During Winter and Spring.—A field which has been kept fallow during a whole season and cultivated either once per week or once in two weeks had the nitrates determined in it on August 25 and again the next spring, April 30. The field was divided into nine plots and the nitric nitrogen was determined in each one to a depth of four feet on both dates. The results are given in the next table.

Table showing the amount of nitric nitrogen found in fallow ground after the leaching of winter and early spring. Pounds per million of dry soil.

	No. of plot.	1.	2.	3.	4.	5.	6.	7.	8.	9.
1st foot	Apr. 30, 1900	75.90	58.31	58.08	55.22	51.66	51.25	38.02	44.34	48.26
	Aug. 22, 1899	16.81	13.58	26.67	26.80	19.09	16.82	5.50	24 07	19.60
2d foot	Apr. 30, 1900	15.81	16.75	7.97	6.51	13 06	15.66	17.33	18.56	14.85
	Aug. 22, 1899	4.34	7.75	1.81	9.07	5.74	2.76	1.43	6.06	6 61
3d foot	Apr. 30, 1900	2.46	4.75	4.93	4.89	3.94	7.35	6.04	8.24	6.71
	Aug. 22, 1899	.70	.54	2.45	.80	0.54	1.37	0.95	0.54	3 01
4th foot	Apr. 30, 1900	2.95	2.37	3.05	2.35	2.01	2.36	3.65	5.60	5 08
	Aug. 22, 1899	.80	......	1.04	......	1.55	0.32	0.26	0.53	3.51

It is clear from this table that however large the leaching may have been it was not enough to prevent the nitrates

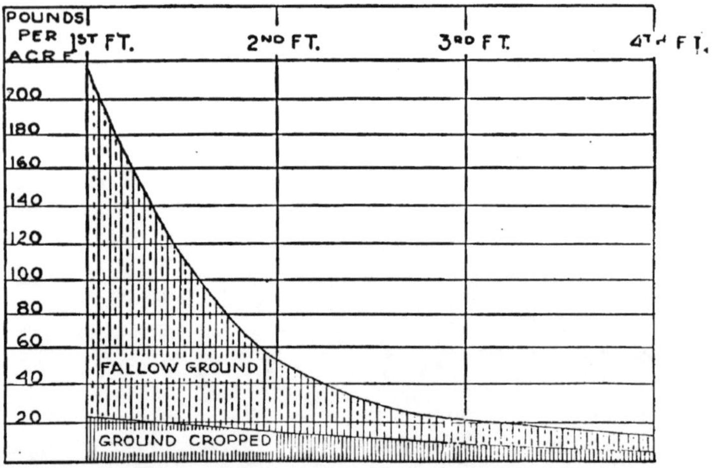

FIG. 33.—Showing the difference in the amount of nitrates in the surface four feet of fallow ground, the succeeding spring, and that upon which crops had been grown.

being higher the following May than they were August 22 before.

121. Nitrates on Fallow Ground in Spring Compared with That not Fallow.—Comparing the mean amount of nitric nitrogen in nine field plots bearing crops in 1899 with that of the nine fallow plots of the same year, as found in the spring of 1900, the amounts are as stated in the table below and represented graphically in Fig. 33.

Table showing the differences in the amounts of nitric nitrogen after the winter and early spring rains in ground kept fallow and free from weeds the previous season and that bearing crops.

Depth.	1st foot.	2d foot.	3rd foot.	4th foot.
Fallow plots, pounds per acre..	212.00	56.22	21.91	13.11
Plots not fallow, pounds per acre	25.24	15 08	10.00	7.24
Difference....................	186.76	41.14	11 91	5.87

From this it is clear that the crops on the fallow ground start out in the spring under conditions very superior to those on the fields which had not been fallow, there being 245.68 lbs. of nitrates more per acre in the surface four feet.

122. Development of Nitrates Influenced by Depth and Frequency of Cultivation.—When a series of cylinders like those represented in Fig. 58, p. 187, are mulched by stirring at different depths and the stirring is repeated at different intervals the rate of formation of nitrates is materially modified, as shown in the table below:

Difference in the amount of nitric nitrogen, after 258 days, due to differences in depth and frequency of cultivation.

Depth of cultivati'n.	Cultivated once per week.	Cultivated once in two weeks.
	Lbs. per acre.	Lbs. per acre.
1 inch deep..	217.69	213.29
2 inches deep........	323.44	199.00
3 inches deep........	441.24	401.68
4 inches deep........	387.96	245.26

It can be seen that the nitric nitrogen has increased in both series to a depth of 3-inch cultivation and it has increased with the frequency of the cultivation.

123. Soluble Salts Affect the Movement of Soil Moisture.—The varying strength of salt solutions in soil moisture modify both the movement of moisture in the soil and its rate of loss from the surface. These movements are influenced (1) by changes in the intensity of surface tension; (2) by changes in the internal friction of the soil moisture or its viscosity; and (3) by modifications of the surface of the soil due to deposits of salts upon and within it, where evaporation is taking place.

124. Modification of Surface Tension by Soluble Salts.—As a general rule the surface tension of a strong soil solution is greater than that of a weaker one, or of pure water, and in so far as this influence is operative it tends to increase the rate of capillary movement toward the surface or toward the roots of plants.

125. Salts in Solution Lessen Rate of Evaporation.—When water has been brought to the surface of the soil by capillarity it has yet to evaporate and unless this takes place the surface soil would become capillarily saturated with water and remain so. Since salts in solution increase the surface tension it will require a greater energy—a higher temperature—to throw the water molecules off into the air than would be required to do so from the surface of pure water and hence the evaporation from soil solutions rich in salts is slower than it is from weaker ones under otherwise like conditions. As the salts become concentrated at the surface by evaporation the moisture becomes a stronger and stronger solution and hence the rate of evaporation becomes less and less so far as it can be influenced by this factor, in this way.

126. Viscosity of Soil Water Modified by Soluble Salts.—The internal friction of soil moisture is made greater by the presence of salts in solution and the more concentrated the soil solution is the greater is the internal friction, and hence the slower must be the rate of flow, and it may be that the much slower rate of capillary movement in a comparatively dry soil is to a considerable extent due to this increased viscosity or internal friction. But as one effect of the salt in solution is to increase the surface tension, while the other decreases the flow by increasing the friction, the two influences work against each other, making the combined result less than it would be could either act alone.

127. Deposits of Salts after Evaporation May Lessen Loss of Soil Moisture.—Where water rich in salts is being evaporated from a soil these salts may accumulate upon the surface and form a sort of mulch more or less effective according to its texture; or they may be deposited as a crust upon, over and between the soil grains, which may nearly close the capillary pores and in this way lessen the loss of water by evaporation. Such a closing of the pores is likely to be more harmful in shutting out the air and in lessening the freedom of entrance of water after rains than it can render assistance in conserving soil moisture.

CHAPTER IV.

PHYSICAL NATURE OF SOILS.

128. Texture of Soils.—The size of soil grains and the way they are grouped in composite clusters forming *kernels* or *crumbs* has a very great influence in determining the physical properties of soils and their agricultural value, and as soils vary quite as widely in the size and arrangement of their grains as they do in their chemical composition it is clear that this phase of soil problems must take at least equal rank with those considered in the last chapter.

In all agricultural soils except the very coarse and sandy ones there is a composite granular structure which renders them much more open and porous than they could otherwise be, and when a soil is puddled this structure or texture is destroyed in a large measure and the separate grains are then brought into the closest possible arrangement, and they become nearly or quite impervious to both water and air, approaching the condition of brick and potter's clays.

129. Size of Soil Grains.—When the fragments of rock are so coarse that very few are smaller than .01 of an inch in diameter we have a sand rather than a soil. Most plastering sands are made up of grains ranging from .01 up to .08 of an inch in diameter.

In the table which follows is given the mechanical analyses of three types of soil:

It will be seen from this table that only .8 per cent. of either soil is made up of grains having diameters so great that only 23 are required to span a linear inch, while the heavy clay soil has nearly one-half of its weight made up of grains so small that 25,000 of them must be placed side by side to span a linear inch.

Sandy Soil.			Loess Soil.			Heavy Clay Soil.		
Diam. m. m.	Number of grains per linear inch.	Per cent.	Diam m. m.	Number of grains per linear inch.	Per cent.	Diam. m m.	Number of grains per linear inch	Per cent.
1 to 3	23.1	.4	1 to 3	23.1	.2	1 to 3	23.1	.8
.5 to 1	31.7	3.0	.5 to 1	31.7		.5 to 1	31.7	1.2
.4	63.5	6.9	.4	63.5	.4	.4	63.5	2.0
.3	84.7	8.1	.3	84.7	.6	.3	84.7	1.6
.16	163.9	3.0	.16	163.9	.9	.16	163.9	.9
.12	211.9	1.6	12	211.9	1.7	.12	211.9	.8
.072	353.4	1.2	.072	353.4	2.0	.072	353.4	.2
.047	540.1	3.6	.047	540.1	14.3	.047	240.1	2.5
.036	704.3	6.8	.036	704.3	16.2	.036	704.3	3.7
.025	1,020.	14.6	.025	1,020.	20.1	.025	1,020.	5.6
015	1,695.	14.8	015	1,695.	5.6	.015	1,695.	10.6
.008	3,226.	30.7	.008	3,226.	33.6	.008	3,226.	24.7
.001	25,000.	4.6	.001	25,000.	2.5	.001.	25,000.	48.0

130. Number of Grains of Soil in a Cubic Inch.—If soii grains were perfect spheres like shot and in a given soil they were all of a single size it would be a simple matter to

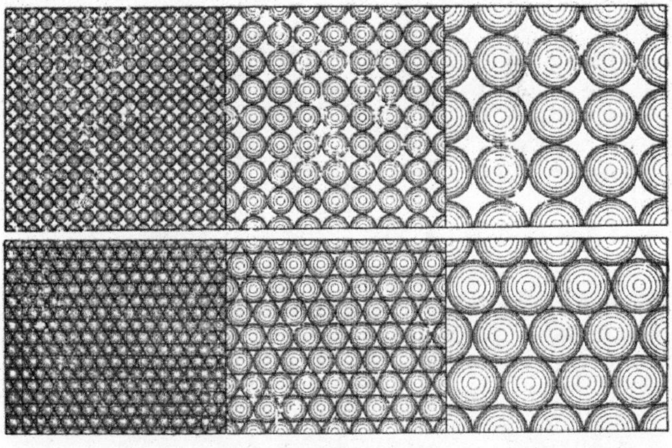

Fig. 34.—Showing the effect of size and arrangement of soil grains on the pore space and upon the movement of air and water through a soil.

determine the number in a cubic inch. If a soil were made up entirely of the largest size given in the last table, then 23 would build one edge of a cube an inch on a side and the number in a cubic inch arranged in the manner represented in the upper part of Fig. 34 would be

$$23^3 = 23 \times 23 \times 23 = 12,167.$$

On the other hand, if they were all the size of the smallest grain in the table then the number would be

$$25,000^3 = 15,625,000,000,000,$$

or enough to form three and a third continuous lines of grains in contact from Boston to San Francisco.

131. The Size of Soil Kernels.—It must be kept in mind that while it is true that the heavy clay soils are made up largely of soil grains of the extremely small size considered in (130) these minute grains are generally bound together in groups or *kernels* of various sizes and it is only by long boiling in water or thorough pestling that these can be broken down. The writer has found that when air-dry samples of the heaviest clay soils are thoroughly pestled in the dry condition it is difficult to reduce their texture to a finer degree than kernels averaging .01 to .005 m. m. in diameter or such that from 2,500 to 5,000 are required to span a linear inch; but even this degree of closeness of texture is too fine to allow of proper drainage and soil ventilation and to permit roots to make their way through the soil with the freedom required for good crops.

132. Specific Gravity of Soil Grains.—The specific gravity of soil grains, or the number of times they are heavier than an equal volume of water, varies somewhat, as does that of the minerals which compose them. As there are not many common minerals more than three times as heavy as water and not many lighter than 2.5 times as heavy, the specific gravity of soil grains will lie between these two figures and it is usually found to be near 2.65.

133. The Pore Space of Soils.—When the weight of a cubic foot of dry soil is known the amount of pore space or space not occupied by the soil grains may be computed from the specific gravity. Taking the weight of a cubic foot of water at 62.42 lbs., a cubic foot of dry soil, if there were no open spaces in it, should be

$$2.65 \times 62.42 = 165.4 \text{ lbs.}$$

With this value and the data given in (**149**) the pore space of those soils may be calculated. Thus, for the surface foot we have

$$\text{Pore space} = \frac{165.4 - 79}{165.4} = 52.23 \text{ per cent.}$$

That is, in this soil the surface foot is more than half open space. The pore space for the six feet will be as given below:

	Weight of soil.	Pore space.
	Lbs.	Per cent.
First foot	79.0	52.23
Second foot	92.62	44.00
Third foot	104.59	36.76
Fourth foot	106 21	35.78
Fifth foot	111.06	32.85
Sixth foot	111.06	32.85

Thus it is seen that the unoccupied space in a soil varies from more than half to less than one-third of its volume, the finest grained soils having the largest pore space and the sandy soils and sands the smallest.

134. Pore Space Between Spherical Grains.—It can be shown mathematically that when a space is filled with spheres all of one size and these are given the closest possible packing, having the arrangement shown in the lower part of Fig. 34 and in Fig. 35, the pore space must be 25.95 per cent.; but when the spheres are given the closest possible packing and the arrangement represented in the upper part of Fig. 34, then the pore space must be as large as 47.64 per cent. In the first case the water capacity of such a soil with the pores entirely filled would

FIG. 35.—Showing the closest packing of spherical soil grains, the element of volume and the direction of lines of flow. Face angles 60° and 120°. (After Slichter.)

be 3.114 acre-inches per acre-foot and with the second arrangement the maximum water capacity would be 5.7168 acre-inches per acre-foot.

Neither of these arrangements would be likely to occur throughout a mass, and hence the general tendency will be to form a pore space between these two extremes, and Fig. 37 shows what the observed pore space is in soils, sand, crushed rock and crushed glass. It will be observed that

Fig. 36.—Showing the closest packing of spherical grains, the element of volume, and the direction of lines of flow when the face angles are 90°, 60° and 120°. (After Slichter.)

the finest clay soils, and indeed the finest grained materials, have the largest pore space. It will also be noted that the largest observed pore space exceeds the largest theoretical pore space and that the smallest observed pore space also falls below the smallest theoretical limit for spherical grains of a single size.

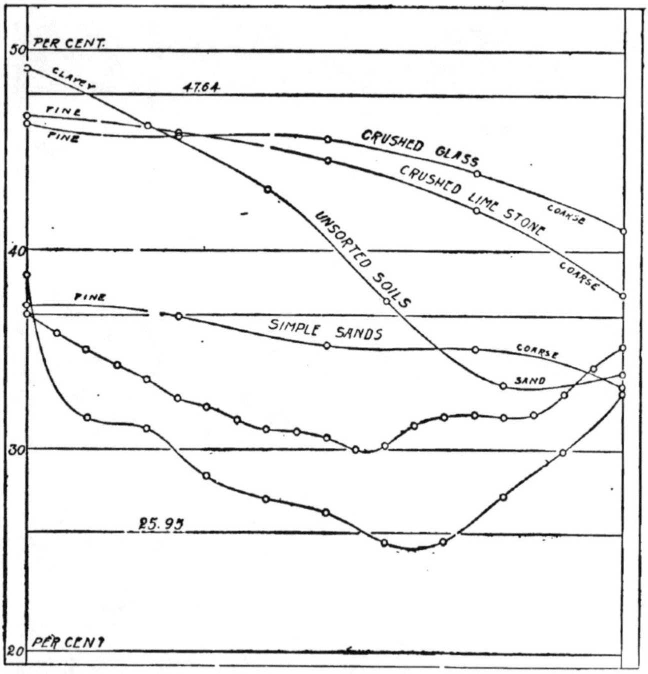

FIG. 37.—Showing the observed pore space of different kinds of soils and sands and their relation to the theoretical pore space of spheres of a simple diameter.

135. Amount of Pore Space Determines Maximum Water Capacity of Soil.—The amount of water a soil may contain when below the level of the ground water surface is measured by the pore space. So too in the case of heavy and protracted rains the pore space determines the number of inches of water which may enter the ground before it becomes so filled that surface drainage must carry away that which is falling, and it will be readily understood that in the clay soils, where the pore space is so high, very large amounts of water may be stored in them to drain away gradually in the underflow.

136. Subdivision of Pore Space Determines the Rate of Percolation and Drainage.—If reference is again made to Fig 34 it will be clear at a glance that water must flow through spaces filled with these different sizes of spheres at very different rates. Where the spheres are largest there are 16 passage-ways for the movement of air or of water; but in the middle section where the spheres have one-half the diameter, the number of passages is 4 times as great, while in the last section with spheres of one-quarter the size the number of passages is 16 times as great.

The aggregate area of the cross-sections of the pores is exactly the same in the three cases, and from this it follows that the areas of the cross-sections of single pores are to each other as 16 : 4 : 1.

The coarse spheres divide the column of water into 16 streams, the medium ones divide it into 64 streams, while the smallest spheres divide the column into 256 streams, each having only one-sixteenth the sectional area of the first. But to subdivide the column into 256 streams instead of 16 means that the friction must be much greater in the aggregate on the smaller streams, and hence that the flow must be slower.

137. Method of Determining the Pore Space of Soil.—The simplest method of determining the pore space of soil is to pack the dry material into a cylindrical vessel containing 100 c. c. until it is even full, and then weigh and compute the per cent. of pore space from the volume, weight and specific gravity, using the formula

$$\frac{Vd - W}{Vd} = P$$

where V is the volume of the vessel in c. c., d is the specific gravity and W is the weight of the soil in grams.

To determine the pore space in undisturbed field soil

the simplest method is to use a soil tube, represented in Fig. 38, taking a number of cores of the desired depth,

FIG. 38.—Showing soil tube for taking samples of soil.

drying them, and then compute the pore space with the formula above.

138. Largest Possible Pore Space.—The largest possible pore space in soils will be found in the cases where the compound or kernel-structure is most marked. Referring again to Fig. 34, imagine each sphere there represented to be made up of other very much smaller spheres having the same general arrangement. Were this the case it is clear that in consequence of the compound spheres the soil must have a pore space not less than 25.95 per cent. with one arrangement and 47.64 per cent. with the other. But in addition to this pore space there must be a like pore space within each compound sphere so that in the first case the total pore space would be

$$25.95 + [25.95 \text{ per cent. of } (100 - 25.95)] = 45.17$$

and in the second case

$$47.64 + [47.64 \text{ per cent. of } (100 - 47.64)] = 72\ 58 \text{ per cent.}$$

The first pore space, 45.17, it will be seen, lies close to that possessed by the finer soils but the latter is larger than anything ever found except it be in the loose mulches.

The smallest pore spaces result when grains of different sizes are so related that the small ones fall into the pores formed by the large ones without at the same time crowding them farther apart. Referring again to Fig. 34, it will be seen that if small spheres are packed into the pores there shown, with the same arrangement that the large ones have, the original 25.95 per cent. and 47.64 per cent.

of pore space would be occupied to the extent of 74.05 per cent. in the first case and of 52.36 per cent. in the second case. Such a condition would leave only about 6.73 per cent. of pore space for the closest packing.

Such arrangements as this are not likely of course to occur in nature but in the construction of macadam roads and in all concrete work a definite effort is made to reduce the pore space to the smallest possible limit by using crushed rock, gravel, sand and finally cement to fill all pores as completely as possible.

139. Number of Soil Grains per Unit Weight.—If soil grains were all spheres and in a given case they were all of the same size the number in a gram could be found by the equation

$$\text{No. of grains} = \frac{\text{Weight of soil}}{\dfrac{\pi d^3 \times \text{sp. gr.}}{6}}$$

where the weight of the soil is in grams and the diameter of the soil grains, d, is in c. m.

In the table below are given in round numbers the number of grains in one gram and in one pound of soil, supposing the grains all spheres and to have a specific gravity of 2.65.

Diameter.	No. of grains in one gram.	No. of grains in one lb.
1. m. m.	720	326,903
.1 m. m.	720,000	326,903,000
.01 m. m.	720,000,000	326,903,000,000
.001 m. m.	720,000,000,000	326,903,000,000,000
.0001 m. m.	720,000,000,000,000	326,903,000,000,000,000

That is to say, 720 multiplied by 10 used as a factor 3, 6, 9 and 12 times gives the number of grains in a gram of soil in round numbers and the number in a pound may be found by using 10 as a factor in the same way and the number 326,903.

If the soil were made up of some grains of all the sizes

in the table, then to find the total number in a gram or pound it would be necessary to multiply those numbers by the per cent. of each size found in a gram of the soil and add the several products. If the soil were made up of 20 per cent. of each size in the table the number would be as follows:

Diameter.	Per cent	No. of grains per gram.
1 m. m..	20	144
.1 m. m.	20	144,000
.01 m. m....	20	144,000,000
.001 m. m.	20	144,000,000,000
.0001 m. m..	20	144,000,000,000,000
Total..	....	144,144,144,144,144

140. Amount of Soil Surface Possessed by a Gram of Soil.—Much of the water retained by soils is held there in the form of thin films surrounding the grains and the larger this surface is the more water may be retained. So, too, the solution of plant food from the grains takes place at their surfaces and the larger the amount of surface the more rapidly the solution may take place.

The extent of soil-surface in a gram of soil can be found by multiplying the number of grains by the surface of one grain or by introducing πd^2 into the equation of (139), thus:

$$\frac{\text{Weight} \times \pi d^2}{\frac{\pi d^3 \times \text{sp. gr.}}{6}} = \frac{6 \times \text{weight}}{d \times \text{sp. gr.}} = \text{soil surface}$$

expressed in square c. m.

Using this formula to compute the surface in one gram of soil grains having the sizes given in the table of (139) the results below are obtained:

Diameter in grains.	Surface per gram sq. cm.	Surface per pound sq. feet.
1. m. m..	22.64	11.05
.1 m. m.	226.41	110.54
.01 m. m....	2,264.15	1,105.38
.001 m. m.	22,641.51	11,053.81
.0001 m. m..	226,415.14	110,538.16

It will be seen from this table that the internal surface of an ideal soil increases in the same ratio that the diameter of the grains decreases, that is, reducing the diameter one-half doubles the surface to which water may adhere and upon which it may act.

141. Difficulties in Determining the Surface of a Soil Accurately.—While it is possible to determine accurately the surface in a given weight of spheres of known dimensions the case is quite different with true soils. Indeed, it is not practicable to determine with much accuracy the surface in a soil. This will be clear from a consideration of a simple problem.

Take a soil composed of grains, (a) .009 and (b) .00015 m. m. in diameter and let these be mixed in the proportions of

A. 90 per cent. of (a) with 10 per cent. of (b).
B. 10 per cent. of (a) with 90 per cent. of (b).
C. 50 per cent. of (a) with 50 per cent. of (b).

Under these conditions the surface of one gram of such mixtures of soil having a specific gravity of 2.65 is

For A.

	Surface.
90 per cent. of grains (a) .009 m. m. diameter	2,264 sq. cm.
10 per cent. of grains (b) .00015 m. m. diameter	15,094 sq. cm.
Total surface	17,358 sq. cm.

For B.

10 per cent. of grains (a) .009 m. m. diameter	251.6 sq. cm.
90 per cent. of grains (b) .00015 m. m. diameter	135,848.9 sq. cm.
Total surface	136,100.5 sq. cm.

For C.

50 per cent. of grains (a) .009 m. m. diameter	1,258.0 sq. cm.
50 per cent. of grains (b) .00015 m. m. diameter	75,481.7 sq. cm.
Total surface	76,739.7 sq. cm.

The number of grains in one gram of each of these mixtures would be as given below:

	A.	B.	C.
(a)	889,753,061	98,861,363	494,306,818
(b)	21,334,187,192,118	192,188,053,097,345	106,770,833,333,333
Total............	21,335,076,945,179	192,188,151,958,708	106,771,327,640,151

It is the custom to find the diameter of soil grains either by direct measurement or else by counting and weighing a given number of grains and then computing the diameter of the mean grain from the weight and specific gravity. If the diameter of the mean grain in the above three problems is computed by each of these methods the results will be as below:

If the surface of a gram of soil is computed from each of these diameters the results given below will be found:

	A.	B.	C.
	sq. cm.	sq. cm.	sq. cm.
Actual surface per gram of soil....................	17,358	136,101	76,740
Surface computed from the grain of mean diameter	150,570	150,939	130,902
Surface computed from the grain of mean weight..	10,053	145,734	119,804

These results are very different and differ so much from the actual as to make them of little value in determining the actual surface a given soil may possess.

It has been the practice to take as the mean diameter of the soil grain the average between the diameter of the largest grain in the group and the smallest, which in the above problem would give .004575 as the mean value.

But to use this to compute the surface in a gram of soil would give the results below:

Computed from the mean of the two extreme diameters.	Computed from the true diameters in true proportions.		
	A.	B.	C.
4,949 sq. cm.	17,358 sq. cm.	136,101 sq. cm.	76,740 sq. cm.

Here it is seen that the computed surface, 4,949, is very far indeed from either of the true values given under A, B and C.

142. Effective Diameter of Soil Grains.—While it is not possible to determine either the mean diameter of the grains in an ordinary soil or the amount of surface a given weight of soil may possess with even approximate accuracy, it is possible for the simple sands, at least, to determine the diameter of *a grain* which, if substituted for the actual ones, would permit, under like conditions, the same amount of air or of water to flow through.

The method is based upon the laws of flow of fluids through capillary tubes and aims to compute from the observed rate of flow of air through a given column of soil the effective diameter of the capillary pores and from this the size of spherical grains which would be required to form such capillary tubes as those computed. The theory of the method is fully set forth in Prof. C. S. Slichter's paper.[1]

143. Description of the Method.—The apparatus used to determine the effective size of soil grains is represented in Fig. 39, and consists of a cylinder in which a sample of soil is carefully packed and weighed to determine the per cent. of pore space. When this has been done the tube is connected with the aspirator and the rate at which air will flow through it under a measured temperature and pressure found. When these data have been obtained, then the formula below, used with the table given, enables the effective diameter to be computed when the flow has been measured at the temperature of 20° C.

[1] Nineteenth Annual Report of the U. S. Geol. Survey, Part II.

$$d^2 = k \frac{h}{spt} \qquad [8.9434 - 10]$$

where

d = diameter of grain in c. m.

h = length of sand column in c. m.

s = area of cross-section of sand column in sq. c. m.

p = pressure in c. m. of water at 20° C.

t = time in sec. for 5,000 c. c. of air to flow through at a temperature of 20° C.

[8.9434 — 10] is a logarithm of a constant

k is a constant taken from the following table.

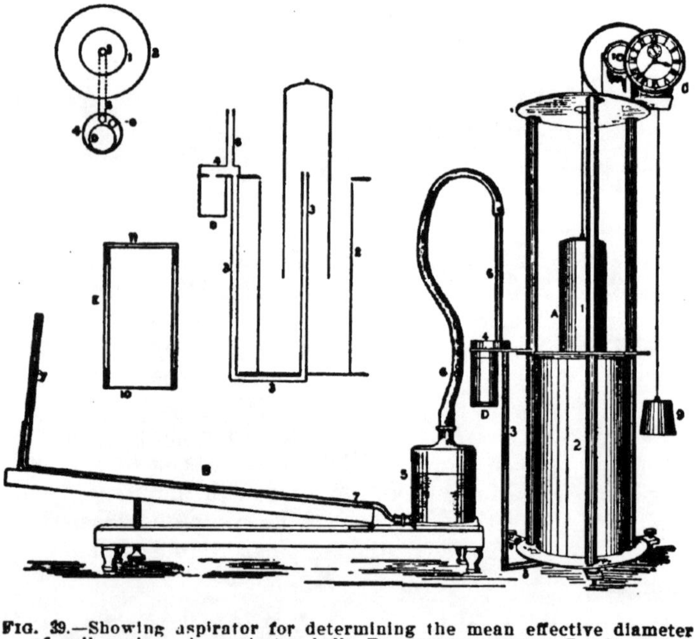

FIG. 39.—Showing aspirator for determining the mean effective diameter of soil grains. A, aspirator bell; B, pressure gauge; C, air meter; D, aspirator tube for samples.

Per cent. of pore space.	Log. k.	d.	Per cent. of pore space.	Log. k.	d.
26	1.9258	563	37	1.4193	877
27	1.8695	500	38	1.3816	871
28	1.8195	490	39	1.3445	867
29	1.7701	502	40	1.3078	853
30	1.7199	467	41	1.2725	851
31	1.6732	455	42	1.2374	845
32	1.6277	430	43	1.2024	839
33	1.5817	438	44	1.1690	820
34	1.5103	410	45	1.1370	812
35	1.4993	407	46	1.1058	829
36	1.4592	400	47	1.0729	....

144. Observed Flow of Water Through Sand Compared With That Computed From the Effective Diameter.—The accuracy of the method described in (**143**) is best shown by computing from the effective diameter of the soil grains what the flow of water ought to be and then measuring the flow of water to see how it corresponds. This has been done and the results are given in the table below:

Grade of sand.	Effective diameter of grain.	Computed flow of water.	Observed flow of water
	m. m.	Gms.	Gms.
8	2.54	2,277	2,296
7	1.808	1,132	1,080
6	1.451	757	756
5½	1.217	522	542
5	1.095	453 2	501.6
4	.9149	297.5	329.2
3	.7988	193	210.0
2	.7146	122	138.6
1	.6006	80.6	94.8
0	.5169	66.8	72.3

When it is observed that the effective diameter of the grains in these sands was found by measuring the flow of air through one sample in one piece of apparatus and the flow of water was measured through another sample and in another piece of apparatus, and that the flow varies as the squares of the diameters of the soil grains, it is clear that the effective diameter has a very exact value so far as the flow of fluids is concerned.

145. The Effective Diameters of Soil Grains and the Amount of Surface Computed From Them.—We have no means of knowing yet how accurately the computed surface of soil grains in a given weight of sample compares with that which is possessed by it. We do know, however, that the comparison is accurate enough to furnish a valuable basis for comparing different types of soils, and in the table which follows is given the effective diameters of several kinds of soils, together with the pore space and the computed amount of soil surface per cubic foot of dry soil.

Table of computed surface of soil grains in different types of soil.

Kind of soil.	Effective diameter of soil grains.	Per cent. of pore space.	Surface of soil grains in one cubic foot.
	m. m.		Sq. Ft.
Finest clay soil	.004956	52.94	173,700
Fine clay soil	.007657	45.69	129,100
Fine clay soil	.008612	48.00	110,500
Heavy red clay soil	.01111	44.15	91,960
Loamy clay soil	.01402	45.32	71,316
Clayey loam	.01810	47.10	53,490
Loam	.02197	44.15	46,510
Loam	.02619	34.49	45,760
Sandy loam	.03035	38.83	36,880
Sandy soil	.07555	34.45	15,870
Sandy soil	.1119	32.49	11,030
Coarse sandy soil	.1432	34.91	8,318

It will be seen from this table that the amount of surface in the true soils is indeed very great, ranging from a little less than a quarter to more than a third of an acre in the sandy soils, through more than an acre in the loams to as much as four acres per cubic foot in the finest clay soils. The amount of soil surface in the upper four feet of every cultivated field ranges from not less than one acre to more than 16 acres per each square foot of surface cultivated.

146. Relation of the Surface of Soil Grains to the Water Capacity.—A large portion of the water held by a soil is spread out as a thin film surrounding the soil grains and it is generally true that the larger the surface of the soil grains the more water the soil will retain.

If a marble is lifted out of water it retains a film surrounding it and its surface is wet; so if rains fall upon a sand or soil surface until percolation takes place, there is held back upon the grains a certain amount of water which is characteristic of or peculiar to each type. It is clear that a soil whose internal surface is 4 acres per cubic foot may contain a large amount of water even though the film is extremely thin. In an acre there are 43,560 sq. ft. and in four acres 174,240 sq. ft. The thickness of a water film on this surface sufficient to equal 4 inches on the level per square foot of soil would be

$$\frac{4}{174,240} = \frac{1}{43,560} \text{ of an inch}$$

or one-half the thickness of the film of a soap bubble when it becomes yellow just before appearing black and breaking, from thinning out. This thickness is also about $\frac{1}{8}$ the diameter of the soil grain itself.

In the case of a fine sand having grains .08188 m. m., which retains, after complete drainage 8 feet above standing water, 3.44 per cent. of water, the film would have to have a thickness of only about $\frac{1}{65}$ of the diameter of the grain, and when containing 20 per cent. of its dry weight then the film need have a thickness of only about $\frac{1}{11}$ of the diameter of the sand grains, that is, .0072 m. m.

It is clear, therefore, from these considerations that the surface of soil grains has much to do in determining the water-holding power of a soil and that the films may be very thin and yet on account of their great extent represent a high per cent. of the soil itself.

147. Movement of Air Through Soil.—There is perhaps nothing which shows how physically different the fine and the coarse grained soils are as clearly as the rates at which air will pass through them when dry, and in the next table some of these are given.

It will be seen from this table that when the grains are so large that 10 of them will span a linear inch only 37 seconds are required for a pressure of .1 foot of water to force 5,000 c. c., 5.3 quarts, of air through a column a foot long and .01 of a square foot in cross section; but in the finest clay soil, which makes the best grass land, where 5,125 grains must be set in line to measure a linear inch, then the time required is 2,933,000 seconds for the same amount of air under the same conditions to be forced through, a ratio of 37 seconds to nearly 34 days.

Table showing the differences in the rate of movement of air through gravel, sand and soils of different types when the columns are 1 foot long, .01 ft. in cross section and under a pressure of .1 ft. of water.

Description of material.	No. of grains per linear inch.	Per cent. of pore space.	No. of seconds for 5,000 c. c. of air to flow through.
Fine gravel, grade No. 8	10.	37.60	37
Fine gravel, grade No. 7	14.0	38.44	67
Fine gravel, grade No. 6	17.5	38.85	99
Fine gravel, grade No. 5.5	20.6	39.20	138
Coarse sand, grade No. 5	24.3	39.88	184
Coarse sand, grade No. 4	27.8	38.53	260
Coarse sand, grade No. 3	31.8	36.26	418
Coarse sand, grade No. 2	35.5	34.66	612
Medium sand, grade No. 1	42.3	34.43	869
Medium sand, grade No. 0	49.1	34.42	1,178
Fine sand, grade No. 60	143	34.20	10,370
Fine sand, grade No. 100	310	35.32	44,310
Coarse sandy soil	177	34.91	14,580
Sandy soil	227	32.49	30,460
Sandy soil	336	34.45	54,910
Sandy loam	837	38.83	227,400
Coarse loam	970	34.49	45,750
Loam	1,156	44.15	2[illegible]2,200
Clayey loam	1,403	47.10	476,600
Loamy clay	1,647	40.19	804,600
Heavy red clay soil	2,286.0	44.15	1,129,000
Clay soil	2,949.0	48.00	1,412,000
Fine clay soil	3,310.0	45.96	2,057,000
Finest clay soil	5,125.0	52.94	2,933,000

It should be understood that this slow rate of movement of air through the finest clay soils was observed when the air-dry soil had been pulverized in a mortar and made as fine as practicable before packing into the aspirator. Under field conditions, as has been pointed out, a good clay soil has its clusters of various sizes and there are passageways of various sizes and forms which allow both air and water to move much more freely than has been recorded in the table and if it were not so plants could not thrive in them.

148. Permeability to Air of Undisturbed Field Soils.—The rate at which air may flow through soils in their natural condition, in place in the field, may be readily studied with an apparatus such as is shown in Fig. 40. When the soil tube A is driven into the ground to near the depth at which the flow of air is to be measured it is recovered, the core of soil removed and the tube returned to its place, when the aspirator is connected as shown in the cut, and the time required for a given volume of air to be drawn through determined. In these field studies it will be found that the dryer the soils are the more freely air passes through them but that when they are saturated with water, as just after heavy rains, little or no air will pass through them even under a pressure of 12 inches of water.

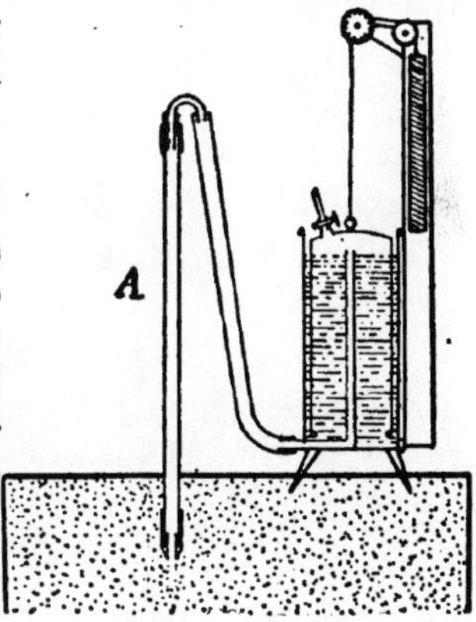

FIG. 40—Showing apparatus for measuring the permeability to air of soils in the field. A core of soil is removed to the desired depth and the soil tube replaced.

149. Weight of a Cubic Foot of Dry Soil.—A cubic foot of undisturbed air-dry soil varies in weight between quite wide limits, the humus soils being the lightest, and the coarse sandy soils the heaviest. The writer has found a dry soil to have the weight per cubic foot given in the table below:

	1st foot.	2d foot.	3d foot.	4th foot.	5th foot.	6th foot.
Pounds per cubic foot	79	92.62	104.59	106.21	111.06	111.06
Pounds per acre......	2,740,000	4,034,000	4,557,000	4,637,000	4,840,000	4,840,000

Shubler gives the weight of a cubic foot of dry soil as follows:

Dry calareous or siliceous sand	110 lbs.
Half sand and half clay......................	96 lbs.
Common arable soil........................	80 to 90 lbs.
Heavy clay	75 lbs.
Garden mould rich in vegetable matter	70 lbs.
Peat soil...................................	30 to 50 lbs.

As a number easy to remember it may be taken as a safe figure that the mean weight of the surface four feet of field soils is, in round numbers, 4,000,000 lbs. per acre-foot.

150. Heavy and Light Soils.—These terms are used more with reference to the ease with which soils may be worked than to their weight per cubic foot. A soil that is naturally mellow and easily stirred is called a light soil, while one that becomes hard when dry and which tends to form clods is often called heavy. Sandy soils, as shown in (**149**) are among the heaviest we have while the clayey varieties are the lightest by weight except the humus types. The prairie loams which contain much humus and the black swamp soils when drained are among the most mellow of all soils, the large amount of humus preventing the soil grains from adhering and baking.

CHAPTER V.

SOIL MOISTURE.

151. Occurrence of Moisture in the Soil.—For purposes of discussing the cultural relations of soil moisture water may be said to occur in the soil under three conditions:

(1) That which fills the pore spaces between the soil grains and is free to move under gravitational or hydrostatic pressure and may be called *gravitational* or *hydrostatic water.*

(2) That which adheres to the surfaces of soil grains and to the roots of plants in films thick enough to allow surface tension to move it slowly from place to place, and which may be called *capillary water.*

(3) That still retained on the surfaces of soil grains when they become air-dry; whose chief movements are those of evaporation and condensation and which has been designated *hygroscopic moisture.*

152. Gravitational Water.—When water in a soil increases in quantity sufficiently to move readily under the pull of gravity it may be harmful in three ways: (1) by washing out the soluble plant foods, thus leaving the soil poor; (2) by excluding the air and thus causing suffocation of the roots of plants and micro-organisms living in the soil; (3) by preventing surface tension and by dissolving cementing materials, thus destroying or reducing the granulation of soils, injuring their texture. It may be helpful in two ways: (1) by replenishing the capillary moisture when this has become too small to enable crops to supply themselves, and (2) by washing out and carrying away soluble substances which, if allowed to accumulate, become injurious, such as black alkalies and possibly toxic principles developed by the roots of plants or soil organisms or during their decay.

153. Capillary Water.—It is in this condition or quantity in the soil from which crops and soil organisms chiefly derive their supply of water, and the right amount at all times is therefore very important. It is in the capillary water, too, that most of the plant foods derived from the soil are held in solution and with it moved to the plants as needed. When the texture of the soil is right the capillary water simply surrounds the soil grains and soil granules as a thin sheet which is continuous where the grains are nearly or quite in contact, but there are always open spaces through which the air may circulate and supply the needs of roots and soil bacteria.

If the soil is puddled and the granules broken down then the surface films on the smaller soil grains come so nearly in complete contact that there is insufficient room for air to diffuse and plants cannot thrive in it.

154. Hygroscopic Water.—Moisture in this form possibly plays an important part in the actual solution of plant food from the soil and fertilizer grains because it is this portion which lies in immediate contact where the action must take place; but if this is true it can only do its work rapidly when capillary water is also present to carry away from the dissolving surfaces the products which are being formed.

Polished surfaces do not as readily rust as those which have become tarnished or otherwise roughened. When a steel knife blade has become a little rusty the rusting then goes on much more rapidly, possibly because each particle of rust becomes invested with its film of hygroscopic moisture, and when these lie against the fresh metal the water can have a greater thickness and permit a more rapid movement of the compounds formed, away from the corroding surface.

It is not probable, however, that the hygroscopic moisture of a soil can in any direct way aid plant growth.

155. Ways of Expressing the Water Content of Soils.—The amount of water a soil will or may contain has been expressed in different ways: (1) As a per cent. of the wet weight of the soil, (2) as a per cent. of the dry weight of the soil, (3) as a per cent. of the volume of the soil, (4) in pounds per cubic foot, (5) in inches per cubic foot. The amount of moisture a soil does contain may be most readily and precisely stated as per cents. of the wet or dry weight, but for agricultural purposes it is best to state the amount in per cent. of the volume or in inches per cubic foot.

156. The Maximum Water Capacity of Soils.—The largest amount of water a soil may contain is expressed by its per cent. of pore space and if reference is made to the table in (**145**) it will be seen that this ranges from about 32 to more than 52 per cent., that is from 4 to 6 acre-inches per acre-foot of soil, and from 20 to 32 lbs. per cubic foot. These amounts of water, however, are never found in soils under field conditions.

157. Water Capacity of Soils Under Field Conditions.—The amount of water which may be retained by soils under field conditions is extremely variable and depends upon a number of factors. In the table below are given the amounts of water which were found in three types of soil with the undisturbed field texture, when they contained as much as they would retain after a few days of drainage following heavy rains.

Capacity of field soils for moisture.

Depth.	Sandy loam.	Clay loam.	Humus soil.
	Per cent.	Per cent.	Per cent.
First foot	17.65	22.67	44.72
Second foot	14.59	19.78	31.24
Third foot	10.67	18.16	21.29

In this table the third foot in each case is more or less sandy and for this reason shows percentagely less water than the soil above. It will be seen that the surface foot of sandy loam contains the smallest per cent. of water and the humus soil the largest, but on account of the differences in dry weight of these soils their water contents are more nearly equal than they appear, the sandy loam containing about 16 lbs., the clay loam 18 lbs. and the humus soil 26 lbs. per cubic foot. Expressed in inches the amounts stand 3, 3.5 and 5 inches nearly.

158. Maximum Capacity of Undisturbed Field Soil.—In the table below are given the amounts of water which completely filled the first five feet of undisturbed field soil, as determined by driving 6-inch metal cylinders one foot long into the soil and, recovering them, covering the bottoms with perforated covers and then placing the cylinders under water until the pores became completely filled.

Table showing maximum capacity of undisturbed field soil for water.

Kind of soil.	Depth.	Per cent. of water.	Inches of water.
Clayey loam	1st foot	41.3	5.88
Reddish clay	2d foot	28.1	5.03
Reddish clay	3d foot	28.4	5.07
Clay with sand	4th foot	24.8	4.67
Very fine sand	5th foot	17.4	3.76
Total			24.41

In this case it is seen that two feet out of five feet of the soil was open space which could be occupied with water.

159. Maximum Capillary Capacity of Soils for Water.—The amount of water which may be retained in soils by capillarity is greatly influenced by the distance of the soil above standing water in the ground and by the frequency and amount of rainfall. The cylinders of soil referred to in (158) when thoroughly dried and then placed in one inch of water in a chamber where no evaporation could take place, took up and retained by capillarity the following amounts of water:

Table showing the maximum capillary capacity for water of field soils with the surface 11 inches above standing water.

	Per cent. of water.	Lbs. of water per cu. ft.	Inches of water.
Surface foot of clay loam contained..........	32 2	23.9	4.59
Second foot of reddish clay contained.........	23 8	22 2	4.26
Third foot of reddish clay contained..........	24 5	22 7	4 37
Fourth foot of clay and sand contained.......	22.6	22 1	4.25
Fifth foot of fine sand contained	17.5	19.6	3.77
Total............	....	110.5	21.24

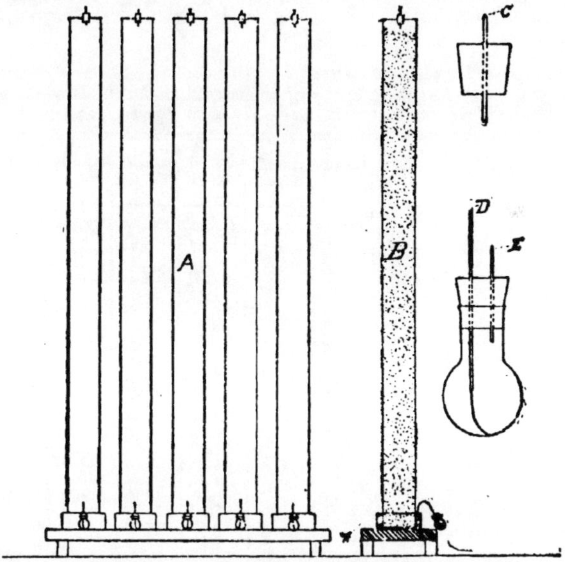

FIG. 41.—Apparatus for measuring the capillary capacity of long columns of sand.

It is clear from this table and the last that much of the pore space in the clayey soils cannot be maintained full of water by capillarity even when the surface is only 11 inches above standing water.

100. Influence of Distance Above Standing Water on the Water Capacity of Soils.—When the distance to the ground-water is considerable the force of surface tension is not great enough to maintain as much water in the soil as when the distance is less, and the table which follows shows how the amount of water retained varies with the distance. The sands and soils were placed in an apparatus represented in Fig. 41, arranged so as to permit free percolation but allowing very little evaporation from the surface. The sand columns were 8 feet long and percolation was allowed to continue nearly 2.5 years. The soil columns were 7 feet long and percolation from them was continued during 60 days, at the end of which time the tubes were cut into short sections and the amount of water still retained determined by drying.

Percentage distribution of water left in columns of sand, sandy loam and clay loam after percolation had continued two and one-half years with the sand and 60 days with the soils.

Height of section above ground water.	Sand No. 20.	Sand No. 40.	Sand No. 60.	Sand No. 80.	Sand No. 100.	Sandy loam.	Clay loam.
Inches.	Pr. ct.	Pr. ct.	Pr. ct.	Pr. ct.	Pr. ct.	Pr. ct.	Pr. ct.
96-93	0.27	0.17	0.22	1.26	3.44	....	....
93-90	.22	.17	.23	1.16	3.44	....	....
90-87	.23	.16	.29	1.34	3.82	....	....
87-84	.22	.15	.32	1.61	3.83	....	....
84-81	.23	.18	.61	1.98	3.93	16.16	21.16
81-78	.29	.19	1.07	2.32	4.19	....	....
78-75	.44	.26	1.33	2.61	4.38	16.08	30.70
75-72	.89	.58	1.57	2.90	4.92	....	....
72-69	1.18	1.16	1.80	3.12	4.94	16.55	31.03
69-66	1.48	1.45	1.85	3.36	5.70	....	....
66-63	1.71	1.67	2.03	3.56	5.91	16.97	31.11
63-60	1.80	1.80	2.18	3.92	6.43	....	....
60-57	1.83	1.86	2.26	4.22	6.77	17.59	31.21
57-54	1.93	1.87	2.27	4.53	7.72	....	....
54-51	1.98	1.98	2.30	4.88	8.59	17.99	31.94
51-48	2.02	1.92	3.38	5.42	9.42	....	....
48-45	2.03	2.12	2.46	6.03	10.50	18.70	31.99
45-42	2.02	2.07	2.71	6.99	11.34	....	....
42-39	2.06	2.18	3.08	7.47	12.58	19.44	32.18
39-36	2.17	2.29	3.46	8.71	13.00	....	....
36-33	2.31	2.48	4.10	10.54	14.95	20.90	32.45
33-30	2.36	2.65	5.09	11.77	15.90	....	....
30-27	2.63	3.14	6.36	12.95	17.20	21.71	33.31
27-24	2.86	3.63	8.74	15.05	17.96	....	....
24-21	3.42	4.71	13.52	17.24	18.92	21.46	34.40
21-18	4.26	6.76	23.57	19.08	20.49	....	....
18-15	6.41	9.38	27.93	19.37	21.34	22.17	35.54
15-12	9.77	14.66	23.61	21.44	21.63	....	....
12- 9	16.08	21.31	22.46	22.69	22.68	22.68	35.97
9- 6	19.33	22.39	22.76	23.20	23.39	....	....
6- 3	20.96	23.52	22.88	24.22	30.28	27.69	37.19
3- 0	21.58	24.61	23.54	25.07	24.08	....	....

This table shows very clearly that the amount of water a soil can retain by capillarity is very materially influenced by the distance it is above the zone of complete saturation or of standing water in the ground. The decrease of water upward is most rapid in the coarsest sand and it is least rapid in the finest soil.

It is remarkable that in sands so coarse as those used water should continue to drain away during more than two years from so short a vertical column and that so small an amount of water was retained in the upper sections of the columns. It is not probable that drainage had become complete from the two soils although it may possibly have been, as there was no percolation during the last five days of the trial.

161. Proportion of Soil-Water Available to Crops.—Not all the water which soils will retain is available to plants. A certain amount must be left overspreading the soil grains which the roots of plants are unable to use. The amount found in one field soil, when corn and clover ceased to grow and when the leaves curled early in the day, is given in the table below. In the same table is also given the moisture of adjacent fallow ground determined at the same time and which contains the least amount of water which, for this soil, will permit maximum crops.

Soil moisture relations when growth is brought to a standstill.

Depth of sample.	Clover.	Maize.	Fallow ground.
	Per cent.	Per cent.	Per cent.
0–6 inch clay loam	8.39	6 97	16.28
6–12 inch clay loam	8.48	7.8	17.74
12–18 inch reddish clay	12.42	11.6	19.88
18–24 inch reddish clay	13.27	11.98	19.84
24–30 inch sandy clay	13.52	10.84	18.56
40–43 inch sand	9.53	4.17	15.9

The moisture contained in the fallow ground shows how much this type of soil may retain, against evaporation and percolation, during a dry season, and it happens to stand just at the under limit for most vigorous growth while the upper limit is given in the next table.

Showing upper and lower limits of best amount of soil moisture for one type of soil.

Kind and depth of soil.	Lower limit of soil moisture.	Upper limit of soil moisture.	Available soil moisture.
	Per cent.	Per cent.	Lbs. per cu. ft.
Clay loam, first foot	17.01	25.77	6.92
Reddish clay, second foot	19.86	24.3	4.112
Sandy clay, third foot	18.53	24.03	5.722
Sand, fourth foot	15.9	22.29	6.786
Total			23.54

It will be seen from this table that, to bring the surface four feet of soil from the lower limit of the best productive stage of water content to the upper limit, requires an application of 23.54 pounds per square foot, or a depth of rainfall equal to 4.527 inches. This therefore represents the available moisture in this type of soil and is about one-third of its full capillary capacity.

162. Kinds of Soil Which Yield Their Moisture to Crops Most Completely.—When the roots of plants come to draw upon the supply of soil-water those soils yield their moisture to the plant most completely whose grains have the largest diameter or, more precisely, which have the smallest internal surface to which the moisture may adhere and over which it is spread.

Referring to the table in (**161**), giving the per cents. of moisture which were too low to permit the plants to supply their needs, it will be seen that under the corn the water in the sand had been drawn down to about 4 per cent.; in the surface loam to 7 and 8 per cent.; while in the intervening more clayey portion only to 11 and 12 per cent. The fundamental truth which should be grasped here is that all these soils are equally dry so far as the needs of the corn crop are concerned, and one of the reasons why they are so is because the thickness of the water film surrounding the grains is nearly the same in all the cases.

The truth of this statement will be evident if we compute the per cent. of moisture in a soil which a given thickness of film surrounding the grains will produce.

163. Relation of Thickness of Moisture Films to Per Cent. of Soil Moisture.—If the data in the table of (**145**) is used the per cent. of soil moisture a given thickness of film will produce may be computed approximately from the formula

$$P = K\frac{ST}{Q}$$

where P = the per cent. of moisture in the soil.

K = a factor, Log. $\bar{2}.497532 = .0314355$

S = surface of soil per cu. ft. taken from (**145**)

T = thickness of film of moisture

Q = per cent. of dry soil obtained by subtracting the pore space in (**145**) from 100.

Using this formula and the data in (**145**) it will be found that the per cents. of moisture stand as given below: With thickness of film $\frac{275}{10000000}$ inch the per cent. of water will be, in the

Heavy red clay	14.24 per cent.
Loamy clay	11.28 per cent.
Loamy clay	8.74 per cent.
Loam	7.20 per cent.
Sandy loam	5.21 per cent.
Sandy soil	2.09 per cent.
Sandy soil	1.41 per cent.
Coarse sandy soil	1.11 per cent.

From this table it will be seen that the coarse sandy soil contains only 1.1 per cent. of its dry weight of moisture when the heavy red clay contains 14 per cent. with the same thickness of film surrounding the soil grains.

Comparing these per cents. of moisture with those contained in the soil in which the corn wilted, it will be seen that the sand of that soil was really the wettest soil there, so far as the available moisture is concerned, there being at least 2 per cent. of moisture yet available. The loamy clay of (**145**), and given in the table, has about the same texture as that of the reddish clay in the table of (**161**) and it will be seen that its per cent. of moisture under the corn was also about the same as that computed.

164. Available Soil-Moisture Affected by Jointed Structure in Clay Subsoils.—The tendency of clay subsoils to shrink and become divided into small cube-like blocks greatly diminishes the available moisture in them. This shrinkage not only often results in breaking rootlets in two but when new rootlets form they advance most readily through the fissure planes and are not able to place themselves in the most favorable relations with the soil to permit capillarity to bring the moisture to the rootlets. It is because the sandy soils and loams seldom develop the structure referred to and because the rootlets and root hairs are able to secure a more uniform distribution throughout them as well as because of the larger size of their grains that plants are able to drain their moisture down to so low a per cent.

165. Available Soil-Moisture Increased by Open Structure.—When soils are in any way left with a loose open structure, as happens with deep plowing and especially with good subsoiling, not only is the ability of the loose soil to retain moisture increased but a larger proportion of this retained water becomes available to the crop. A larger amount of water is retained because when perfect capillary connection with the unstirred soil below, is broken, surface tension opposes rather than aids gravity in producing percolation and spaces too large to remain full of water otherwise are able to retain it.

When the soil is open and loose the case is quite different from that resulting from shrinkage referred to in (**164**), for in this case the roots and root hairs are better able to enter the separated portions and, as the moisture films are thicker, the moisture is more readily gathered.

166. Drainage May Increase the Available Soil-Moisture.—When the subsoil is too close and too fully saturated with water to permit the roots of crops to penetrate it, as is the case where drainage is needed, the roots of plants are forced to develop in so limited an amount of soil that when a drying time comes, and when the demands of the crops for moisture are large because of rapid growth, capillarity from below is not able to supply the moisture as fast as needed, and the result is the zone of soil occupied by the roots becomes so dry that growth is impeded.

On the other hand, where a field is well drained the roots are extended through much larger volumes of soil; the local demands are thus less urgent and the water need not move so far by capillarity before the plant comes in possession of it. Under these conditions the moisture of the surface four feet of soil is in close reach of the roots and capillarity may still add to this supply from below.

167. The Amount of Water Required by Crops.—It has been determined by careful and extended observations in this country and in Europe that almost any one of the cultivated crops withdraws from 300 to 500 tons of water from the soil for each ton of dry matter produced. In Wisconsin the amounts of water lost from the soil by evaporation during the growing season and through the plant are given in the table below:

Table showing the mean amount of water used by various plants in Wisconsin in producing a ton of dry matter.

	No. of trials.	Water used per ton of dry matter.	Water used.	Dry matter per acre.	Acre-in. of water per ton of dry matter.
		Tons.	Inches.	Tons.	
Barley	5	464.1	20 69	5.05	4.096
Oats	20	503.9	39 53	8.89	4.447
Maize	52	270.9	15 76	6.59	2.391
Clover	46	576.6	22 34	4.39	5.089
Peas	1	477.2	16.89	4 009	4.212
Potatoes	14	385.1	23 78	6.995	3.399
	138	Av. 446.3	23,165	5.987	3.939

From this table it is seen that the amount of water used ranges from 270 tons of water with corn to 576 tons with clover per ton of dry matter; or when expressed in acre-inches from 2.4 to 5.1 inches nearly, the average for the six crops being nearly 450 tons or 4 acre-inches per ton of dry matter.

When the yields per acre are 2, 3 and 4 tons the numbers given above must be multiplied by the same factors.

168. Amounts of Water Required for Different Yields of Wheat.—In order to express the data of the last section in terms which it is more customary to use, there is given in the next table the amount of water required by a crop of wheat when the yields per acre range from 15 to 40 bushels.

Observations made by Hellriegel in Germany show that wheat uses about 453 tons or 3.998 acre-inches of water for a ton of dry matter. Using this ratio and one pound of grain to 1.5 pounds of straw the water required will stand as below:

Table showing the least amount of water required to produce different yields of wheat per acre when the ratio of grain to straw is 1 to 1.5.

Yield per acre.				Water used.
Number of bushels.	Weight of grain.	Weight of straw.	Total weight	
	Tons.	Tons.	Tons.	Acre-inch.
15	.45	.675	1.125	4 498
20	.60	90	1.500	5.998
25	.75	1 125	1.875	7.497
30	.90	1.350	2.250	8.997
35	1.05	1.575	2.625	10.495
40	1.20	1.800	3.	12.

This table shows that 12 inches of effective rain during the growing season of wheat, starting with the soil moisture in good condition, should enable a yield of 40 bushels per acre to be produced.

169. Least Amount of Water Which Will Permit Yields of Different Amounts.—In the next table there is given the least amount of water taken from the soil which can be expected to give the yields for the different crops there stated:

This table must be regarded as showing the minimum amounts of water which will bring the crops named to full maturity so as to produce the yields specified under conditions of absolutely no loss by surface or under-drainage, and where the evaporation from the soil itself is as small as it can well be. It must be farther understood that the soil at seeding time already possesses the needful amount of water for the best conditions, and that at the end of the growing season it is yet so moist that no check to vigorous, normal growth has occurred.

Table showing the highest probable duty of water for different yields per acre of different crops.

Bushels per acre	15	20	30	40	50	60	70	80	100	200	300	400
Name of crop	Least number of acre-inches of water.											
Wheat	4.5	6	9	12	15	18	.....	.....	....	....	.....	.....
Barley	3 21	4 28	6.42	8 56	10 7	12.84	19 98	.....	....	....	.....	.....
Oats	2.3[illegible]	4.136	5 701	6 272	7 84	9 40	10 98	12.51	15 6[illegible]	....	.....	.....
Maize	2.52	4 36	5.04	6.72	8.4	10 08	11.75	13 43	16.77	...	.....	.....
Potatoes	....	.41	.62	.83	1.03	1.24	1.45	1.65	2.07	4.14	6.2	8.27

Tons per acre	1	2	3	4	6	8	10	12	14	16	18	20
	Least number of acre-inches of water.											
Clover hay, 15 per cent. water	4.43	8 85	13 28	17.7	26.55	35.4	14.25	.....	.....	.....	.....	.....
Corn with ears 15 per ct. water	2.08	4.16	6.24	8.32	12.47	16.61	20.72	24.95	29.1	33.26	37.42	41.58
Corn silage, 70 per cent. water	1.41	2.82	4.23	5.64	8.46	11.28	14.1	16.92	19.74	22.56	25.38	28.2

CHAPTER VI.

PHYSICS OF PLANT BREATHING AND ROOT ACTION.

MECHANISM AND METHOD OF TRANSPIRATION IN PLANTS.

170. Breathing of Plants and Animals.—The transpiration of plants and the respiration of animals are processes which have much in common. Both plants and animals are provided with internal cavities into which air may enter. They both breath air. While breathing air both give off large quantities of moisture. The primary object of the lungs is to supply the body of the animal with oxygen and to remove carbon dioxide. The corresponding structure in the leaves of plants is to supply it with carbon dioxide and to throw off oxygen. In both cases the breathing surface has a very delicate texture and is situated where it can always be kept wet; the chief function of the water escaping from the breathing surface is to keep it moist.

If the lining of the lungs were to become dry and parched the gases would not as readily pass through and there would be like difficulty in the case of leaves, if their breathing surfaces were not kept moist. In both plants and animals the breathing surfaces are carefully guarded from the intense sun and strong drying winds.

171. Respiratory Organs in Plants.—The air passages or breathing chambers of plants are chiefly located in the leaves, but they are also found to greater or less extent in all the green parts. They are simply irregular chambers left between the cellular tissue and are represented in the lower portion of Fig. 42, which shows a section of barley leaf with the epidermis removed and much magnified.

172. Breathing Pores.—Leading into the air chambers are many breathing pores through which the air enters. Eight of these are represented in Fig. 42. They are most numerous on the under sides of leaves where evaporation may be least.

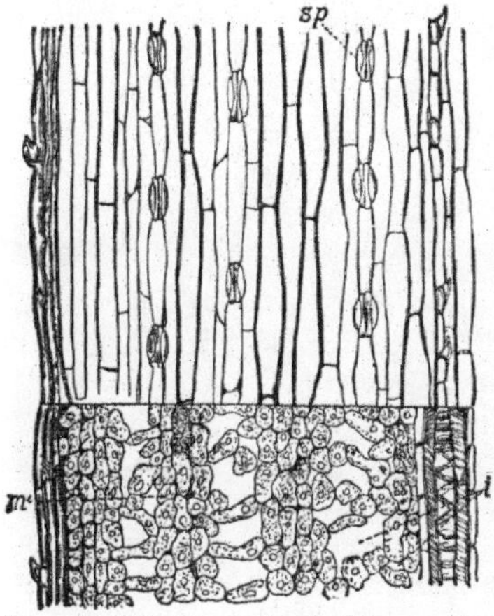

FIG. 42—Structure of barley leaf. (After Sorauer) sp is a breathing pore; m, chlorophyll cells; i, respiratory chambers.

The breathing pores or stomata are very small and numerous, Weiss estimating, from an average of 40 plants, as many as 209,000 in each square centimeter of surface, an area equal to the square shown in Fig. 43. In the case of a corn leaf 21 per cent. of the surface is occupied by the doorways to the breathing chambers.

173. Chlorophyll Cells.—Surrounding the air chambers in every leaf there are multitudes of tender, thin-walled cells in which are found the green chlorophyll grains, giving color to the leaf, which absorb the sunshine and use it in breaking down the carbon dioxide for the carbon, which is one of the chief constituents of plant tissues, and of the starches, sugars and most other compounds.

174. Guard Cells.—In order that the loss of water may be as little as possible each breathing pore is surrounded by a pair of guard cells, represented in Fig. 42, and on a much larger scale in Fig. 43. These guard cells have for their function the regulation of the amount of evaporation from the plant. The chlorophyll grains can be effective in breaking down the carbon dioxide only in comparatively

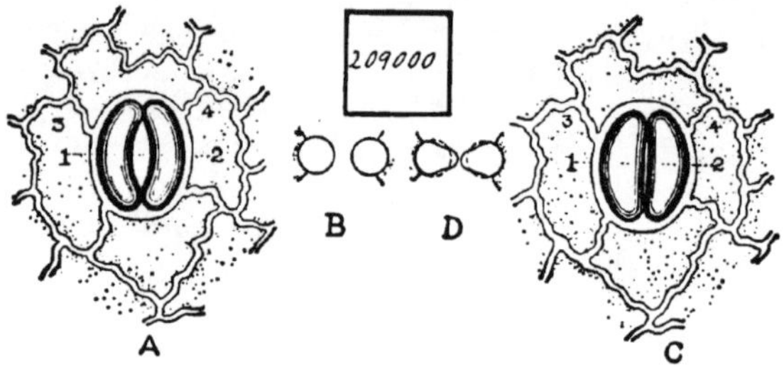

Fig. 43.—Diagram showing the mechanical action of guard cells in opening and closing breathing pores. The square shows the area of under side of leaf containing an average of 209,000 breathing pores or stomata. (From Irrigation and Drainage.)

bright light and so, during cloudy days and at night, the guard cells automatically change their form and close the doors, reducing evaporation. Indeed they remain open only when there is light enough to utilize it in decomposing the carbon dioxide.

175. Action of the Guard Cells.—The opening and closing of the guard cells is brought about by their peculiar shape and changes in the amount of material they contain.

Unlike the other cells in the epidermis of the leaf these contain chlorophyll grains and are thus able to carry on the process of developing plant food. The advantage of having this work done here is to increase the osmotic pressure through the rendering of the sap denser when the sun is shining, thus distending the cells and changing their shape so as to open the doors widest when the sun shines brightest, as represented at A, Fig. 43. When night comes or it is cloudy then the osmotic pressure forces the assimilated material out of the guard cells faster than it is produced and the walls collapse, taking the attitude represented at C and in cross section at D, closing the opening. B and D are cross sections of a pair of guard cells along the lines 1–2 and B shows how a full cell must pull the edges apart while

D shows how the limp condition will permit the walls to fall together.

176. Loss of Water Through the Guard Cells.—The epidermis of the leaf is so close in texture and often so waterproofed that when the guard cells close there is but little loss of moisture. But when the sun shines and there is moisture enough in the soil to keep the leaves from wilting the guard cells open wide and great evaporation may take place even in a saturated atmosphere.

By admitting live steam into our plant house on bright sunny days, keeping the air highly saturated, we have found corn to lose nearly as much moisture as in the dryer condition of the air with the sun also shining. The reason this is possible is that the epidermis acts like the glass of the hot bed, permitting the sunshine to enter but preventing the longer dark heat waves from escaping. In this way the air saturated outside is not so inside on account of the higher temperature. This remarkable provision of the plant to save moisture should teach how important it is to assist, in every way practicable, the conservation of soil moisture.

STRUCTURE AND MODE OF ROOT ACTION.

There is scarcely a better illustration anywhere in Nature of the adaptation of living organisms to their environments than is furnished by the mechanism by which the higher land plants supply themselves with moisture; and one of the most remarkable of remarkable tasks is that of a corn plant pumping into its stem and leaves, from a comparatively dry soil, 2.896 pounds of water daily for 13 consecutive days.

177. Functions of Roots.—The roots of ordinary land plants have three distinct functions to perform: First, to gather from the soil its moisture and the salts dissolved in it for the use of the plant; second, to convey and deliver into the stem and leaves the water absorbed; and third, to act as an anchor or support, holding the plant upright in the soil, air and sunshine.

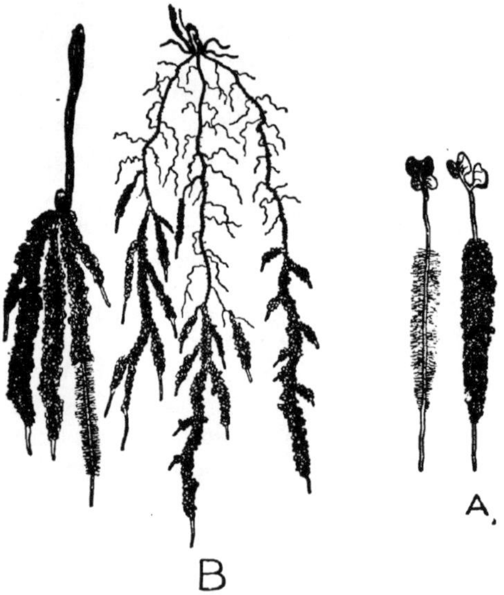

FIG. 44.—A, Root-hairs of mustard plants, with soil adhering, and with soil removed. B, root-hairs of wheat, when very young, and four weeks later. (After Sachs.)

178. The Absorbing Portion of Roots.—It is the general belief of plant physiologists that the active portion of roots—that which is immediately concerned in gathering the water from the soil—is what are known as root hairs, represented at the left of A, Fig. 44, and at A buried in the soil grains in the same figure. In Fig. 46 is a much enlarged view of a single root hair which has worked its way in among the soil grains where it is in place to absorb soil moisture and soluble salts. The appearance of root hairs in relation to soil grains can be clearly demonstrated by growing plants in rather coarse sand between glass plates as represented in the apparatus shown in Fig. 45.

179. Structure of Root Hairs.—Root hairs are extremely thin walled and greatly lengthened single cells, having lengths ranging up to an eighth or quarter of an inch and a diameter of $\frac{1}{500}$ of an inch. They stand out about the main root like the pile of velvet, forming a brush-like appearance as shown in Fig. 44. The object of this form is to secure a large area around which surface tension may force the water in the same way that it does about the soil grains. Indeed root hairs have forms adapted to drawing upon themselves a portion of the water film investing the soil grains.

FIG. 45.—Apparatus for observing the growth of roots and their relation to soil grains. The sides of the apparatus are two panes of glass, 1.5 inches apart.

180. Relation of Root Hairs to Soil Grains.—The manner in which root hairs place themselves among the soil grains is clearly shown in the form of a diagram in Fig. 46 where h h is a root hair; e is the main root, 2 a soil granule, and 1 an air space; while the concentric lines represent the films of capillary moisture which surround both the granules and the root hairs. In Fig. 47 is represented the tip of a young growing root advancing into fresh soil and having five root hairs developed in place among the soil grains ready for work.

181. Method by which Root Hairs Gather Water.—As the root hairs force their way through the pore spaces among the soil granules they bring their walls into close touch with them in such a way that in form and position they make up a part of the soil mass. In this relation the force of adhesion draws the capillary water out over their walls so

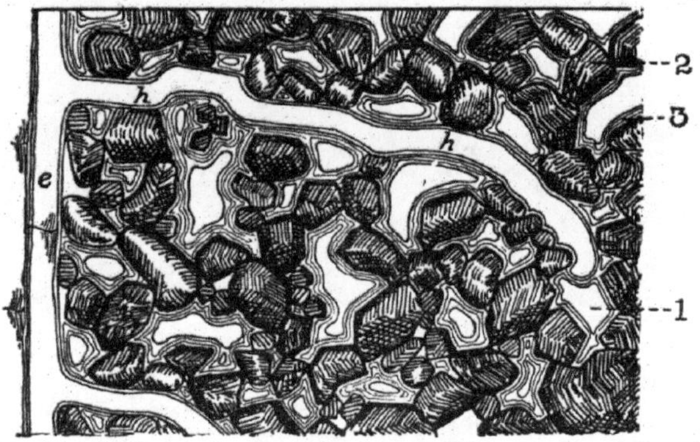

FIG. 46.—Distribution of water on the surfaces of soil grains and of root-hairs. e, main root; 1, air space; 2, soil grain; 3, film of water; h h, root-hairs. (After Sachs.)

as to leave them and the soil granules surrounded by the water film. Each root hair is or should be in a sense under water, that is invested in a film of greater or less thickness. When a portion of this water enters the root hair and passes on into the root and up to the leaves, the water layer surrounding the root hair is left thinner; but no sooner does this thinning out occur than the equilibrium is destroyed and surface tension at once squeezes more water onto the surface from the surrounding soil. In this way capillarity keeps the water moving to the root hairs as they pass it on to the plant.

182. Advance of Roots through the Soil.—Until the method by which roots advance through the soil is understood it is difficult to realize how it is possible for such delicate structures to set the heavy soil aside sufficiently to reach the great depths they do. Nature's method of overcoming the difficulty is simple enough and it is as effective as it is simple. The large amount of open space there is in the surface four to six feet of soil makes it easier to set the soil aside, and the setting of fence posts proves how large this space is. A 6-inch post set in the hole dug for it seldom occupies so much of the space but that all of the soil removed may be returned by thorough ramming. It is the existence of such large amounts of open space in the soil which makes the movements of water, air and roots through it possible and the absence of it which makes a puddled soil so uncongenial to plant growth.

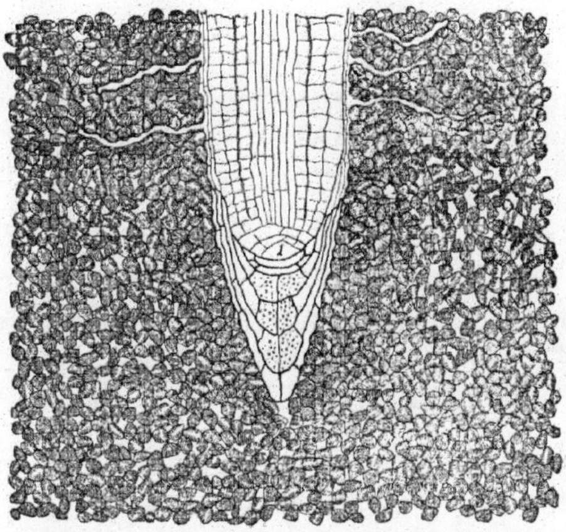

FIG. 47.—Method by which root-hairs advance through the soil. (Adapted from Sachs.)

In Fig. 47 is represented a section of the tip of a root growing and advancing through the soil. It has been found that at 1, a short way back from the tip, there is a center of growth. Here new cells are forming by division and subsequent enlargement. On the forward side of this cell the new ones build the root cap, which acts as a shield and wedge, while those in the rear are finally transformed to make the various structures found in the root.

At the center of growth new cells are forming and expanding under the intense power of osmotic pressure and, as the root is anchored behind, the root cap is pushed forward and wedged sidewise, setting the soil aside and thus making room for itself. The root cap does not slide forward past soil grains but is anchored rigidly to them; the tip entering existing cavities is enlarged by growing forward under and through the cap, the rear cells of which die after the root has grown past them, the root cap being a sort of point continually renewed as the root advances.

183. The Extent of Root Development of Corn.—It is only by careful study that the extent of root development in a soil can be learned. In Figs. 48 and 49 are shown the amount and distribution of corn roots at two stages of growth. When the corn was 30 inches high the whole of the soil to a depth of two feet was as full of roots as the engraving shows between the two hills; when the corn was coming into tassel the roots had penetrated to a depth of three feet and had come closer to the surface; and at maturity the roots had reached four feet in depth, making their way through a fairly heavy clay loam and clay subsoil, the fourth foot only being sandy.

It should be understood that the roots here shown grew in undisturbed field soil and were obtained by going into the field at the stage of growth shown and digging a trench around a block of soil a foot through and the length of the width of the row. The cage was then set down over the block; wires run through the block of soil to hold the roots in place and then the soil washed away by pumping water in a fine spray upon the block. Three days' work for two men were required to secure the sample in Fig. 49.

184. Extent of Root Development of Grain.—In Fig. 50 is represented the depth to which the roots of winter wheat, barley and oats penetrated a heavy clay soil and subsoil. The roots are what were found in a cylinder of soil just one foot in diameter and were obtained by driving a cylinder of metal four feet long its full depth into the soil and then washing the dirt out of it. It will be seen that in each case the roots have reached a depth of fully four feet.

FIG. 48.—Showing amount and distribution of corn roots under natural field conditions.

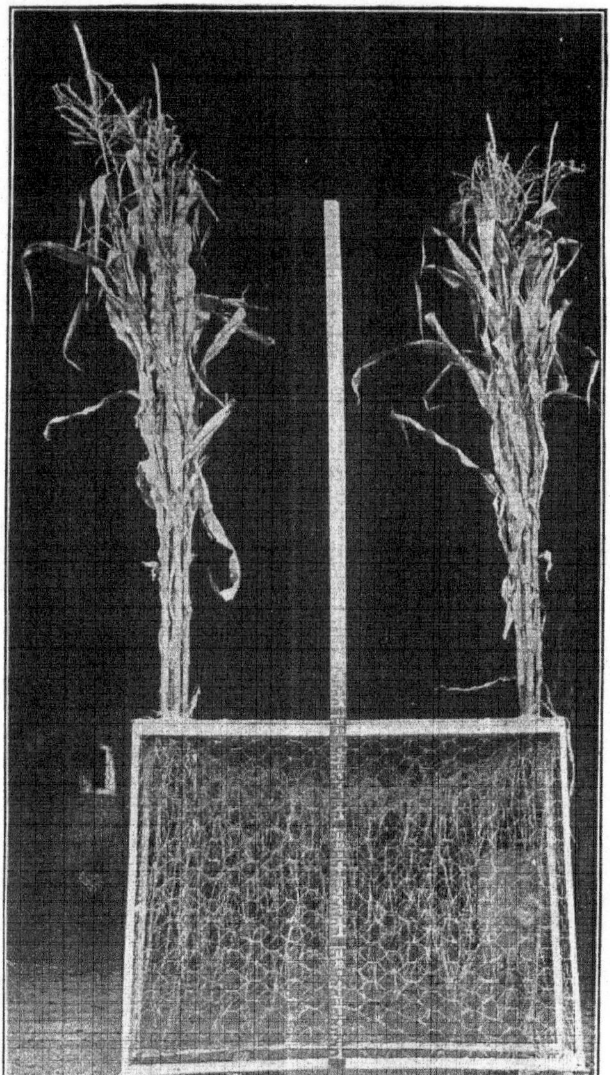

FIG. 49.—Showing amount and distribution of corn roots under natural field conditions.

Wheat. Barley. Oats.

Fig. 50.—Showing amount of roots found in the field in cylinder of soil one foot in diameter, extending to a depth of four feet.

FIG. 51.—Showing the total root of one hill of corn.

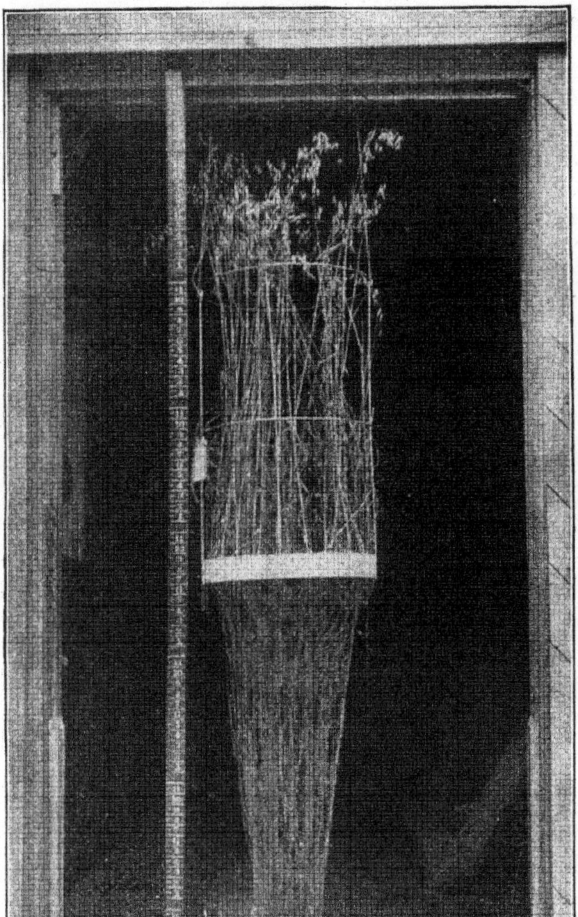

FIG. 52.—Showing total roots of oats.

FIG. 53.—Showing total roots of medium clover.

The coarse branches shown with the winter wheat roots are the roots of a red oak tree which was growing in a pasture 33 feet away, and they serve to show how far forest trees send their roots foraging through the soil for water and food, and through what long lines the water must be pumped after it has been gathered.

185. The Total Root of Plants.—In the preceding sections the samples simply show the amount of root found in a given volume of field soil. In Fig. 51 is shown the total root of four stalks of corn, while Figs. 52 and 53 show the same thing for oats and medium clover. These were secured by growing the plants in cylinders 42 inches deep and 18 inches in diameter, filled with soil. When the crops were mature the cylinders were cut down and the soil washed away.

In each case the roots extended to the bottoms of the cylinders, forming a dense mat there, as the engravings show.

The roots shown with the clover, and which gathered the moisture for the top, forced from the soil water enough to cover the space to a depth of 29 inches. It will be seen that the stand of clover is very close, fully three times as heavy as a good clover crop in the field. This was made possible by having a rich soil and supplying all the water the plant could use at just the right time.

The length of all these roots is less than it would have been had the cylinders been deeper, as proven by the matting at the bottom.

CHAPTER VII.

MOVEMENTS OF SOIL MOISTURE.

186. Types of Soil Moisture Movement.—The moisture which is found in the soil above the surface of the ground water is continually subjected to three types of movement: (1) Gravitational, (2) Capillary and (3) Thermal; the first due to the action of gravity, the second to surface tension and the third to heat.

When rain falls upon the soil one portion of it begins to flow vertically downward through the pore spaces, urged to do so by the pull of gravity; a second portion increases the thickness of the water film surrounding the soil grains and root hairs and is made to do so by surface tension; while a third portion is returned to the atmosphere through evaporation, caused by heat.

GRAVITATIONAL MOVEMENTS.

187. Percolation of Soil Moisture.—The direct gravitational flow of soil moisture, which occurs during and after rains, is nearly always vertically downward until the ground-water surface is reached. The movement takes place chiefly through the shrinkage cracks and passageways left by the decay of roots and the burrowing of animals, but also through the capillary pores formed by the grains of the coarser soils and by the granules of the finer types.

The rate of movement is most rapid following heavy rains when the soil is already well saturated. After prolonged periods of drought, when the soil has become very dry, there is so much air in the pore spaces that it greatly impedes percolation except in those cases where wide shrinkage checks and cracks have resulted.

Where percolation is influenced chiefly by soil texture it is most rapid through the sandy soils and the more granulated clay types. It is least rapid through the puddled clays.

188. Rate of Percolation Through Sands.—When the simple sands are once completely filled with water the percolation from them is quite rapid but decreases with the size of the sand grains. In the table below is given the amount of water which percolated from the columns of sand referred to in (160).

Table giving the rate of percolation from sands under the gravitational head of the inclosed water.

Grade of Sand.	Effective diameter of grain.	Per cent of pore space.	Weight of sand per 8 cubic feet.	Amount of Water Percolated in— First 30 min.		Second 30 min.	
	m. m.		Pounds.	Lbs.	Inches.	Lbs.	Inches.
No. 20	0.4745	38.86	809.28	53.33	10.25	24.26	4.683
No. 40	.1818	40.07	793.28	39.27	7.519	27.35	5.258
No. 60	.1551	40.76	784.00	29.99	5.674	23.52	4.522
No. 80	.1183	40.57	786.64	7.86	1.512	6.73	1.294
No. 100	.08265	39.73	797.76	6.31	1.213	4.40	.845

It will be seen from the above table that the rate at which the water moved downward through the coarsest or No. 20 sand was such as to average during the first thirty minutes 492 inches per twenty-four hours, while for the finest or No. 100 sand the mean rate was 58.16 inches, the flow from the first being nearly 8.5 times as fast, with grains not quite 6 times as large.

After the end of the first nine days of percolation these coarse sands lost about 1.7 per cent. of their dry weight in each case, or only about .33 of an inch.

189. Rate of Percolation from Soils.—The percolation of water from the sandy loam and from the clay soil, given in the table of (**160**), when the eight-foot columns were completely full of water at the start, took place at a much slower rate than from the sands, as indicated in (**188**), the rates being

	Sandy loam. inches.	Clay loam inches.
First 21 hours	2.640	....
First 23 hours	....	1.958
First 10 days following the above	5.072	2.111
Second 10 days following the above	.905	.493
Total in about 21 days	8.617	4.562

The rates in these cases were such that more water percolated from the three coarsest sands during the first 30 minutes than from the clay loam in as many days; and yet the loam contained at the start the largest amount of water. It is clear from these differences in the rate of percolation why the sand could not be productive under ordinary conditions of rainfall, no matter how much plant food it might contain. It is clear also that fineness or closeness of texture is one of the most important qualities of a good soil, for without this the water drains away so rapidly that, with the ordinary intervals between rains, not enough could be retained for the needs of crops.

190. Percolation Through Dry Soil.—When soils have become relatively dry, as happens especially during the middle and later summer, water does not percolate into them as readily as it does in the spring when the pores are more nearly filled. When the volume of air in the soil is large, and when the films of water surrounding the soil grains are very thin, the flow downward past the air is very slow. It is on this account, in part, that the lighter rains are less effective in midsummer than they are in the spring, the water being retained close to the surface where it is quickly lost by evaporation.

CAPILLARY MOVEMENTS OF SOIL MOISTURE.

The capillary movements of soil moisture are relatively slow, when compared with those of percolation, and are slower in dry than in wet soil.

The general tendency of capillarity is to bring water to the surface from varying depths, but its movements may occur in any other direction, the flow being always from a soil where the water films are relatively thick toward those where they are thinner, or from the wetter toward the dryer soils.

If the roots of plants have made the soil dryer in their immediate neighborhood capillarity may carry water to them from below, above or from either side. When heavy rains follow a dry spell then capillarity will assist gravity in carrying the water more deeply into the ground; and when water is applied by the furrow method in irrigation capillarity carries it laterally away from the furrows.

191. The Rise of Water in Capillary Tubes.—When a clean glass tube whose bore is small and wet is held vertically in water the liquid rises to a certain height above the level outside, the amount varying with the diameter of the tube, as given in the table below:

In a tube	1.	inch in diameter the water raises	.054 inches.
In a tube	.1	inch in diameter the water raises	.545 inches.
In a tube	.01	inch in diameter the water raises	5.456 inches.
In a tube	.001	inch in diameter the water raises	54.56 inches.

That is to say, reducing the diameter of the tube one-half doubles the height the water may be raised by capillarity, and reducing the diameter to one-hundredth enables the water to rise 100 times as high. The results in the table above will be true only when the walls of the tube are very clean, the water pure and the temperature 32° F.

192. Cause of the Variation in Height to Which Water Is Raised in Capillary Tubes.—The reason for the differences in height to which water may be raised in capillary tubes by surface tension is found in the relation existing between the volume of the tube and its internal circumference at the level of the water surface. Quinke has shown that the force of cohesion is exerted over a distance of $\frac{1}{500000}$ inch; so that when a glass tube is thrust into water the molecules in the surface of the wall just above the water draw upward upon the rows of molecules in the surface lying nearest, raising them above the natural water level. But as the edge of the surface film is raised the whole water column is carried upward also until the weight lifted above the hydrostatic level is equal to the cohesive attraction between the glass and the water.

As each molecule of glass has a fixed power to pull, the tube of large diameter will be able to lift as much more water than the small one, as the number of molecules in its circumference is greater. But the circumferences of tubes increase in the same ratio as their diameters, and hence a tube whose diameter is .1 inch will lift above the water level 10 times as much water as the one .01 inch in diameter. But, as the weight of water lifted increases as the squares of the diameters of the tubes, the first tube will only lift its column one-tenth as high as the second tube, for then its load becomes 10 times as great, and this is the limit of its power, as expressed in the table below:

Diameter of tube.	Relative area of cross-section of tube.	Height to which water is lifted.	Relative amount of water lifted.
1.0 inch	1,000,000 ×	.05456 inches =	54,560.00
.1 inch	10,000 ×	.5456 inches =	5,456.00
.01 inch	100 ×	5.456 inches =	546.00
.001 inch	1 ×	54.560 inches =	546.50

The actual amount of water lifted by the surface film stretched across the tube and carried upward by the pull of the glass molecules just above its edge is as follows.

In the 1.0	inch tube.................	.01285	cubic inch.
In the .1	inch tube.................	.001285	cubic inch.
In the .01	inch tube.................	.0001285	cubic inch.
In the .001	inch tube.................	.00001285	cubic inch.

193. Capillary Rise of Water in Soils.—The spaces left between the soil grains form more or less triangular capillary tubes whose cross-section, formed by four spherical grains, placed as closely together as possible, is represented at the left in Fig. 54; and these tubes extend in all directions through a soil.

The effective diameters of these capillary tubes are somewhat nearly proportional to the diameters of the soil grains so that for soils with spherical grains having the closest packing, doubling the diameters of the grains would also double the effective diameters of the capillary tubes through which the water must be moved.

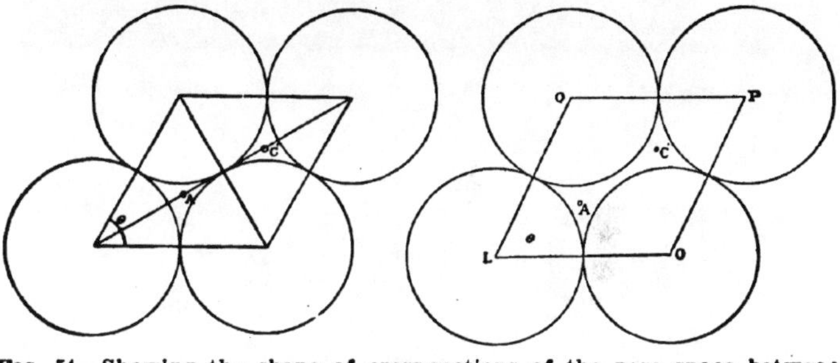

FIG. 54.—Showing the shape of cross-sections of the pore space between soil grains.

The area of cross section of the two capillary pores shown in Fig. 54 is equal to the area of the rhombus connecting the centers of the four grains minus the area of a circle having the diameter of the soil grains, so that dividing this area by two gives the area of the section of the pore.

Where the pore has the smallest section its area is given by the equation

$$\text{Area} = \left(\sqrt{3} - \frac{\pi}{2}\right) \times r^2 = .1613\ r^2$$

where r is the radius of the soil grain.

The capillary pores in an ideal soil do not have a uniform diameter but are shaped like the cast shown in Fig.

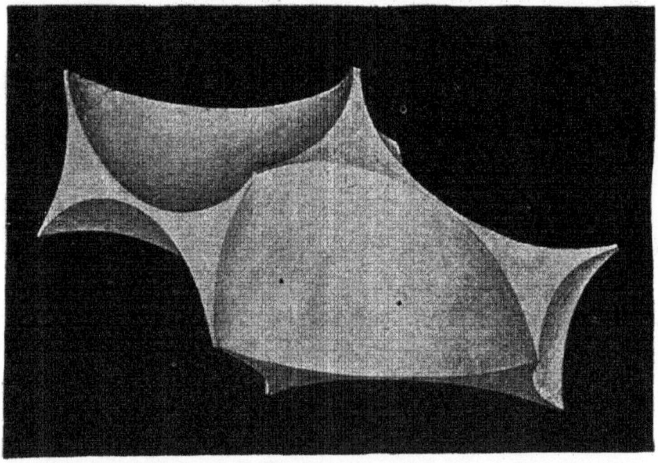

FIG. 55.—Showing a cast of the pore space between spherical grains, much enlarged.

55, largest at one place and decreasing in either direction to the area given by the equation above. The mean area of the section of the pore, is given by Slichter,* as

$$\text{mean area of section of pore} = 0.2118\ r^2$$

which would make the largest or effective cross section of the capillary pore not far from

$$(.2118 \times 2) - .1613 = .2623\ r^2$$

From this the effective diameter of the capillary tubes may be found, using the formula

$$D = 2\sqrt{\frac{.2623\ r^2}{}}$$

where r is the radius of the soil grain
and D is the diameter of the capillary pore.

* Theoretical Investigation of the Motion of Ground Waters, 19th annual report of the Geological Survey, part II, p. 316.

On this basis spherical soil grains of one size and the closest packing, having diameters of

m. m.	m. m.	m m.	m. m.	m. m.
1.	.5	.1	.05	.01

would form capillary tubes whose largest cross sections are nearly equivalent in area to circles having diameters of

m. m.	m. m.	m. m.	m. m.	m. m.
.289	.1445	.0289	.01445	.00289

Did such soil grains have the attractive power of glass for water and were their triangular pores capable of raising water to the height of circular tubes of equivalent cross sections they should be able to lift water at 32° F. to very nearly the height of

.4 ft. .8 ft. 4 ft. 8 ft. and 40 ft. respectively.

194. Observed Height of Capillary Rise of Soil Moisture.—To measure the rise of water by capillarity in ordinary soils four cylinders, 10 feet long and .04611 sq. ft. in section, were filled, two with a sandy loam and two with a clay loam, the first containing 18.88 per cent., and the second 32.63 per cent. of water uniformly distributed throughout the columns. On one of each set of tubes a soil mulch was developed 3 inches deep, when they were all placed in front of a ventilator where a current of air was maintained across their tops during 314 days. At the end of this time the tubes were cut into 6-inch sections and the water content of the soil determined, with the results given in the table which follows:

It is clear from this table that there has been an upward movement of water and loss through the surface even from the bottom layers of soil in the case of the medium clay, and probably also from the sandy loam. This follows from the fact that the clay soil contained, when put into the cylinders, 32.63 per cent., whereas the lower six inches is 1.38 per cent. drier in the mulched cylinder and 3.17 per cent. drier in the cylinder not mulched.

Table showing the loss of water by surface evaporation from columns of soil 10 feet long, mulched and not mulched.

	SANDY LOAM.		CLAY SOIL.	
	Mulched 8 inches.	Not mulched.	Mulched 8 inches.	Not mulched.
	Per cent.	Per cent.	Per cent.	Per cent.
Surface 6 inches	8.33	7 41	17 66	7.79
6 inches to 12 inches	12 97	14.48	24.59	18.30
12 inches to 18 inches	14.59	14.70	26.58	21 46
18 inches to 24 inches	15.25	14.96	26.95	26 28
24 inches to 30 inches	15.55	15 53	27.45	26.39
30 inches to 36 inches	15.89	16.17	27.92	27.16
36 inches to 42 inches	16.22	16.33	27.94	27.61
42 inches to 48 inches	16.29	16.33	28.24	27.64
48 inches to 54 inches	16 58	18 10	28.46	27.28
54 inches to 60 inches	17.07	16.76	28.47	28.28
60 inches to 66 inches	17.05	17.31	28.87	27.79
66 inches to 72 inches	17.26	17.43	28.70	28.05
72 inches to 78 inches	17.56	17.79	29.24	28.93
78 inches to 84 inches	17.73	17 88	29.28	28.31
84 inches to 90 inches	17 94	17.85	29.35	28.32
90 inches to 96 inches	17.96	17.67	29.79	28.80
96 inches to 102 inches	18.25	18.05	30.32	29.14
102 inches to 108 inches	18.67	18.09	31.15	29.16
108 inches to 114 inches	18.53	18.63	30.47	29.33
114 inches to 120 inches	19.21	19.95	31.25	29.46

In the case of the sandy loam the lower six inches in each case is wetter than when it went in, showing that at first percolation downward had taken place, and as this soil when allowed to drain freely only retained 19.44 per cent. of water at a depth of 36–42 inches, it is quite probable that at some time the lower soil 10 feet below the surface may have been wetter than found at the end of the trials, and if this is true then even the sandy loam has lost water upward from a depth of ten feet below the surface.

It is quite certain that a drying of these soils has taken place through a depth of ten feet, and hence that moisture ten feet below the surface of the ground may become available for vegetation purposes at or near the surface.

The effective diameter of the soil grains in these two cases was found to be, for the sandy loam, about .01635 m. m., and for the medium clay loam, .01254 m. m.; this would indicate that there might be a capillary rise of 23.6 and 30.8 feet respectively.

195. Capillary Rise of Water in Sand.—In the case of a sorted sand with grains .4743 m. m. in diameter, when saturated with water in an apparatus represented in Fig. 56, it was found that water was raised through a column 6.75 inches above the level of water in the reservoir at the rate of 44.09 inches of water on the level per 24 hours, but that when the column was made 11.75 inches long no water was raised to the surface.

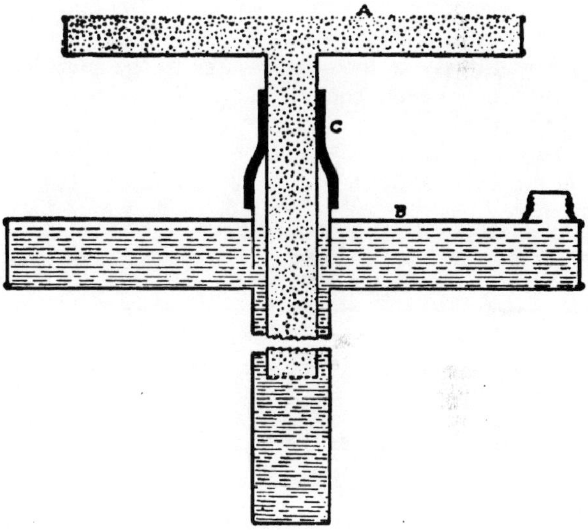

FIG. 56.—Apparatus for measuring the maximum rate and height of capillary rise of water in sands. A, evaporating reservoir; B, water reservoir; C, rubber tube.

From the formula in (193) a glass sand with grains the size of this one should be able to lift water by capillarity to a height of 10.11 inches and, since the quartz sand used did lift water at the rate of 44.09 inches in depth in 24 hours through a height of 6.75 inches, and failed to lift any water to a height of 11.75 inches, it is clear that its maximum limit must lie very close to that computed for the glass sand.

196. Rate of Capillary Rise of Water in Wet Soil.—There is yet no very satisfactory data as to just how rapidly water may be moved by capillarity through wet soils. It is probable that the case cited in (**195**) represents about the maximum rate in that coarse quartz sand, through that height, namely, 44.09 inches in depth per 24 hours. This is an enormous quantity of water to be raised by capillarity and was rendered possible only by expanding the column of sand at the top, as shown in the figure, so as to increase the rate of evaporation until it exceeded the ability of capillarity to bring the water to the surface.

Experiments have shown that with a strong current of air passing across the wet surface of the soil, water was lifted by capillarity in a square foot of soil, through the different distances and at the rates given in the table below:

	1 foot.	2 feet.	3 feet.	4 feet.
	lbs. per day.	lbs. per day.	lbs. per day.	lbs. per day.
Fine quartz sand	2.37	2.07	1.23	.91
Clay loam	2.05	1.82	1.00	.90

It is quite certain that these figures do not represent the maximum rate of capillary rise through these soils; because, as the surface of the soil had no greater area than the section of the soil column, the rate of rise could not exceed the rate of evaporation.

197. Rate of Capillary Movement of Water in Dry Soil.—The movement of water through a thoroughly dry soil, by capillarity, is not as rapid as it is through the same soil when wet; the case being analogous to the much slower absorption of water by a dry cloth or sponge than by a similar one when damp.

In the table which follows is given the rate at which water entered 5 cylinders of water-free soil, 6 inches in diameter and 12 inches long, standing in one inch of water and possessing the undisturbed field texture. The cylinders stood in a saturated atmosphere and the amount of water absorbed was determined by weighing every third day, the samples being the same ones used in (**158**) and (**159**).

Table showing the mean daily absorption of capillary water by undisturbed field soil. Cylinders 6 inches in diameter, 12 inches long, standing 11 inches out of water.

	POUNDS PER CUBIC FOOT.				
	First foot.	Second foot.	Third foot.	Fourth foot.	Fifth foot.
Water absorbed during 1st 3 days......	12.50	12.42	9.61	13.50	10.73
Water absorbed during 2nd 3 days......	2.57	2.18	2.33	3.58	2.93
Water absorbed during 3rd 3 days......	1.74	1.02	1.56	1.71	2.15
Water absorbed during 4th 3 days......	1.33	.79	1.28	.51	.61
Water absorbed during 5th 3 days......	.96	.59	1.16	.23	.16
Water absorbed during 6th 3 days......	.44	.46	1.00	.17	.06
Water absorbed during 7th 3 days......	.12	.32	.69	.10	.01
Water absorbed during 8th 3 days......	.07	.25	.48	.03	.02
Water absorbed during 24 days	19.73	18.03	18.32	19.83	16.67
	Per ct.	Per ct.	Per ct.	Per ct.	Per ct.
Complete saturation....................	32.2	23.8	24.5	22.6	17.5
Degree of saturation attained...........	28.28	20.43	20.39	21.30	15.72
Difference	3.92	3.37	4.11	1.30	1.78

From this table it is seen that the amount of water absorbed during the first three days was only at the mean daily rate of 4.16, 4.13, 3.20, 4.5 and 3.58 lbs. respectively; after the first period the rate of rise was much less rapid and did not equal the rate at which an almost identical soil (**196**) raised water through 4 feet as measured by the daily evaporation; and yet the daily rise of water of .91 and .90 lbs. per sq. ft. would have been greater had the evaporation only been more rapid. In the case of the sand of (**195**) the water was lifted by capillarity at the enormous rate of 228.6 lbs. per sq. ft. in 24 hours while the sandy loam of (**194**), placed under the conditions of (**195**), using the same piece of apparatus, lifted water at the rate of 26.62 lbs. per sq. ft. in the same 24 hours.

In the case of the 6-inch cylinders of soil above, with their tops only 11 inches out of water, the length of time required for the surface of the soil to begin to appear damp was

2 days for the fine sand or 5th foot.
6 days for the sand and clay or 4th foot.
6 days for the clay loam or 1st foot.
18 days for the reddish clay or 3rd foot.
22 days for the reddish clay or 2nd foot.

It is clear from the data presented that the rate of capillary movement of soil moisture is greatly influenced by the water content of the soil.

198. Capillarity Is Stronger in Wet than in Dry Soils.—It follows from (196) and (197) that capillary action in a given soil is stronger when the soil contains a certain amount of moisture than it is when that amount is much reduced. When soils have their water content so much reduced that they begin to look dry, and especially after they become air-dry, they act as effective mulches and water will neither rise through them so rapidly nor so high the dryer they become, and if, under these conditions, a light shower should fall it might have the effect of leaving the surface soil with a greater increase of moisture than is represented by the rain which fell.

199. Rain May Cause a Capillary Rise of the Deeper Soil Moisture.—It was observed in 1889, when determining the water content of soils at different depths in the field, just before and immediately after rains, that frequently the lower soil showed a smaller amount of moisture than it had before the rain, while the surface layers had gained in water more than that represented by the rainfall. It was later shown that, by applying a known amount of water to a section of a field, the lower soil became dryer while the surface layers had gained more water than was added, as shown in the table.

Table showing the translocation of soil moisture due to wetting the surface.

DEPTH.	PER CENT. OF WATER.		DIFFERENCE.	
	Before wetting.	After wetting.	In per cent.	In pounds per cub. ft.
0–6 inches....	14.	22.23	+ 8.23	+2.873
6 inches to 12 inches	15.14	15.71	+ .57	+ .199
12 inches to 18 inches	16.23	15.75	− .48	− .218
18 inches to 24 inches	17.70	16.92	− .78	− .347
24 inches to 30 inches	16.76	14.41	− 2.35	−1.032
30 inches to 36 inches	15.51	15.21	− .30	− .132

The amount of water applied to the surface in this experiment was 2 lbs. per sq. ft. but when samples of soil were taken 26 hours later there had been an increase of 3.072 lbs. in the surface foot and a loss of 1.724 lbs. from the second and third feet. Observation showed that a tray of soil, on a pair of scales at the place, lost, by evaporation during the same time, .428 lbs. per sq. ft.; and, assuming that the field soil lost water at the same rate, makes the water to be accounted for

$$3.072 + .428 = 3.5 \text{ lbs.},$$

while the total loss from the lower two feet plus the water added was

$$2 + 1.724 = 3.724 \text{ lbs.}$$

an amount as nearly equal to the 3.5 lbs. as could be expected.

In another trial, adding 1.33 lbs. of water to the surface produced the gain, by translocation upward into the upper four feet, shown in the next table.

Depth.	Water Content of the Soil.		
	Before wetting.	After wetting.	Change.
	Pounds per cu. ft.	Pounds per cu. ft.	Pounds per cu. ft.
First foot	11.78	14 06	2.28
Second foot	15 79	17 52	1.73
Third foot	14 73	15.58	.85
Fourth foot	14.03	15.40	1.37
Gain	....	....	6.23

The interval during this experiment was one of very little evaporation and the adjacent untreated ground gained 1.21 lbs. per sq. ft. in the same depth. This amount and the water added deducted from the gain in the treated area leaves the translocation

$$6.23 - (1.21 + 1.33) = 3.69 \text{ lbs. per sq. ft.}$$

200. Farmyard Manure May Strengthen Capillary Rise of Soil Moisture.—When a soil is treated with farmyard manure which has become well incorporated with it, it has the effect of causing a stronger rise of the deeper soil moisture into the surface three feet, where it is most needed in the production of crops. The table which follows shows the mean results of experiments aiming to measure this effect during three years.

Table showing effect of farmyard manure in strengthening the capillary rise of soil moisture.

	1st foot.	2nd foot.	3rd foot.	4th foot.	5th foot.	6th foot.
	Per cent. of water.	Per cent. of water.	Per cent. of water.	Per cent. of water.	Per cent. of water.	Per cent. of water.
Manured	19 88	19.79	18.88	17.29	14.35	16.98
Not manured	18.79	19.33	18.60	17 32	14.63	17.13
Difference	+1.09	+.46	+ 28	—.03	—.28	—.15

It is seen here that the surface three feet have in some way been maintained more moist, and apparently by the manure, at the expense of moisture from below.

201. Heavy Soil Mulches May Strengthen the Capillary Rise of Soil Moisture.—Since capillary action is not as strong in a dry as in a well moistened soil it should be anticipated that any condition which would maintain a fair degree of saturation in the surface one to three feet of soil would permit it to bring up from below, for the use of crops, a larger supply of capillary water.

On three different kinds of soil, where the ground had been cultivated during the season in alternate groups of four rows 3 inches deep and 1.5 inches deep, the distribution of moisture, on July 16, was found to be as follows:

Table showing the effect of mulches in strengthening the capillary rise of soil moisture.

	1st foot.	2nd foot.	3rd foot	4th foot.
	Per ct. of water.	Per ct. of water.	Per ct. of water.	Per ct. of water.
Field No. 1 cultivated 3 inches deep ..	11.30	15.57	10.54	11.37
Field No. 1 cultivated 1.5 inches deep	9.92	15.43	11.56	13.99
Difference	1.38	.14	—1.02	—1.62
Field No. 2 cultivated 3 inches deep	13.96	22.74	23.39	19.47
Field No. 2 cultivated 1.5 inches deep	12.98	20.44	24.02	21.34
Difference	.98	2.30	—.63	—1.87
Field No 3 cultivated 3 inches deep	11.65	17.47	16.44	13.03
Field No. 3 cultivated 1.5 inches deep	10.65	16.85	17.81	13.32
Difference	1.00	.62	—1.37	—.29

This table indicates that the 3-inch mulch, by maintaining the surface soil more moist, enabled capillarity to bring up from below a larger supply of water; that is, the maintaining of a relatively high per cent. of moisture in the upper two feet of soil makes it possible, through capillarity, for crops to utilize a larger amount of the soil moisture which is stored in the deeper layers. This view is confirmed by the fact that, in the fields of the table above, the largest yields of corn were in all cases taken from the ground cultivated 3 inches deep, where the upper two feet of soil contained, in spite of the larger crop, much more moisture, but at the expense of that deeper in the ground, as shown by the fact that in every case these soils were dryest in the 3d and 4th feet.

202. Firming the Soil May Strengthen the Capillary Rise of Soil Moisture.—When soils have been rendered open and loose by plowing or other deep stirring the first effect is to permit the loose and open soil to become dry, because this soil is less perfectly in contact with that below. If, after such soil has become dry, it is firmed again the moisture films will then increase in thickness over the surface of the soil grains and, as a result of this, moisture will be raised from depths as great as four feet to saturate the firmed dryer soil. In the table below are shown the changes which occurred in the deeper and superficial soil layers as the result of rolling.

Table showing how rolling may strengthen the capillary rise of soil moisture.

Depth of sample.	No. of trials.	Rolled ground.	Unrolled ground.	Change produced.
		Per cent. of water.	Per cent. of water.	Per cent. of water.
Surface 2 to 18 inches	62	15.85	15.64	+.21
Surface 24 inches	61	19.49	19.85	—.36
Surface 36 to 54 inches	24	18.72	19.43	—.71

From this table it is seen that the first effect of rolling is to increase the amount of moisture in the upper 18 inches of soil, but that when samples are taken deeper than 18 inches the total amount in the soil is decreased. In other words, the first effect is to concentrate the deeper soil moisture toward the surface.

If, however, the soil is left firmed very long then the whole column, to the surface, becomes dryer, until it has lost so much moisture that it begins to act as a mulch.

THERMAL MOVEMENTS OF SOIL MOISTURE.

Besides the gravitational and capillary movements of soil moisture there are others due to the molecular vibrations set up in the soil-air and water by the absorbed solar energy.

203. Hygroscopic Soil Moisture.—It is seldom if ever true that any solid surface, even when in the dryest air, can be found which is not invested with a film of moisture of greater or less thickness. It is also true that even when all moisture has been driven from the surface of a solid by drying at the high heat of 200° C., the same body will again become coated with moisture when exposed to a moisture-bearing atmosphere. Water thus collected on the surface of solids is called *hygroscopic* moisture.

204. The Movements of Hygroscopic Moisture.—It will be seen that the movements of hygroscopic moisture are the same as those of evaporation. The same molecular attraction which causes the capillary rise of water in a glass tube tends to collect the water molecules, which may be moving about in the air, upon solid surfaces. So when a dry soil is exposed to a damp atmosphere some of the moving water molecules are brought in contact with, and retained by, the surfaces of the soil grains. The moisture will go on accumulating upon the soil grains until the rate of evaporation from them equals the rate of condensation. Since the water molecules are attracted to the soil grains more strongly than they are attracted to one another the water in immediate contact with the soil grains cannot evaporate as readily as that which is further removed when the water films are thick, as they are in a well saturated soil.

Neither can the innermost layers of molecules adhering to the soil grains escape to enter the root hairs of plants by osmotic pressure as readily as those from the layers farther removed, and hence there must always be a certain quantity of water upon the surfaces of soil grains which neither evaporates readily nor becomes easily available to plants. and this may be regarded as the hygroscopic moisture.

205. Relation of the Diameter of Soil Grains to the Hygroscopic Moisture.—It was shown in (**163**) that with the same thickness of water surrounding the soil grains the per cent. of water was necessarily much higher in the soils having the smallest soil grains. In (**192**) is given Quincke's observation of the distance across which the force of cohesion is sensible, or $\frac{1}{500000}$ inch. Since this attraction of the soil for water is stronger than that of the water for the water it appears likely that a layer of water surrounding the soil grains, at least as thick as this, would not be as free to evaporate or to otherwise move about as that much farther removed from this cohesive attraction, and if so it is important to know what per cents of soil moisture a water-film of such a thickness would represent. This may be computed for spherical soil grains with the formula

$$\text{Per cent. of water} = \frac{\frac{\pi (d + 2t)^3}{6} - \frac{\pi d^3}{6}}{\frac{\pi d^3 \text{ sp. gr.}}{6}}$$

where d = diameter of soil grain in c. m.

t = thickness of water film.

sp. gr. = the specific gravity of the soil.

Taking a very fine soil having grains with a diameter of .00508 m. m. and a coarse one with a diameter of .1 m. m., a film of moisture on each, having the thickness of the range of sensible cohesive attraction, as given by Quincke, would make the per cent. for the finest soil 2.31 but for the coarse soil only .1153. No crop can survive in soils as dry as these; and air-dry soils whose grains range between those given will generally contain more than these amounts of moisture. It follows from these considerations, therefore, that what has been regarded as the hygroscopic moisture is more than that held within the range of sensible cohesive attraction. It appears clear also that no hard and fast line can be drawn between capillary and hygroscopic moisture, nor indeed between either of these and the gravitational water; each must shade by insensible degrees into the other.

206. The Amount of Moisture a Soil May Absorb from the Air.—The amount of so-called hygroscopic moisture a given soil may absorb from the air depends primarily upon the relative temperature of the soil and of the air and its degree of saturation. If the temperature of a soil could be maintained continually below that of a saturated atmosphere above, it would in time become so fully charged with water as to result not only in capillary saturation but in percolation as well; and it frequently occurs on clear nights in summer, when dews are heavy, that a thick, loose, dry dust blanket will cool down so much that moisture condenses upon it in sufficient quantity to make it appear damp. Indeed dew, wherever it forms, is a demonstration of the truth of the statement made; when it evaporates with the rising of the sun the loss of moisture from the blades of grass may carry the amount all the way from the drops, too heavy to be retained upon the blades, through the thick adhering films, to those which become invisible and are called hygroscopic.

207. Observed Absorption of Moisture from the Air.—The rate and amount of moisture which may be absorbed from the air is influenced by many factors. Hilgard has studied the rate and amount of absorption of moisture by soils when spread out in layers about 1 m. m. thick in a fully saturated and a half saturated atmosphere, maintained at a uniform temperature. He finds that fully 7 hours are required for an equilibrium to be reached in so thin a layer. In the table which follows are given some of his observations.

Table showing the absorptive power of soils spread out in thin layers.

KIND OF SOIL.	SATURATED ATMOSPHERE.			HALF SATURATED ATMOSPHERE.		
	Temp. Far.°	Time, hrs.	Per cent. of water absorbed.	Temp. Far.°	Time, hrs.	Per cent. of water absorbed
Dark alluvial loam, Putah Valley, Solano county...	58	19	11.745	57	43	6 547
	59	19	11.886	....	....	....
	61	18	11 408	....	....	....
	72	7	12.013	70	7	6.424
	77	7	12.233	77	7.5	6.305
	88	7	13.141	88	7	6.356
	100	6	13.481	100	6	6.209
Black adobe soil, University grounds, Alameda county	55	19	7.144	61	18	4.008
	57	19	7.880	61	7 5	4 122
	70	7	7.696	80	6	4.024
	80.5	17	8.681	83	7.5	3 926
	82.5	7.5	8.948	89.5	7.5	3.910
	100	7	9.569	100	7	3.885
Calcareous silt soil, Fresno county..................	61	18	2.133	59	18	0.987
	79	6	2.983	79	6	0.959
	84	7	3.396	84	7	0 858
	95	6	4.211	95	6	0.821

It will be seen that in the saturated atmosphere the largest amount of moisture was absorbed at the highest temperature, while the reverse was true in the half saturated atmosphere. Under the high temperature the rate of molecular movement is so rapid that the rate at which the water from the air falls upon and enters the soil is so much increased that more water must have accumulated in the soil before the number of molecules which can leave its surface in a unit of time equals that which falls upon it. In the dryer atmosphere, on the other hand, where there are less molecules to fall upon the soil and increase its amount, the higher temperature favors the rapid escape as much as when the saturation was high and, since less water is condensing, a lower per cent. is finally present when an equilibrium of interchange has been reached.

208. Internal Evaporation of Soil Moisture.—It is likely that under certain conditions the thermal movements of soil moisture may be considerable and perhaps of sufficient importance to materially influence vegetation, directly or indirectly. When the per cent. of unoccupied pore space in a soil has been materially increased by the loss of water and when the moisture films have become so thin that capillarity is much enfeebled it is possible that *internal* evaporation of soil moisture may result in a considerable change of its position. If, for example, when the soil has become quite dry, to considerable depths, the surface six inches should become cooler than that below, the tendency to continual diffusion of water vapor under the impulse of heat would produce more internal evaporation of moisture where the soil is warmest and most moist, and a larger condensation of moisture where the soil is dryer and cooler. Even where there is little difference in temperature between adjacent layers of soil there must be, if they are not equally saturated, a tendency for diffusion to take place more rapidly from the wettest layer of soil toward that which is least moist. It is possible that during dry times and in dry climates during the dry season some moisture, too far below the root zone to be made available through capillarity, may be carried upward by these thermal or evaporation movements so as to become helpful to crops in a measure. We are yet lacking in experimental data to form any just conception as to the magnitude of such a movement.

209. Temperature Influence of Hygroscopic Moisture.—It is Hilgard's view that, in dry climates and during droughty periods in humid climates, the moisture still retained by soils when capillarity has become very feeble may exert an important influence in preventing the soil from becoming overheated during dry soil conditions, by the cooling effect of internal evaporation. It must be observed, however, that in order that this influence may become effective the moisture evaporated must have left the soil and not have been replaced by an equal amount through condensation from some other place.

It appears to the writer possible that the ability of such soils to withstand drought may perhaps be partly due to a more rapid evaporation from the soil grains and condensation of moisture on the root hairs, the thermal movement, in this way, tending to supplement the enfeebled capillarity.

CHAPTER VIII.

CONSERVATION OF SOIL MOISTURE.

There are very few fields upon which crops of any kind, in any climate, can be brought to maturity with the maximum yields the soils are capable of producing without adopting means of saving the soil moisture. There are fields, it is true, where, at times, the moisture in the soil is too great, and drainage becomes necessary; but even under these conditions it will usually be found advisable to adopt measures for conserving the water not so removed.

210. Modes of Controlling Soil Moisture.—In aiming to control soil moisture three distinct lines of operation are followed, based upon as many different aims. These are:

(1) To conserve the moisture already in the soil (a) by different modes, times and frequencies of tillage, (b) by the application of mulches, and (c) by establishing wind breaks.

(2) To reduce the quantity of water in a soil (a) by frequent stirring, (b) by ridging or firming the surface, (c) by decreasing the water capacity, and (d) by surface or under drainage.

(3) To increase the amount of water in a soil (a) by increasing its water capacity, (b) by strengthening the capillary movement upward and (c) by irrigation.

211. Late Fall Plowing to Conserve Moisture.—There is no method of developing so effective a soil mulch as that furnished by a tool which, like the plow, completely cuts off a layer of surface soil and returns it loosely, bottom up, to place again.

When ground is plowed late in the fall, just before freezing, it then acts during the winter and early spring as a mulch, diminishing the loss of water by surface evaporation, and at the same time the roughened surface tends to hold the snows and to permit winter and early spring rains to penetrate more deeply into the soil, leaving the ground more moist at seeding time than would be the case if it were left unplowed. Determinations of the moisture in the spring, as late as May 14, have proved that late fall plowed ground may contain fully 6 pounds per square foot more water in the upper four feet than similar adjacent ground not plowed. This difference represents a rainfall of 1.15 inches and is a very important saving in climates of deficient water supply for crops.

212. Late Tillage for Orchards and Small Fruits.—Late fall plowing and deep cultivation in orchards of fruit trees and in vineyards of small fruits, after the wood is fully matured and growth arrested by the cold weather, will do very much toward giving the soil better moisture relations the next spring, tending to secure such results as are cited in (**211**). In cases where injury from deep freezing is liable to occur the late plowing will lessen this danger because the loose soil blanket will help to retain the heat in the ground as well as the soil moisture.

In the late plowing and deep tillage, advised in this and the last section, there is little danger of increasing the loss of plant food by leaching because the season is too late and the temperature of the soil too low to stimulate the formation of nitrates.

213. Early Fall Plowing to Save Soil Moisture.—In those cases where winter grain is to be sowed, the early plowing of the ground, or plowing as soon as the field has been freed from the preceding crop, is in the direction of economy of soil moisture. So too in sub-humid climates, even where winter grain is not to be sowed, it will often be desirable to plow as early as possible in order to retain soil moisture and to facilitate the entrance of the fall rains more deeply into the ground. The early plowing or disking in these cases may also be helpful in hastening nitrification in the soil.

It is the strong tendency of early fall plowing, in climates where there is plenty of soil moisture to develop nitrates and where there is much rain in the late fall and early spring, which has led to the sowing of "cover crops" having for their primary object the locking up of the soluble plant foods to prevent them from being lost by soil leaching; and the tendency of early fall plowing to diminish surface evaporation and thus, in wet climates, to increase percolation and the loss of plant food may sometimes make this practice undesirable in such cases.

214. Early Spring Plowing to Save Soil Moisture.—In all climates where there is a tendency of the soil to become too dry the earliest stirring in the spring, which is practicable without injuring the soil texture, is in the direction of economy in most cases because, at this season of the year, the effectiveness of tillage in conserving soil moisture is greater than at almost any other time. This statement follows from (**198**), where it is shown that a wet soil carries water to the surface much more rapidly and from a greater depth than a dry soil can. In the spring the soil at the surface is usually not only wet but also well compacted, two of the most important conditions for the rapid movement of water to the surface, and it is because of these that early and deep spring tillage is so important as a means of saving soil moisture.

In one instance, where two immediately adjacent pieces of ground, in every way alike, were plowed in the spring, 7 days apart, it was found that the earliest plowed ground contained, at the time the second piece was plowed, a little more moisture in the upper four feet than it had 7 days before, while the ground which had not been plowed had lost, in the same interval of time, an amount of moisture from the surface four feet equal to 1.75 inches, a full

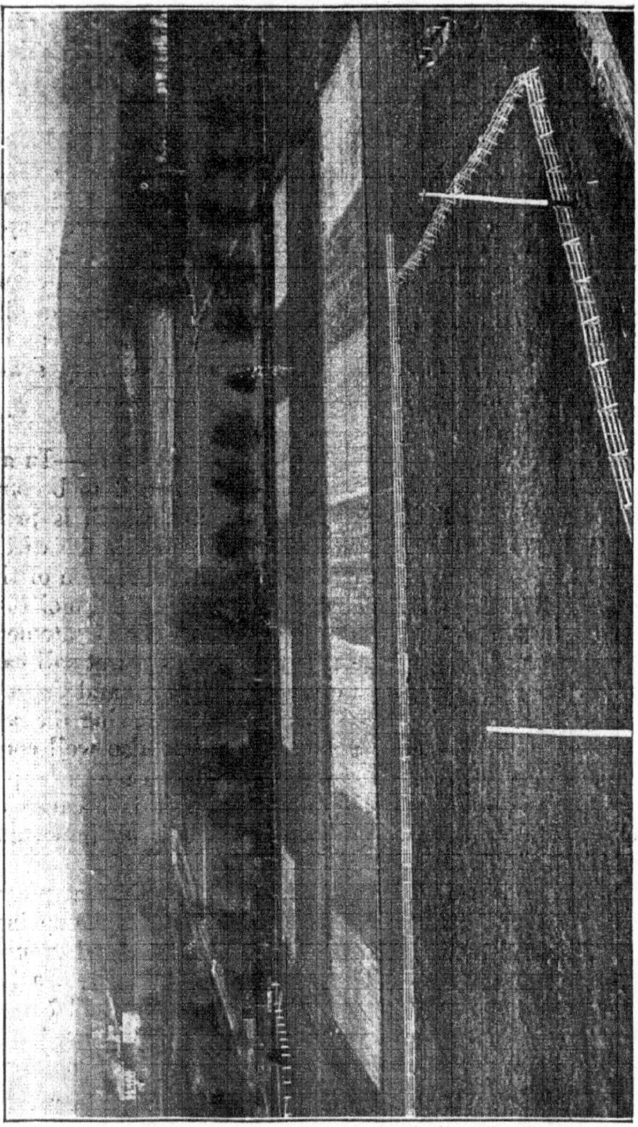

FIG. 57.—Showing the influence of early and late spring plowing on soil moisture. The light colored areas were plowed last and are dryest.

eighth of the rainfall of the growing season of that locality.

Nor was the saving of moisture the only advantage gained by the early plowing, for the soil plowed last had dried so extensively as to become very hard and lumpy, thus greatly increasing the labor necessary to fit it for planting.

In another experiment to study the effectiveness of early as compared with late spring plowing in conserving soil moisture Fig. 57 shows how evident the effects were to the eye.

215. Disking or Harrowing Where There is Not Time to Plow.—It often happens in the spring that hot dry winds come on when there is not opportunity to get the ground plowed in time to save the needed moisture and prevent the development of clods. In such cases the use of the disk harrow, or even the ordinary spike tooth harrow, will do very much to save the moisture and preserve the tilth of the soil, if only the fields are gone over with these. The disk harrow is one of the best of tools for early use in the spring to work the soil and develop mulches.

216. Corn and Potato Ground, Orchards and Gardens Plowed Early in the Spring.—Ground to be planted to corn or potatoes, as well as the orchard and garden, should generally be plowed quite early in the spring and a considerable time before it is intended to plant them. By doing this, not only will moisture be saved but the development of nitrates in the soil will be hastened and thus larger crops secured on this account. It is only in the event of long, frequent and heavy rains, following such early tillage, that loss can result from such a practice.

217. Effectiveness of Soil Mulches.—The effectiveness of soil mulches as means for diminishing evaporation varies (1) with the size of the soil grains, (2) with the coarseness of the crumb structure, (3) with the thickness of the mulch and (4) with the frequency with which the soil is stirred. Soils which maintain a strong capillary rise of water through them will, when converted into mulches, still permit the water to waste through their mulches faster than it will be lost through the mulches of soils which permit only slow capillary movements. That is, the sandy soils form more effective mulches than do the clayey ones and this greater effectiveness of the sandy soils, as mulches, goes a long way toward making the smaller amount of water they are able to retain effective in crop production.

In Fig. 58 is shown an apparatus for measuring the relative effectiveness of mulches and in the table which follows are given the results of a series of trials with three types of soil. The cylinders in this series, however, stood out in the open air of the field rather than in the case shown in the cut.

Table showing the effectiveness of soil mulches of different kinds and different thicknesses.

	No mulch, water lost per 100 days.	Mulch 1 in. deep, water lost per 100 days.	Mulch 2 in. deep, water lost per 100 days.	Mulch 3 in. deep, water lost per 100 days.	Mulch 4 in. deep, water lost per 100 days.
Black marsh soil:					
Tons per acre	588.0	355.0	270.0	256.4	252 5
Inches of water	5.193	3.12	2.381	2.265	2.230
Per cent. saved by mulches	....	39.54	54.03	56.39	57.06
Sandy loam:					
Tons per acre	741.5	373.7	329.3	287.3	315 4
Inches of water	6.548	3.300	2.996	2.539	2.785
Per cent. saved by mulches	....	49.69	54 24	61.22	57.47
Virgin clay loam:					
Tons per acre	2,414.	1.260.	979 7	859.2	883.9
Inches of water	21.31	11.13	8.652	7.852	7.805
Per cent. saved by mulches	....	47.76	59.38	63.13	63.34

From this table it will be seen that the soil mulches have exerted a very great influence in saving soil moisture.

It should be understood, however, that if the water reservoirs had been much farther below the surface of the soil, and below the mulch, the mulches would have been more effective as well as less water would have been lost from the unmulched cylinders.

218. Frequency of Cultivation May Make Mulches More Effective.—When a fresh mulch is formed upon the surface of a well moistened soil the first effect of the stirring is

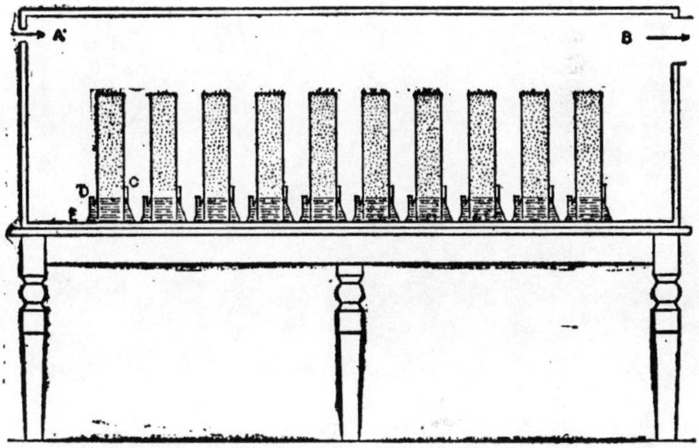

FIG. 53.—Apparatus for measuring the relative effectiveness of mulches.

to increase the rate of evaporation from the field, on account of the much larger surface of wet soil which is exposed to the air. This greater loss of water, however, is largely from the *stirred soil.* If dry winds and sunny weather follow the formation of the soil mulch it soon becomes so dry that but a relatively small amount of water can pass up through it. On the other hand if a series of cloudy days follow, when the rate of evaporation must be small even from firm wet soil, and if at the same time the soil below the mulch is quite moist, so much water may pass up into the mulch as to nearly saturate the lower portion of it and to cause the kernels to be drawn together and again compacted and reunited with the unstirred soil below. If this change does take place the mulch is rendered less effective and a second stirring is needed.

FIG. 59.—Showing large cylinders for studying soil problems.

The relative effectiveness of mulches stirred twice per week, once per week, and once in two weeks, for a virgin clay loam, in cylinders 52 inches deep and 18 inches in diameter, standing in our plant house, as shown in Fig. 59, is given in the table which follows.

Table showing the relative effectiveness of soil mulches of different depths and different frequencies of cultivation.

	Not cultivated. Per acre.	Once in 2 weeks. Per acre.	Once per week Per acre.	Twice per week. Per acre.
Cultivated one inch deep:				
The loss in tons per 100 days was	724.1	551.2	545.0	527.8
The loss in inches per 100 days was	6.394	4.867	4.812	4.662
The percentage of water saved was	....	23.88	24.73	27.10
Cultivated two inches deep:				
The loss in tons per 100 days was	724.1	609.2	552.1	515.4
The loss in inches per 100 days was	6.394	5.380	4 875	4.552
The percentage of water saved was	....	15.88	23.76	28.81
Cultivated three inches deep:				
The loss in tons per 100 days was	724.1	612.0	531 5	495.0
The loss in inches per 100 days was	6.394	5.402	4.694	4.371
The percentage of water saved was	....	15.49	26 60	31.64

It will be seen that with each of the three depths of cultivation the percentage of moisture saved, over that which was lost from the ground not cultivated, increased with the frequency of cultivation.

219. Too Frequent Cultivation Undesirable.—When a soil mulch is well loosened and thoroughly separated from the firm ground beneath, and especially after the mulch has become quite dry, little can be gained by stirring the soil. Indeed it must ever be kept in mind that it costs to cultivate a field and when this is done without need the work is a dead loss. Further than this, late in the season, when the surface of the ground has become relatively dry, positive harm may be done by unnecessary cultivation because at this season many plants have put up, very close to the surface, great numbers of fine roots in order to avail themselves of the moisture from light showers and from the dew which may be condensed in the surface layer of soil on the coolest nights. To destroy these roots will, in most cases, cause a greater loss by root pruning than can be gained by saving moisture. It is possible also, by too frequent tillage, to make the texture of the mulch so fine that its effectiveness is decreased.

220. Cultivations Should Be Most Frequent in the Spring. In the early part of the season when the aeration of the soil, the warming of it and the killing of weeds are other important objects to be attained it is more important to cultivate frequently. This is the season of the year when the effectiveness of mulches decreases most rapidly, it is the season when there is least danger of destroying the roots of the crop, and it is the time when cultivation is needed to help develop plant food.

221. Cultivation After Heavy Rains.—Whenever a rain has occurred which has thoroughly united the soil crumbs to one another, and with the soil below, it is time to cultivate again if this can possibly be done without too heavy root pruning, and the cultivation should be done just as quickly as the soil will permit. In the early part of the season there is little danger of root pruning if the cultivator teeth do not go too close to the plants and not more than 3 inches deep.

A rain which does not wet down more than 3 inches cannot be saved by cultivation; all that can be done in this case is to permit the surface roots to get as much of it as possible and to stir, if it appears expedient, when the wetting is likely to strengthen the upward movement too much. It must be remembered in this connection, however, that if, late in the season, the roots of the crop have spread horizontally through the whole soil, anything which strengthens the rise of the deeper water, causing it to come nearer the surface, at the same time brings it to the roots where it is needed, and hence it will seldom happen that a crop like corn or potatoes can be helped by cultivation after the corn is in tassel or the vines begin to well cover the ground.

222. Depth of Cultivation to Save Moisture.—In regard to this point it must be kept in mind that the soils out of which mulches are made are the richest on the farm and that when they are converted into perfect mulches they are practically useless so far as direct plant feeding is concerned. The general rule must then be to make the mulch just as thin as it can be and not permit too heavy a waste of the deeper soil water.

On the lighter and coarser grained soils the mulches may be shallower than on those of the clayey type.

In Wisconsin we have found that with the ordinary narrow pointed tooth cultivators a depth of about three inches saves more moisture and permits larger yields of corn in about 15 cases out of 20 than less depth of cultivation. Where the tool is of such a character that it shaves off the whole surface of the ground and leaves the stirred soil spread in a blanket of uniform thickness the stirring may be shallower than if the surface of the ground is left in either narrow or wide ridges.

223. Depth and Frequency of Cultivation Should Vary With the Season and Crop.—From what has been said in the preceding paragraphs it follows that the soil may to advantage be cultivated more deeply and more frequently during the early part of the season when the soil temperatures tend to be low, when the moisture may be overabundant, and when weed seeds are germinating. Later in the season, however, when there is not as great need to encourage the development of nitrates by tillage, when the roots have come closer to the surface, and the maintenance of a soil mulch is the chief or only object, the cultivation may evidently be less deep and not so frequent. The general practice then should be to gradually make the cultivation both less deep and less frequent. It should also be kept in mind that cultivation may generally be a little deeper in the middle of the space between rows, than close to the hills, because of less danger of root pruning.

224. Best Time to Cultivate Corn and Potatoes.—The best time to till land for corn, potatoes and similar crops, where intertillage is practiced, is before the ground is planted and just as the crop is coming up. When the ground is plowed two or three weeks before the crop is to be planted there is opportunity to develop the nitrates, to kill one or two crops of weeds, and to store in the upper five feet of soil the largest reserve of soil moisture from the spring rains. Besides these advantages there is no period in the growth of the crop when the ground can be stirred so rapidly and so cheaply. Before planting the disk or spring-tooth harrow may be used and afterward the different weights of spike-tooth harrows, which enable a larger area of ground to be covered in a day by a man and team. The harrowing of corn and potatoes should be continued until the plants are well out of the ground and if care is taken to do the work during the hot portion of the day, when from slight wilting the plants do not break off readily, there need be but little serious injury to them.

The different types of mulch producing tools are discussed in Chapter XI.

225. Harrowing and Rolling Small Grain After It Is Up.—It sometimes happens in humid climates, when drying weather follows a wet period, that a crust forms on the surface of fields sowed to the small grains, which may be injurious to the plants by preventing sufficient aeration and increasing the loss of moisture. In such cases the difficulties may be partly corrected by using either the roller or the light harrow with teeth sloping backward.

If the grain is large, and especially if the surface of the field has been left narrowly ridged and somewhat lumpy, the use of the roller *when the surface soil is dry* will break up the crust by crumbling down the ridges and lumps and at the same time develop a true and effective mulch. The light harrow, when driven across the ridges, may be effective in breaking up the crust and in developing a mulch.

In sub-humid climates, such as that of western Kansas, fields seeded permanently to alfalfa have been, in the very early spring, gone over with the disk harrow and then crossed with the spike-tooth harrow, thus developing a very effective mulch which materially increases the yield.

226. Mulches Not Made From Soil.—While it is true that most conservation of moisture must be through earth mulches it should be understood that all vegetation growing upon the ground, whether it completely covers the surface or not, exerts a protective influence and diminishes the loss of moisture directly from the soil itself. This protection comes partly from shading, partly from diminishing the wind velocity and partly from the saturation of the air with moisture by the transpiration from the growing plants.

Even in pastures where the grass is short, but close, the mulching effect is strong and hence it is not in the direction of economy to allow the feeding to be too close, not only because the growth of the grass is slower from too severe destruction of the foliage, but because there is a greater loss of soil moisture besides that passing through the grass.

The surface dressing of meadows with farmyard manure, thoroughly harrowed to spread it evenly over the ground, is extremely beneficial through its mulching effect as well as in the plant food it brings to the soil. When such dressings are applied in the winter and early spring and spread over the surface while the soil is yet wet beneath, the saving in soil moisture is greatest and in the case of meadows where the clover has disappeared, for any reason, such a dressing may make it possible to get a new seeding, by sowing the clover broadcast before the frost is out in the spring, so that the thawing and freezing will tend to cover the seed and the thin mulch protect the ground from too rapid drying until the young plants are well rooted.

The use of straw and other coarse litter and coarse sand for mulching will generally only be practicable in gardens and orchards and for the protection of shade trees and the like.

227. Ridged and Flat Cultivation.—It used to be a common practice to "lay by" the corn and potato crop with a strong hilling of the rows. This practice, however, except for potatoes, is now generally abandoned unless in localities where surface drainage is needed. The general abandonment of the practice rests in part upon the belief that the evaporation from the soil is appreciably increased by this process on account of the greater amount of surface exposed to the air.

In making a practical test during the season of 1899 the results recorded in the following table were secured.

These plots, each seven rows wide, alternated across a field of nearly uniform soil and samples were taken under and between every row. It will be seen that the soil receiving the flat cultivation contained at the end of the growing season a little less water than the ridged plots, which is contrary to the accepted belief. Since the ridges are all shaded by the potato vines and since the wind currents may be supposed to be less strong between the furrows, perhaps this is as should be expected. It is true, however, that the plots cultivated flat produced a little larger yield per acre and on this account the soil should have lost more moisture. It may be that the flat cultivation did really make a larger saving of water and that this saving was the cause of the larger yield.

Table showing the water content of soil, Sept. 19, under and between rows of potatoes hilled and left flat when laid by.

Depth of Sample.	Nos. of sub-plots.	Hilled.		Flat.	
		In row.	Between row.	In row.	Between row.
		Per cent.	Per cent.	Per cent.	Per cent.
First foot	1	12.83	14.11	11.85	14.23
	2	12.01	13.61	12.18	13.54
	3				
	Mean	12.42	13.86	12.02	13.89
Second foot	1	16.71	18.56	15.38	17.69
	2	15.84	17.85	16.03	17.84
	3				
	Mean	16.28	18.21	15.71	17.77
Third foot	1	18.00	18.61	16.41	18.03
	2	17.09	17.55	16.13	17.97
	3				
	Mean	17.55	18.08	16.27	18.00
Fourth foot	1	15.78	16.95	9.79	11.75
	2	14.41	13.98	13.08	14.01
	3				
	Mean	15.06	15.46	11.44	12.88
Mean of four feet		15.33	16.40	13.86	15.64

228. Subsoiling to Save Soil Moisture.—The deep plowing or stirring of the soil, to which this name has been applied, has the effect of making a larger per cent. of the rainfall available in producing crops, but it will never have the wide applicability that is possible for surface tillage. In sub-humid climates where the subsoils are less liable to be puddled and where there is the greatest need of economy this method of conserving soil moisture will find its widest usefulness.

A piece of ground when subsoiled, as represented in Fig. 60 and given, with an adjacent area, a like amount of water, and protected from surface evaporation, was found to have retained not only the water given it but to have gained an additional supply through capillarity from below; while the ground not subsoiled lost a large per cent. of that given to it through percolation and capillary creeping. From the subsoiled area 8 inches of the surface were removed, the subsoil spaded to a depth of 13 inches more, and the soil returned to its place. After taking

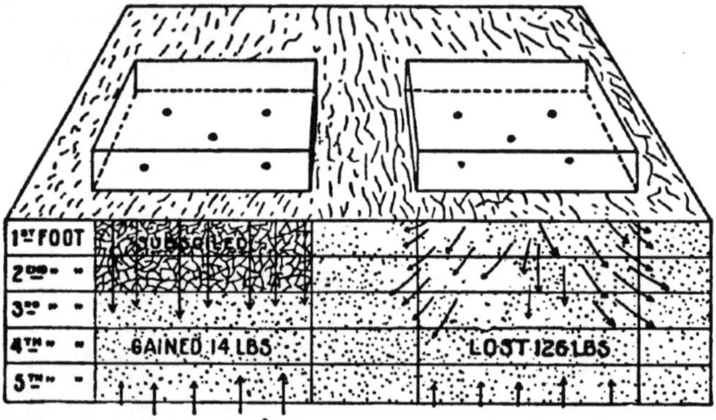

FIG. 60.—Method of demonstrating the influence of subsoiling on soil moisture.

samples from the five places indicated by the dots, 1.36 inches of water were gradually sprinkled over the two areas on June 11th and they were allowed to remain covered until the 15th, when samples were again taken. The changes in the water content of the soil in the two areas are shown in the table which follows:

Table showing the ability of subsoiled ground to hold water against gravity.

	Subsoiled.	Not subsoiled.	Difference.
	Lbs.	Lbs.	Lbs.
The first foot gained	124.6	102.1	+22.5
The second foot gained	72.57	10 34	+62.23
The third foot gained	38.22	12.05	+26.17
The fourth foot gained	33.26	3.82	+29.43
The fifth foot lost	2.29	19.5	−17.21
Total water gained	268.65	128.31	
Total water added	254.41	254.41	
Difference	+14.24	−126.1	

The subsoiled ground had therefore not only retained all the water added but it had gained by capillarity 14.24 lbs. more. It is noteworthy too that the fifth foot in both places had lost water upward by capillarity, 2.29 lbs. in the former and 19.5 lbs. in the latter case.

The effect of subsoiling on the capillary rise of water from below was demonstrated by using the same piece of apparatus in the same way except that the two areas were covered to prevent evaporation, without adding any water, the experiment extending from June 26 until July 2, giving the results shown in the next table.

Table showing the effect of subsoiling on the capillary rise of water from the deeper soil when no evaporation can take place from the surface.

	On Subsoiled Ground.				
	1st foot.	2nd foot.	3rd foot.	4th foot.	5th foot.
	Per ct.	Per ct.	Per ct.	Per ct.	Per ct.
June 26............ Moisture at start	23.29	21.89	17.85	14.14	19.55
July 2............ Moisture at close.	22.66	22.50	17.49	14.45	20.27
Change..................	— .63	+ .61	— .36	+ .31	+ .72
	On Ground not Subsoiled.				
June 26—start.................	22.52	20.67	17.74	15.06	19.34
July 2—close..................	23.97	22.09	18.92	14.62	18.38
Change..................	+1.45	+1.32	+1.18	— .44	— .96

It will be seen that in the subsoiled area there had been but little change in the water condition while the ground not subsoiled had gained a very material amount of water in the surface three feet at the expense of that deeper in the ground, the gain in the upper three feet amounting, on the 36 square feet, to 129.69 lbs., 53.52 lbs. having come from the fourth and fifth feet and the balance probably partly from the sides and partly from the sixth foot.

When the ground was subsoiled in the same manner as

before and allowed to stand exposed under natural conditions, and the surface kept free from weeds by shaving them off close to the surface with a sharp hoe, it was found, after an interval of 75 days from June until September, that the water content of the soil stood as in the next table.

In this case the surface foot of subsoiled ground is dryer than that not so treated, but the second, third and fourth have gained in moisture, over and above that lost from the other two feet, enough to represent a rainfall of 1.64 inches.

	Subsoiled ground	Not subsoiled ground.	Difference.
	Per cent.	Per cent.	Per cent.
First foot	17.07	18.91	−1.84
Second foot	23.29	19.42	+3.87
Third foot	22.76	17.78	+4.98
Fourth foot	16.35	14.19	+2.16
Fifth foot	18.14	19.20	−1.06

229. Moisture Effects of Subsoiling.—The results which have been given in the last section illustrate several very distinct effects produced by subsoiling:

(1) Subsoiling increases the percentage capacity of the soils stirred for moisture.

(2) Subsoiling decreases the capillary conducting power of the soil stirred.

(3) Subsoiling increases percolation through the soil stirred or its gravitational conducting capacity.

230. How Subsoiling Increases the Water Capacity of the Soil Stirred.—When a soil is broken into lumps lying loosely together, and these become filled with water, each one behaves in a measure much as if it were standing by itself and much as a lump of sugar would, plunged into water and then withdrawn, coming forth with its pores practically filled with water. In short columns of soil, like the lumps, the surface films of water which span their capillary pores are strong enough to maintain their whole interior nearly full of water, drainage being largely confined to those passageways and cavities which have larger than capillary dimensions.

If a dozen strands of candle-wicking, two feet long, are twisted loosely together, saturated in a basin of water, and then held horizontally from the two ends to drain, more water will be retained than if it is allowed to sag into a loop and drainage from it will be still more complete when hanging from one end. So it is with long continuous columns of soil; from them the drainage is more complete than from shorter ones.

231. How Subsoiling Decreases the Capillary Conducting Power.—When large open spaces have been formed in a soil, by any means, as is the case in subsoiling, every such cavity cuts off the capillary connection with the unstirred soil below and above and in this way reduces the number of capillary passageways by which water may rise to the surface. This being true, when rains fall upon subsoiled ground, water travels downward quite slowly until after it has become capillarily saturated and, if the rain is not enough to over-saturate the layer, the whole will be retained.

On the other hand, when the subsoiled layer has once become dry, the poor connection with the firmer ground below and its open texture makes it impossible for the moisture to rise through it to the surface as rapidly as it could through a more compact layer.

It is clear, from these relations, that when the root system of a crop once develops through the subsoiled layer it may then act as a mulch of great thickness and increase the yield; but should a crop fail to get its roots below the subsoiled layer before the moisture becomes too scanty then a diminished yield might be the result even with an abundance of water below.

232. How Subsoiling Favors Percolation.—When rain enough has fallen upon an earth mulch or upon subsoiled ground to completely saturate the soil the balance of the water is then free to move rapidly downward through the large non-capillary pores, urged by the strong force of gravity. Not only this, but, since the pores are many of them too large to be filled by the percolating streams, there is left an easy egress for the soil-air, which must escape upward before the water can enter, and this does not retard percolation as it does in a compact soil.

233. A Larger Percentage of the Moisture of Subsoiled Ground Available to Crops.—When a soil has been made more open by subsoiling, and its capacity for holding water thereby increased, this extra amount of water retained becomes wholly available to crops. It was shown in **(161)** and **(162)** that there is a certain per cent. of water in a soil which the roots of plants are unable to remove with sufficient rapidity to meet their needs and as this amount depends upon the size of the soil grains, which subsoiling does not alter, the increased percentage held becomes a clear gain to the crop.

234. Dangers From Subsoiling.—One of the most serious difficulties associated with subsoiling, aside from the expense, is the danger of puddling, and this is particularly great in humid climates where the subsoil, especially in the spring, is liable to be too wet. The danger is intensified on account of the fact that the surface soil may be in good condition for plowing when that below is much too wet. If this work is attempted when the ground is not in condition very great harm may be done and so it is generally much safer to subsoil late in the fall in humid climates, when the deeper ground is generally dryest.

235. Early Seeding.—When the crop is started to growing upon the ground as early as the temperature of the soil and of the air will permit the farmer is conserving soil moisture, by taking advantage of that which otherwise would be lost by surface evaporation, and enabling his crop to use this in growth. Such timely planting may not only save moisture from going to waste, both by evaporation and by percolation, but it may save plant food from loss in the drainage waters.

Yet, while due diligence should be exercised in timely planting and sowing, there is danger of too great haste and it will generally be better to make the mistake of getting the crop in a little late rather than too early. The soil should by all means be warm enough and dry enough to make germination prompt and vigorous, for otherwise weak and sickly plants will result, if the seed does not rot in the ground.

236. Danger From Green Manuring.—In the practice of growing cover-crops, and in green manuring, attention must always be given to the effect these have upon the soil moisture, as related to the crop which is to follow. When either rye or clover is used in green manuring, and the plants are allowed to make a heavy growth before plowing under, the soil will be found very much dryer than if the field had been plowed and tilled early but left naked, or even if not plowed at all. The next table demonstrates the truth of this statement, showing, as it does, the strong drying effect of clover as early as May 13.

Table showing the drying effect upon the soil of a green manure crop.

	1 to 6 inches.	12 to 18 inches.	18 to 24 inches.
	Per cent.	Per cent.	Per cent.
Ground not planted.............. ..	23.33	19.13	16.85
Ground in clover....................	9.59	14.75	13.75
Difference	13.74	4.38	3.10

In such a case as this, with the soil as dry when plowed as that under the clover, not only would there be danger of the seed not germinating properly but the large growth of herbage, when plowed under, would so much cut off the capillary connection with the deeper soil moisture that it could not readily become available until after the roots had penetrated below this level.

Nor is this all; any such crop would have locked up in insoluble form, for the time being, a large portion of the soluble plant food, and unless abundant and timely rains were to follow the plowing speedily to develop a new supply, the next crop would suffer for lack of nitrates and other plant foods.

On soils naturally too wet and in wet seasons the dangers referred to will of course not be so great and the green manure crop might even be an advantage from the soil moisture side by making the over-wet soil more open, thus favoring stronger root action and more rapid nitrification.

237. Wind-breaks and Hedges.—"In* sub-humid climates, especially like those of our western prairies, where there is a high mean wind velocity, and in the level districts of humid climates, where the soils are light and sandy, with a small water capacity, and which are lacking in adhesive quality, the fields may suffer greatly at times, not only from excessive loss of moisture, but the soil itself may be greatly damaged by drifting caused by the winds. Under such conditions, it is a matter of great importance that the wind velocities close to the surface should be reduced as much as possible."

On the lighter sandy lands, wherever broad fields lie unsheltered by any wind-break, strong dry winds frequently sweep entirely away crops of grain after they are four inches high, and at the same time drift away even as much as three or four inches of the surface soil, the best in the field. In such cases wind-breaks and hedge-rows exert a very strong protective influence and greatly lessen such disastrous results.

Not only do trees along line fences and roadsides, under these conditions, prevent such direct injuries to soil and

* Irrigation and Drainage, p. 168.

crops but they materially lessen the evaporation of moisture from the soil and thus help to secure a higher yield of crops. *"The writer has observed that, when the rate of evaporation at 20, 40, and 60 feet to the leeward of a grove of black oak 15 to 20 feet high was 11.5 c. c., 11.6 c. c., and 11.9 c. c., respectively, from a wet surface of 27 square inches, it was 14.5, 14.2 and 14.7 c. c., at 280, 300 and 320 feet distant, or 24 per cent. greater at the three outer stations than at the nearer ones. So, too, a scanty hedge-row produced observed differences in the rate of evaporation as follows, during an interval of one hour:

At 20 feet from the hedge-row the evaporation was	10.3 c. c.
At 150 feet from the hedge-row the evaporation was	12.5 c. c.
At 300 feet from the hedge-row the evaporation was	13.4 c. c.

Here the drying effect of the wind at 300 feet was 30 per cent. greater than at 20 feet, and 7 per cent. greater than at 150 feet from the hedge.

Then, too, when the air came across a clover field 780 feet wide the observed rates of evaporation were:

At 20 feet from clover	9.3 c. c.
At 150 feet from clover	12.1 c. c.
At 300 feet from clover	13 c. c.

Or 40 per cent. greater at 300 feet away than at 20 feet, and 7.4 per cent. greater than at 150 feet."

* Irrigation and Drainage, p. 169.

CHAPTER IX.

RELATION OF AIR TO SOIL.

NEEDS OF SOIL VENTILATION.

Air in the soil in which crops are to be grown is as essential to the life of the plants as the air in a stable is to the life of the animals housed.

Careful observations and lines of experimentation have proved, in many ways, that when oxygen is completely excluded from seeds that are otherwise under good conditions for germination they fail to start. It has been found, too, that even after a seed has begun to grow, if the oxygen supply is cut off, it makes no farther progress. Growth does take place in seeds in a very dilute atmosphere of oxygen, but after the amount has been reduced below $\frac{1}{32}$ of the average in the air the plants advance very slowly and are sickly.

A soil in the best condition for crops must permit of ready entrance of fresh air and an abundant escape of the air once used; in other words, like the stable, it must be well ventilated. This ventilation is needed:

(1) To supply free oxygen to be consumed in the soil.

(2) To supply free nitrogen for the use of the free-nitrogen-fixing germs.

(3) To remove the excess of carbon-dioxide which is set free in the soil.

238. Needs For Free Oxygen in the Soil.—Free oxygen in the soil is required not only by the seeds, when they are germinating, but throughout the active life of the plant in order to permit the roots to live, for they, too, must breathe.

Then in the conversion of the nitrogen of humus, manure,

and decaying organic matter in the soil into nitric acid, large amounts of oxygen are needed, for each of the three known forms of microscopic life which do this work are unable to live in its absence.

239. A Water-logged Soil.—One of the chief reasons for the unproductiveness of a water-logged soil is the deficiency of free atmospheric oxygen in it. When the soil pores are filled with water and this water is stationary, that is, not changing, the free oxygen which it may contain in the air dissolved in it is soon used up and then the rate at which oxygen from the air above the soil is able to make its way downward through the soil-water and around and between the soil grains is much too slow to meet the ordinary needs of the roots of any crop. Not only this, but, as pointed out in (**103**), even the microscopic organisms in the soil find so scanty a supply that they are obliged to decompose the nitric acid for the oxygen it contains in order to supply their needs. The chief need of draining wet lands, then, is to secure to the soil a more rapid change of air.

240. Floating Gardens.—The instances where the Chinese and Mexicans grow crops upon floating rafts of logs anchored in a stream or lake and thinly covered with soil may seem to contradict the statements in the last paragraph regarding a water-logged soil because, in these cases, the soil is very wet in its lower portion and the roots of the plants are continually immersed in a saturated soil or in the water itself beneath. A little reflection, however, will make it clear that the two cases are very different. Both in the lake and in the running stream the water is changing continually so that a new supply, charged with fresh oxygen, is being continually brought to the roots or very near them.

It is the abundance of oxygen which rain water and that used for irrigation contains which prevents it from killing crops when the water entering the soil is excessive. As long as the water is moving through the soil, and a fresh supply from above entering, an abundance of air is carried with it for the needs of the roots.

241. Excessive Soil Ventilation.—The higher temperature of a pile of open horse manure, as compared with that of the closer heap of cow-dung, illustrates how important the free and rapid access of air to the interior is to the formation of the ammonia, for the difference in temperature in the two cases is largely due to a difference in the rate of fermentation, and this to the too rapid entrance of air. In these cases the air is entering too rapidly and a loss of nitrogen is the result. And the same thing may occur in a too open soil. Indeed, the small amount of humus in the sandy soils is in a large measure due to the freer access of air to the interior.

It is for this reason that unusual care must be exercised to keep the supply of humus in these soils up, not only because of its need for plant food, but because it enables the sandy soils to hold more water, and this in turn makes them less readily penetrated by the air and the humus does not waste as rapidly.

242. Return of Carbon-Dioxide to the Air.—It is of course necessary to the continuance of plant life that the vast systems of roots which are developed in the soil should be broken down, first into humus and then into carbon-dioxide, water and free nitrogen, and all of the processes concerned in these changes demand free oxygen taken from the air and the escape of the carbon-dioxide and nitrogen gas set free, and here again is ample soil ventilation necessary.

243. The Fixing of Free Nitrogen.—In the processes of symbiosis discussed in (101), which lead to the removal of the free nitrogen of the air in the soil and soil moisture, and the conversion of it into organic compounds suitable for the food of higher plants, soil ventilation is necessary in order to supply both the oxygen and nitrogen of the air which the micro-organisms are obliged to use in carrying on their life processes.

PROCESSES OF SOIL VENTILATION.

The interchange of gases between the soil and atmosphere is brought about in several ways and by different agencies. Among these are (1) the slow process of diffusion described in **(5)** and **(14)**. (2) The expansion and contraction of soil-air due to changes in temperature. (3) The expansion and compression of the air due to changes in barometric pressure. (4) The suctional effect of the wind, especially when it is gusty. (5) The air absorbed by rainwater is carried into the soil when percolation takes place. (6) When water drains away from a soil or is carried upward and out by capillarity or root action it acts by suction to draw into the soil a volume of air equal to that of the water which flows out.

244. Ventilation of Soil by Diffusion.—The exchange of air between that in the soil and the atmosphere above by diffusion is a very slow process but, because it is all the time taking place, the total exchange during the growing season is considerable. The more open the texture of the soil is and the higher the soil temperature the more rapidly will the interchange by this process take place.

245. Soil Ventilation Due to Changes in Soil Temperature.—When the temperature of air is changed its volume is also altered and in the ratio of $\frac{1}{491}$ for each degree F. or $\frac{1}{273}$ for each degree C.; so that if 491 cubic feet of soil-air were to have its temperature changed 1° F. this would result in one cubic foot of air being forced out of the soil, if the temperature was raised, and a like amount would enter if the temperature were to fall the same amount.

The temperature of the surface three inches of soil often changes as much as 16° to 20° F. and that at 18 inches deep as much as 1.5° F. A soil like the surface foot in **(133)**, containing 18 per cent. of water, would enclose about 5.3 acre-inches of air in the surface 1.5 feet and, with a diurnal change of 16.4° F. in the upper 3 inches and 1.5° F. at a depth of 18 inches, the amount of soil-air which would be forced out and again taken in each 24 hours would be about 14 cubic inches for each square foot of surface. So that the soil ventilation due to diurnal changes in soil temperature will range from 0 up to possibly 20 cu. in. per square foot.

246. Influences of Changes in Barometric Pressure on Soil Ventilation.—Any change which may occur in the pressure of the air above the soil is followed by a change in the volume of the soil-air, causing an escape from the soil, if the pressure above falls, and the entrance of an extra supply whenever the pressure is increased.

With soil like that in (133), having 18 per cent. of water in the first foot, 20 per cent. in the second and 15 per cent. in the third and fourth feet, there would be 7.88 inches in depth of soil-air contained in the four feet and every change in atmospheric pressure amounting to .1 inch would cause the escape or entrance of 3.78 cubic inches for each square foot of surface and 18.9 cubic inches for each change in pressure of .5 inches of barometer.

It is common in the United States for waves of high and low pressure to pass a given locality about twice each week, and the differences in pressure between high and low barometer are generally not far from .5 inch, so that the results stated above give a fair measure of this influence in soil ventilation.

247. Wind Suction and Soil Ventilation.—It is seldom true that the wind blowing across a field has a uniform velocity, the general tendency being for it to blow in gusts. This unsteady action tends at times to increase the pressure on the soil-air and at other times to decrease that pressure and, as a result, there is a nearly constant tendency for air to leave or enter the soil on this account, and it is possible that this factor in soil ventilation may be stronger than any other, on account of the great frequency with which the changes recur.

248. Movements of Water and Soil Ventilation.—The water which enters the soil as rain must displace a volume of air equal to the rainfall which penetrates the soil and then, when this water is again lost by the soil, whether by percolation or by capillary or root action, the same volume of air must again be returned. In a climate where the rainfall, which penetrates the soil, is 24 inches during the growing season, two cubic feet of air per square foot of surface enters the soil in consequence.

WAYS OF INFLUENCING SOIL VENTILATION.

There are important means and methods of controlling and modifying the rate and extent of soil ventilation, which are under the control of the farmer.

249. Soil Ventilation Modified by Tillage.—Nearly all of the operations of surface tillage modify the rate of entrance or escape of air from the soil. Plowing effects a sudden and complete change of air in the soil to the depth stirred and in the spring, when nitrates are deficient, and the pores largely closed with water, this breaking up of the soil may be very beneficial.

The thorough preparation of the seedbed before planting, so strenuously insisted upon by the best practical men, has a portion of its rational basis in the need of soil ventilation; and deep subsoiling, when done at such a time as not to puddle the soil, must always profoundly affect the relation of air to soil, as well as of moisture. Indeed, all of the operations of soil loosening serve, not only to admit air more freely to the soil stirred, but the undisturbed portions beneath will also be better ventilated because of the surface loosening.

250. Rolling and Harrowing For Soil Ventilation.—It frequently happens, especially with small grains in the spring, when the season has been unusually wet and evaporation large, that a crust forms upon the surface, partly by shrinkage, partly by the crumb-structure breaking down and partly by the deposit of soluble salts between the soil grains, thus closing up the pores and greatly impeding the entrance of air. Under such conditions the harrowing or rolling of small grains after they are up owes its advantages in part to the better soil breathing it secures, by breaking the crust.

But it will sometimes happen, when small grains are rolled immediately after seeding, if the ground chances to be a little too moist, that soil ventilation will be so much hindered by the packing as to result in defective germination and sickly plants. In one case a crop of barley was so much affected in this way that a serious reduction of yield was the result and the plants, even when mature, were so evidently influenced, that the rolled strip, between two adjacent areas not rolled, but in other respects the same, showed in strong contrast on account of the smaller plants.

251. Underdraining For Soil Ventilation.—When heavy soils are underdrained they are so much more deeply and better aerated that this is one of the chief advantages of that method of land improvement. In such cases the roots of plants penetrate the subsoil so much farther, and earthworms and ants burrow so much deeper, that with the decay of the roots the more or less vertical galleries formed by these agencies permit much freer and deeper soil ventilation.

Then when the under clays dry out, as they do after draining, great numbers of shrinkage checks form and into these both the roots of plants and the free soil-air penetrate and are brought together.

After this last stage of soil improvement has taken place the bringing in of carbonic acid with the air leads, through its action upon the lime, to the flocculation of the minuter soil particles and thus to a more extensive granulation of the whole subsoil, which in turn extends the soil ventilation still more widely.

But all of these effects upon the soil are only the means which permit the underdrains to render their greatest service in permitting a strong and extensive movement of air into and from the soil; for once the soil is opened up in this way, the air, through the action of the wind, changes in barometric pressure and changes in soil temperature, readily enters the soil, not only through the surface above but throughout the whole length of the underdrains.

When it is seen that changes in soil temperature and in atmospheric pressure make such marked changes in the flow of water from springs and from tile drains as are shown in (337) and (338) it becomes clear that the movements of soil-air into and out of tile drains must be even more marked than the movements of ground water.

252. Influence of Vegetation on Soil Ventilation.—In the case of such crops as clover, which send long and somewhat fleshy roots down deeply into the subsoil, there are very many and important passageways opened up after the roots decay, which greatly facilitate the deeper and more rapid change of soil-air, and, as has been pointed out, the removal of water by the living roots must also draw into the soil a volume of air equal to the amount of water used, except in so far as this is made good by the rise of capillary water from below.

CHAPTER X.

SOIL TEMPERATURE.

253. Importance of Soil Temperature.—None of the chemical, physical or biological changes essential to the development of plant food in the soil and to the action of roots, can take place in the absence of the energy stored up in the soil and indicated by its temperature. When the temperature of the soil falls to 32° F. nearly all the life processes become dormant and for most of the cultivated crops and higher plants these cannot begin until a temperature above 40° F. has been reached. All living bodies must have their temperature maintained between certain limits in order to have growth take place.

254. Soil Temperature at Which Growth Begins.—According to the observations of Ebermayer growth will not begin, with most cultivated crops, until the soil has attained a temperature of 45° to 48° F. and it does not take place most vigorously until after it has reached 68° to 70° F. Neither do the niter germs begin the formation of nitric acid from humus until a temperature above 41° F has been reached and its greatest activity is not attained until the soil temperature has risen to 98° F.

255. Best Soil Temperature for Germination.—There is, for most seeds, a certain range of soil temperature under which germination is most rapid, under which the plants become most vigorous, and which ensures the highest percentage of plants from the seed. This general truth should never be overlooked in the spring when it is possible to plant in a too cold soil. In the table which follows are given the best soil temperatures and the lowest and highest temperatures at which certain seeds have been observed to germinate.

Name of Plant.	Best Soil Temp.		Lowest Soil Temp.		Highest Soil. Temp.	
	Sachs.	Van Tiegham.	Sachs.	Van Tiegham.	Sachs.	Van Tiegham.
Wheat	84° F.	81° F.	41° F.	41° F.	104° F.	99° F.
Barley	84	83	41	41	104	100
Peas	84	80	44.5	44	102	....
Maize	93	93	48	49	115	115
Scarlet bean	79	....	49	....	111	....
Squash	93	....	54	....	115	....
Red clover	....	70	....	42	....	82
Turnips	....	89	....	....	....	108
Mustard	....	81	....	32	....	99
Melon	....	99	....	....	....	....

The two important facts fixed by these data are: (1) The soil temperatures at which the seeds of most cultivated crops germinate best, lie between 70° and 100° F., with an average of about 85° F. (2) The soil temperatures below which germination does not take place are between 41° and 54° F. From these it is clear that seeding should not begin until the thermometer will show the temperature of the soil at the depth of planting, well up toward 70° F. during the warmest portion of the day. These statements should not be understood as advising against the sowing of clover seed early in the spring, while the frost is yet on the ground, under conditions where it might not be possible to get a stand otherwise.

256. Observed Soil Temperatures.—The temperatures which the soil does attain at different depths during the different months of the growing season will be of interest in connection with the statements made in the last two sections. In the two tables which follow are given the mean seasonal variations of soil temperature at two stations, one in this country and the other in Europe.

Table showing the mean monthly soil temperatures, at State College, Pa., by Dr. Frear, and at Munich, Germany, by Ebermayer.

AT STATE COLLEGE, PENNSYLVANIA.

Depth.	April.	May.	June	July.	Aug.	Sept.
	°F.	°F.	°F.	°F.	°F.	°F.
3 inches..............	43.74	55.13	67.29	70.16	63.70	61.32
6 inches..............	43 08	54.72	65.34	69.75	68.49	61.70
12 inches..............	42.69	53.83	65.03	68.89	68 66	62.73
24 inches..............	41.43	51.45	61.90	66.42	67.41	63.59

AT MUNICH, GERMANY.

Depth.	April.	May.	June	July.	Aug.	Sept.
5 9 inches............	44.65	56.79	61.11	67.28	64 09	58.21
11.8 inches............	44.31	57.51	60.06	66.16	63 61	57.38
23.7 inches............	44.40	53.58	59.11	63.12	63.35	58.82
35.4 inches............	43.56	51.24	57.33	62.92	62.26	58.51

It may appear that the temperatures recorded in these tables are too low to be in harmony with the comparatively high temperatures given as the best for germination. It must be understood, however, that the average must be lower than would be found in the soil during the warmest portion of the day. In regard to the minimum temperature at which germination takes place it will be clear enough that the April records for soil temperature are quite in harmony with those given for germination.

257. Influence of Soil Temperature on the Rate of Germination.—The more quickly seeds are permitted to germinate after they are placed in the soil the higher will be the per cent. of seeds growing and, as a rule, the more vigorous will the plants be. Indeed, seeds of low vitality placed in too cold a soil often fail to germinate at all.

Haberlandt found that, when corn would germinate in 3 days at a temperature of 65.3° F., it required 11 days when the soil was as low as 51° F., and Hellriegel showed that when corn was planted under a mean temperature of 48° only 2 out of 10 kernels sprouted in 42 days; that under the same temperature rye germinated in 9 days, winter wheat in 12 days, and barley and oats in 13 days, while cucumbers did not germinate in 42 days.

258. Effect of Soil Temperature on Root Pressure.—The power which sends the soil moisture into the roots of plants and up into the leaves is osmotic pressure, developed by the warmth of the soil, and unless the soil temperature is sufficiently high plants may wilt, as Sachs has shown, where he demonstrated that pumpkin and tobacco plants wilted badly, even at night with an abundance of moisture, as soon as the soil temperature fell much below 55° F., the moisture not rising fast enough to compensate for even the slow evaporation during the night.

259. Influence of Soil Temperature on the Formation of Nitrates.—The nitrates in the soil do not develop until the temperature has risen above 41° F.; the action of the germs is extremely feeble at 54° and they do not attain their maximum activity until a soil temperature of 98° has been reached; but if the earth becomes as warm as 113° F. then the action is nearly stopped, it being as weak as at 54°.

CONDITIONS INFLUENCING SOIL TEMPERATURE.

260. Specific Heat of Dry Soil.—When the same number of heat units are given to like weights of different kinds of soil their temperatures are not raised through the same number of degrees and this is because their specific heats (40) are different.

From the determination of Oemler it appears that the number of heat units required to raise the temperature of 100 lbs. of water and 100 lbs. of soil of different kinds from 32° to 33° F. is as stated in the table which follows:

Table of specific heat of dry soils.

	No. of heat units required to raise 100 lbs. from 32° F. to 33° F.	Temperature of 100 lbs. after the application of 100 heat units.
	Heat units.	°F.
Water	100.00	33.00
Moor earth	22.15	36.51
Humus	20.86	36.79
Sandy humus	14.14	39.07
Loam rich in humus	16.62	38.02
Clayey humus	15.79	38.33
Loam	14.96	38.68
Pure clay	13.73	39.28
Sand	10.08	41.92
Pure chalk	18.48	37.41

It is clear from this table that much more heat is required to raise the temperature of water through one degree than of a like weight of dry soil, and hence that a dry soil will warm in the sunshine more rapidly than a moist soil can.

261. Specific Heat of Wet Soil.—The differences in the weight per cubic foot of dry soils and the differences in their water content greatly affect the specific heat or the rate at which the surface temperatures will rise under the same conditions.

Sand has a small capacity for water and on this account is naturally warm, but its greater weight per cubic foot acts as an offset, tending to make it colder. If a loosely packed clay loam weighs 70 lbs. per cubic foot and a sandy soil 106 lbs. and the two hold 33 per cent. and 18 per cent. of water respectively, when capillarily saturated, then the number of degrees F. that 100 heat units will raise the temperature of a cubic foot of each soil when saturated, half saturated and dry are given below:

	Saturated.	Half saturated.	Dry.
Sandy soil	3.4° F.	5.° F.	9.92° F.
Clay loam	2.98	4.49	6.02
Difference	.42	.51	3.9

One thousand heat units would raise the differences in temperatures to 4.2°, 5.1° and 39°, making it clear that the differences in weight and in water content greatly influence the degree of warmth.

262. Influence of Color on Soil Temperature.—The color of a soil, especially when dry, so that the rate of evaporaiton from its surface is small, has a marked influence on the temperature, even at considerable depths. Wollny made a series of experiments to note the effect of color, using white marble dust and lampblack in different proportions, to secure different shades from light grey to black, in which he placed two thermometers, one with the bulb just beneath the surface and the other 4 inches below. The temperatures were taken every two hours of the 24 and the results are given in the table below, together with those of a similar trial using yellow ocher.

Table showing the influence of color on the temperature of soil.

	At the Surface.				At Four Inches Deep.			
	Black.	Dark grey.	Med'm grey.	Light grey.	Black.	Dark grey.	Med'm grey.	Light grey.
	°F.	°F.	°F.	°F.	°F.	°F.	°F.	°F.
Mean temp..	32.82	32.39	31.98	30.94	28.33	28.46	27.83	27.20
Variations...	31.55	32.90	32.45	30.10	15.20	14.25	12.50	11.85
	Dark brown.	Medium brown.	Light brown.	Faint brown.	Dark brown.	Medium brown.	Light brown.	Faint brown.
	°F.	°F.	°F.	°F.	°F.	°F.	°F.	°F.
Mean temp	31.76	31.65	30.93	30.70	27.29	27.19	27.31	26.40
Variations .	31.95	31.75	29.90	27.65	12.30	12.15	11.80	10.75

From this table it appears that the darkest soil, whether black or brown, was more than a degree warmer than the light soil at four inches deep; and that the black soil had a daily variation in temperature at four inches more than 3° F. greater than the light soil, and the dark brown soil one of 1.55° F.

263. Influence of Topography on Soil Temperature.—The degree of inclination of the land surface and the direction of the slope, whether facing east, west, north or south, may exert a marked influence upon the temperature of the soil and particularly upon its diurnal range. The temperature of a stiff red clay soil, upon a level table, and upon a south exposure sloping about 18°, was found in the surface three feet to be as represented in the table below:

Showing the influence of topography upon soil temperature.

Kind of Soil.	Depth Below the Surface.		
	1st foot.	2nd foot.	3rd foot.
Red clay, south slope	70.3° F.	68.1° F.	66.4° F.
Red clay, level surface	67.2	65.4	63.6
	3.1	2.7	2.8

Here it is seen that the effect of a south exposure is to make a difference in temperature of from a little more than 3° F., in the surface foot, to a little less in the second and third feet.

The reason for these differences will be readily understood from a study of Fig. 61. Suppose A 6 5 B to represent a section of a prism of sunshine falling upon the hill A E B, where A E is the south slope and E B is the north. On account of the sun not being directly vertical over the hill the south slope receives as much more heat in a unit of time than the north slope as the line 4–6 is longer than the line 4–5.

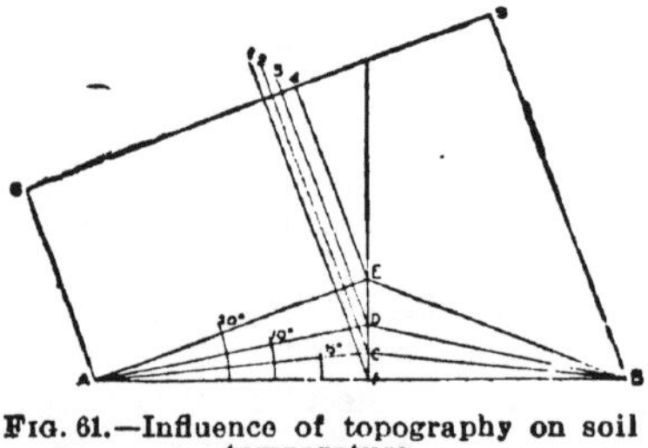

Fig. 61.—Influence of topography on soil temperature.

264. Influence of Looseness and Unevenness of Surface on Soil Temperature.—When a field is left very uneven, and especially if covered with lumps, the large amount of surface exposed to the sky and to the air permits the heat of the surface soil to be lost rapidly in warming the air above and the result is the deeper soil remains at a lower temperature. So, too, if the soil is loose and open, the dry superficial layer becomes warm and heats the air, while the poor conducting capacity of the open soil prevents the heat from being conveyed deeply below the surface and a lower temperature is the result.

265. Influence of Surface Tillage on Soil Temperature.—When corn ground was cultivated 3 inches deep as compared with 1.5, in alternate groups of four rows, the mean temperatures of the soil in the first, second and third feet below the soil stirred was found to be .82° F. warmer in the first foot and .59° F., and .36° F. respectively in the second and third feet on the ground receiving the shallower cultivation.

266. Influence of Chemical and Physical Changes on Soil Temperature.—When heavy dressings of farmyard manure are plowed in, and when heavy crops are turned under for green manure, the fermentation which is set up in these materials results in a measure of heat which warms the soil in the same way that a manure heap heats when fermenting. Indeed all of the steps in the formation of nitrates in the soil result in the evolution of some heat.

Again, when the surfaces of dry soil grains become moistened with water, whether by rain or by capillary movements, surface tension in forcing the water to surround the soil grains generates a small amount of heat, which affects, in so far, the soil temperature.

267. Influence of Rains on Soil Temperature.—Heavy rains which fall upon fields and penetrate the soil may exert very marked effects upon its temperature on account of the relatively high specific heat of the water as compared with that of the soil.

If the atmosphere is warmer than the deeper soil, as

may be the case in the spring, and if rains fall which result in heavy percolation, a large amount of heat is conveyed rapidly and deeply into the soil with the water and the temperature of the ground, two to four feet below the surface, may thus be very materially raised.

268. Influence of Evaporation on Soil Temperature.— There is no factor, except the direct sunshine and the direct radiation of heat away from the earth into space, which exerts so strong an influence on the temperature of the soil as the evaporation of moisture from its surface; and the chief reason why an undrained clay soil is colder than one well drained is the cooling effect associated with the larger evaporation of soil moisture.

To evaporate a pound of water from the surface of a square foot of soil, by means of the heat contained in the soil, makes it imperative that 966.6 heat units be expended to do the work and this, if withdrawn from a cubic foot of saturated clay soil, would lower its temperature some 10.3° F.

The difference in temperature shown by the wet and dry bulb thermometers measures, in one way, the cooling effect of evaporation; the wet bulb often reading as much as 15 or even 20 degrees lower than the dry one, under otherwise identical conditions.

Table showing the influence of rapid evaporation upon the temperature of the soil.

Date.	Time.	Condition of weather.	Temp. of air.	Temp. of drained soil.	Temp. of undrained soil.	Difference.
			°F.	°F.	°F.	°F.
April 24	3.30 to 4 p. m.	Cloudy, with brisk east wind.	60.5	66.5	54.00	12.50
April 25	3 to 3.30 p. m.	Cloudy, with brisk east wind.	64.0	70.0	58.00	12.00
April 26	1.30 to 2 p. m.	Cloudy, rain all the forenoon.	45.0	50.0	44.00	6.00
April 27	1.30 to 2 p. m.	Cloudy and sunshine, wind S. W. brisk.	53.0	55.0	50.75	4.25
April 28	7 to 8.30 a. m.	Cloudy and sunshine, wind N. W. brisk.	45.0	47.0	44.50	2.50

In the table above are given the observed differences in temperature of a well drained sandy loam and an adjacent black marsh soil, not well drained, the observations being taken simultaneously and the differences in temperature being due largely to differences in the rate of evaporation in the two cases.

MEANS OF CONTROLLING SOIL TEMPERATURE.

269. Effect of Rolling on Soil Temperature.—In the spring of the year, when the soil is naturally cold, the first effect of rolling is to cause the soil to warm deeply at a more rapid rate, and Fig. 62 shows how strong this influence may be. In extreme cases the soil temperature, at 1.5 inches below the surface, has been found as much as 10° F. higher than on entirely similar and adjacent ground, not rolled, and 6.5° at 3 inches below the surface. This difference is due to the better conducting power of the soil, on account of its firmer texture, and is in spite of the loss of heat due to greater evaporation which takes place from the rolled surface.

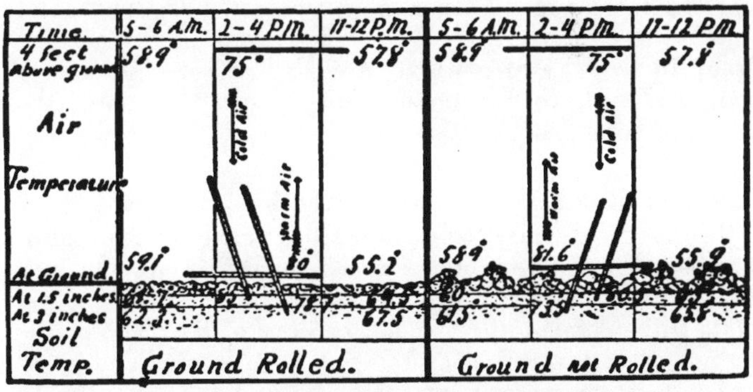

Fig. 62.—Showing the effect of rolling on soil temperature.

The average difference in temperature of soil on eight Wisconsin farms, at the season when oats were germinating, was found to be as given in the table below:

Time.	Mean air temp.	Mean soil temperature at 1.5 inches deep.		Mean soil temperature at 3 inches deep.	
		Rolled.	Unrolled.	Rolled.	Unrolled.
2 to 4 p. m...	65.37° F.	71.69° F.	68.57° F.	67.33° F.	64.39° F.

Here is a mean difference of 3.1° F. at 1.5 inches, and 2.9° F. at 3 inches deep in favor of the rolled surface.

270. Influence of Thorough Preparation of the Seed-bed on Soil Temperature.—It follows, from what has been said in previous paragraphs, that the practice of thoroughly preparing the seed-bed before sowing or planting must have the effect of decreasing the capillary rise of cold water from below and its loss by evaporation from the soil. This then would tend to concentrate the sun's heat in the seed-bed itself, first by lessening its rate of conduction downward, and second by diminishing its loss, by lessening the evaporation. In the spring, then, early and thorough preparation of the seed-bed tends to make the seed-bed warmer; it diminishes the loss of soil moisture; it increases the formation of nitrates, thus making the soil richer; it hastens and makes stronger the germination and it enables one or more crops of weeds to be destroyed before the crop is up in the way of cultivation. Hence there is much to gain and little to lose in the thorough preparation of the seed-bed before planting.

271. Controlling Soil Temperature by Underdraining.—When land naturally too wet for tillage early in the spring has been thoroughly underdrained, the soil is brought into fit condition for seeding much earlier than would be possible without this improvement, and one of the great points gained is the warming of the soil to a greater depth, on account of the removal of the water and the lessening of the loss of heat by evaporation.

CHAPTER XI.

OBJECTS, METHODS AND IMPLEMENTS OF TILLAGE.

Tilling the soil is one of the oldest of agricultural arts, and during its long practice very many methods have been adopted and tools devised for securing the ends sought.

272. Objects of Tillage.—The term "tillage" has been applied to the different methods of working the soil in order to secure the conditions needful for the growth of cultivated crops. The chief objects which tillage aims to secure are:

1. To destroy and prevent the growth of weeds and other vegetation not desired upon the ground.
2. To place beneath the surface manure, stubble and other organic matter where it will not be in the way and where it may be converted rapidly into humus.
3. To develop various degrees of openness of texture and uniformity of soil conditions suitable to the planting of seeds and the setting of plants.
4. In still other cases the object of tillage may be to so modify the movements of soil moisture and of soil air.
5. In still other cases the objects of tillage may be to so change conditions as to make the soil either warmer or colder.

TILLAGE TO DESTROY WEEDS.

It must ever be kept in mind that wherever weeds are allowed to grow they are removing from the soil both available moisture and plant food in the form of soluble salts and, to whatever extent this is permitted, to that extent is the possible yield of any crop lessened. No soil can mature a maximum crop of corn when weeds are permitted to grow with it. Neither is it possible for an orchard of any kind to come into bearing as quickly or to produce as vigorous trees where the soil between and beneath them is occupied by either weeds or grass. It may be thought that so long as the weeds are destroyed upon the ground they return to it whatever they have taken out and therefore cannot leave the soil poorer. To this it must be said that whatever moisture is removed is a positive loss because it is carried away by the winds; the nitric acid that is taken up and the potash, phosphoric acid and other ash ingredients are also largely a positive loss so far as that season is concerned for they are removed from the soil moisture and converted into dry matter in the tissues of the weeds where the crop cannot use them. Even if the weeds are killed while the crop is yet on the ground they cannot furnish food for it for they are likely not to decay soon enough to become at once available.

273. The Best Time to Kill Weeds.—The best time to kill weeds is just as the seeds are germinating or while they are yet very small. When this is done but little moisture is lost through them and they render but little plant food insoluble. In the thorough and early preparation of the seed-bed many weeds are destroyed by killing them just as they are coming up. So, too, in the case of a grain field, which is rolled after being seeded and is then harrowed, the rolling hastens the germination of the weed seeds and the harrowing then throws them out into a dry soil which kills them. If such a field is again harrowed just after the grain is up a second crop of weeds may be destroyed and the yield made greater as a consequence.

In the case of potatoes and corn it is very easy to destroy at least two crops of weeds before the corn or potatoes are large enough to cultivate, by harrowing before and just after the plants are up. This is very important because it not only saves plant food for the crop but it can be done so much more cheaply and rapidly with the broad light harrows and weeders than it can later with the cultivator.

274. Weed Seeds Do not All Germinate at Once.—It must be remembered in handling soils to kill weeds that the seeds do not all germinate at once. The first harrowing which is done to kill weeds may itself bring up from below seeds which were too deep in the ground to grow or it may cover some seeds which were lying upon or too close to the surface to germinate, hence frequent cultivations for hoed crops are needful.

275. The Best Tools for Weed Killing.—The tool which will do the most effective service in killing weeds depends upon the character and condition of the soil and the size of the weeds. When they are not yet fairly out of the ground or are just coming up and before a root system has been developed there is no tool equal to a medium weight or light spike-toothed harrow represented in Fig. 62a. The stiffer and more compact the soil is the heavier should be the harrow or rather the deeper it should be run in the ground.

FIG. 62a.—Tilting harrow, best tool for killing young weeds.

The tilting harrow, constructed so that the teeth may be inclined forward or backward, is one of the best forms as, with this arrangement, it may be made to run deep or shallow as desired.

On sandy soils and other soils when very loose the form of tool represented in Fig. 63 may be used to kill very young weeds before they are well rooted; but this is not an effective tool when weeds have a start nor where the soil is at all hard or heavy.

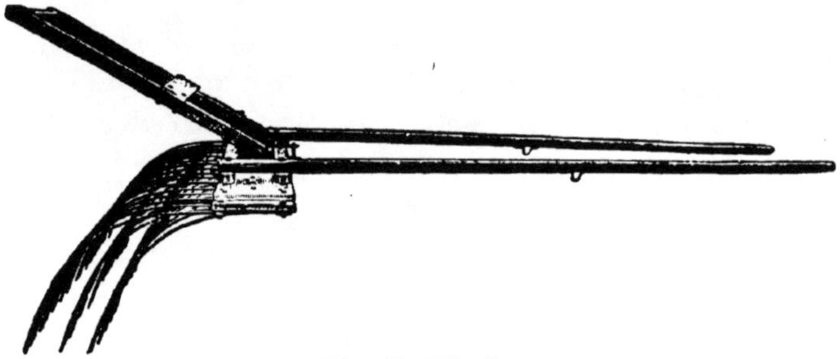

FIG. 63.—Weeder.

276. Cultivation After the Harrowing Stage.—When plants have become too large to permit the harrow or weeder to be used to advantage a tool with broader teeth is needed. Cultivation or intertillage should begin as soon as the first fresh weeds start and great pains should be taken to work so close to the row that all the soil is either stirred or covered with a thin layer of fresh soil. Few realize how close it is possible to work to a row without either covering the plants or seriously injuring the roots, until they have learned to do it. It is early and frequent harrowing and careful close first cultivation that insures scrupulously clean fields and the largest yields the season's rainfall will permit.

277. Cultivators for Intertillage.—When harrowing has been properly practiced intertillage may begin with a tool whose teeth are about 2 inches wide and there should be enough of them to thoroughly stir the whole soil surface to a depth of two and one-half to three inches. Fig. 64 shows a good set of teeth for soils not too heavy, while Fig. 65 shows a tool which should not as a rule find a place in well cared for fields, for the teeth are too wide and too few for good general work. They are wasteful of moisture, wasteful of fertility and liable to do too much root pruning.

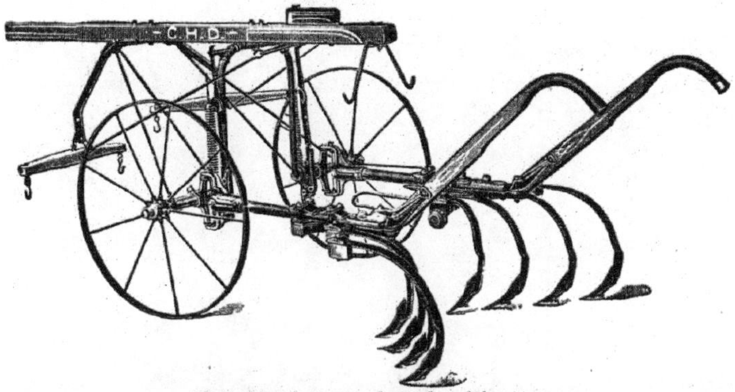

FIG. 64.—A type of good cultivator.

Cultivators with rigid teeth like those of Fig. 66 do better work as a rule than those of the spring tooth type represented in Fig. 64, for the reason that the ground is stirred more completely and to a more uniform depth. On naturally mellow soils the spring tooth is good and where the land is very stony it is safer against breaking.

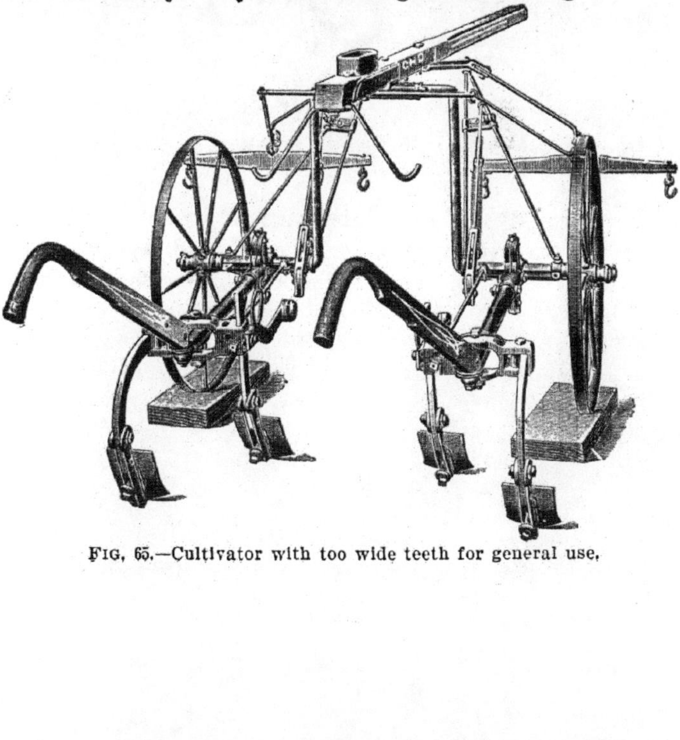

FIG. 65.—Cultivator with too wide teeth for general use.

278. Easy and Quick Movement of Teeth.—A very important feature of a riding or walking sulky cultivator is to have the gangs of teeth so swung from the carriage that a slight effort will produce a quick and certain movement. This is indispensable in order to work close to the rows.

FIG. 66.—Cultivator with rigid teeth; best where soil is heavy and not stony.

279. The Teeth of the Cultivator Adjustable.—Another important feature sulky cultivators should possess is the possibility of tilting the gangs so as to allow them to work more deeply in the soil toward the center of the row in the later stages of cultivation because then the roots near the rows have developed close to the surface, and deeper cultivation in the center, where the soil is more exposed to the sun, is needed for effectiveness as a mulch.

280. Covering Weeds in the Row.—It sometimes happens with the most careful management that weeds will get such a start in the row that either hand hoeing must be resorted to or else a tool must be used which will throw enough

FIG. 67.—Cultivator which can be used to cover weeds in row.

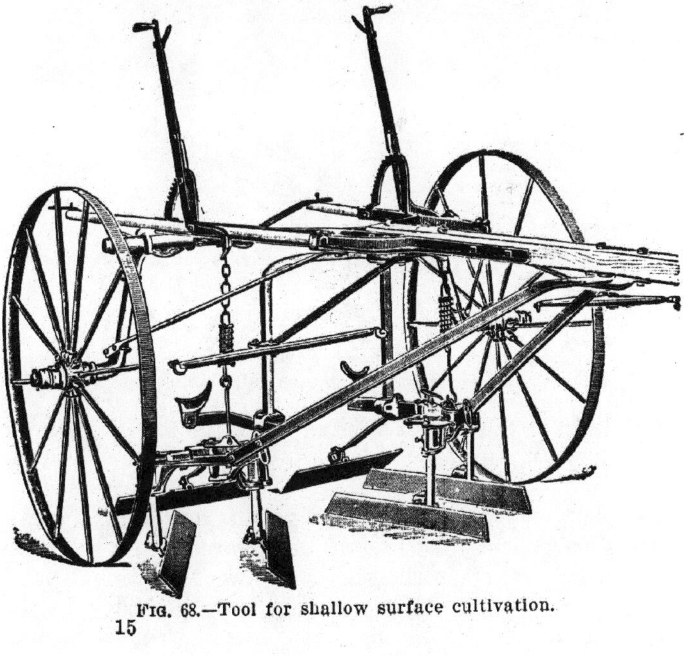

FIG. 68.—Tool for shallow surface cultivation.

earth to cover the weeds in the row. A good cultivator for this kind of work is represented in Fig. 67. The levelers represented in the rear of the discs are intended to throw

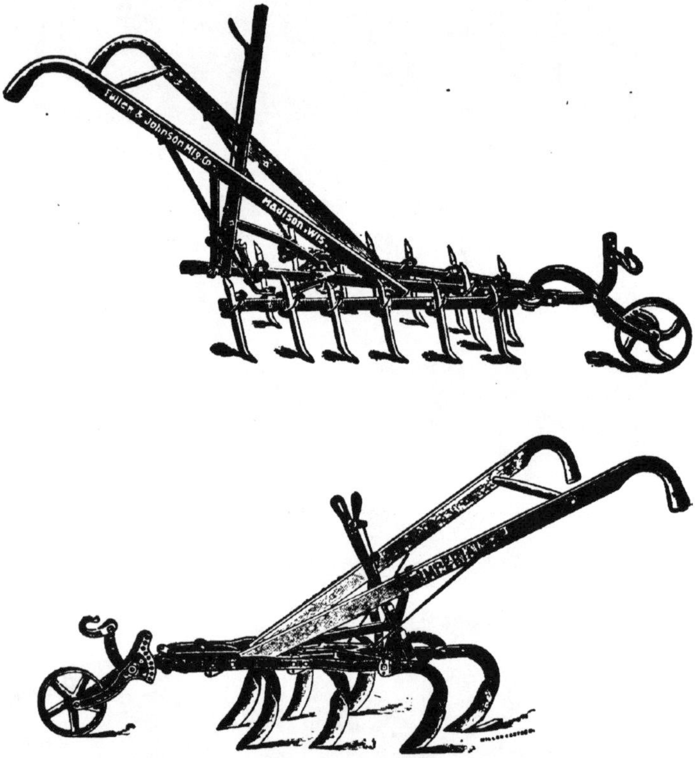

FIG. 69.—Two good garden cultivators.

the earth back to prevent ridging when the tool is used for ordinary cultivation and ridging is not desired.

281. Garden Cultivators.—Two good forms of garden cultivators are represented in Fig. 69, where the upper one is to be used early, when the plants and weeds are small, and the lower one when the harrow-stage has passed. In the garden as in the field the best time to kill weeds is just as the seeds are germinating and emerging from the soil and the harrow-toothed cultivator is very effective in doing this. It stirs the surface thoroughly enough to throw the young weeds out and cause the soil close to the surface to dry sufficiently to kill them. Much worry and hard work will be saved by the timely use of this or a similar tool.

TILLAGE TO MODIFY SOIL TEXTURE.

282. Soil Texture and Tilth.—Texture of soil, like the texture of cloth has reference to the size of the elements which give it its evident structure; and just as the threads of a piece of cotton, a piece of woolen or a piece of silk are

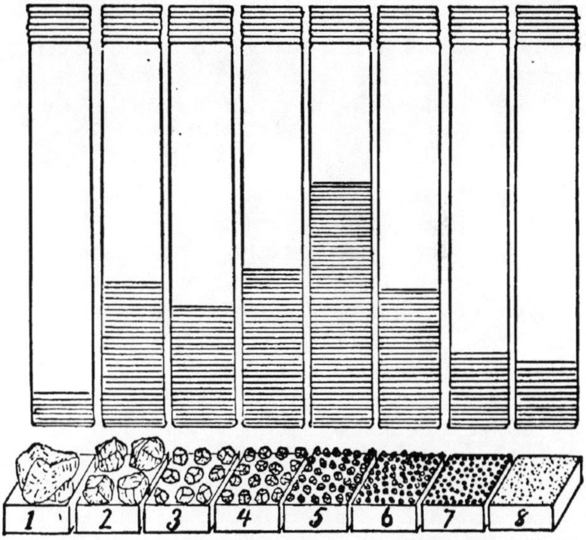

FIG. 70.—Showing the granular character of a soil in good tilth after cultivation.

made by twisting together varying numbers of small fibers, making the threads coarse or fine, so is it with soils; they are composed of granules of varying sizes formed out of ultimate soil grains which are cemented together more or less firmly. Fig. 70 represents the textural elements of a clay loam in pretty good tilth. There are shown seven sizes of granules large enough to be readily distinguished with the naked eye, and each size is composed of fine soil grains cemented together. All are represented natural size and were carefully drawn from an actual sample taken from a three inch mulch as left after the cultivator.

The granules were sorted by means of a series of sieves and the relative amount of each size of granules is represented by the shading in the vials where it is seen that the largest size constitutes the smallest part of this soil, and No. 5 the largest portion. The finest grade, No. 8, is also largely composed of compound grains, many large enough to be clearly distinguished by the unaided eye, but many more of the ultimate grains which were rubbed off from the larger grains by cultivating and during the process of screening.

Just as woolen cloths differ when the threads are of the same size because some are twisted from finer and others from coarser wool, so soils differ in having their granules made of coarser or finer soil particles cemented together.

Then, too, just as one cloth may differ from another in having its threads loosely twisted, while another is hard twisted, so one soil may differ from another in the degree of firmness with which the soil particles are cemented together.

Still again, just as one fabric may be loosely woven while another is firm, so one soil may have its granules more strongly cemented together than another, making it hard to work and heavy while the other is light and mellow.

A sand differs from a soil in being composed of simple separate grains, usually of rather large size, while a clay is composed very largely of extremely fine granules made from the finest of particles.

A soil is in good tilth when its granules are neither too fine nor too coarse, and when they are not too firmly cemented together.

283. Why Good Tilth and Good Tillage Are Important.—It is clear from the rounded form of the granules of soil shown in Fig. 70, that when they are massed together without being crushed a very large amount of unoccupied space must exist; this unoccupied space in a soil is needed for the movement of air and of water; for the spreading out of the root fibers and root hairs, and for the home of micro-organisms which develop the available nitrogen used by all the higher plants.

If the granules are too large and too loosely packed the soil lets the rains fall through it too freely and does not bring it back rapidly enough by capillarity to meet the needs of crops. If the granules are too small and too close then the water moves too slowly, too much is retained by capillarity and there is too little air. If the granules are bound together too strongly, the soil is too hard and the roots are unable to set it aside in making their advance and this lack of freedom reduces the yield.

284. How Texture and Tilth Are Developed.—The soil particles are drawn together into the rounded granules by the tension of the soil water in the same way that water forms itself into spheres when sprinkled on a dust covered floor. As long as there are large open spaces in the soil not filled with water the water is all the time drawing itself together, tending to form spheres, and in this system of pulls the soil particles become involved and are drawn together also. As the water is lost by evaporation and the salts dissolved become too strong to remain in solution they are deposited upon and between the grains and granules tending to cement them together.

285. Difference Between Soil and Potter's Clay.—When the granules of a fine soil are all broken down and separated into their ultimate grains we have the puddled condition so fatal to crops, but the one the potter strives to secure to make his wares close in texture and strong. In the puddled soil and potter's clay enough of the granules have been broken down to fill the spaces between the larger simple grains and finer granules not yet broken down to make a close textured, impervious material in which no plant can thrive, and through which neither water nor air can move.

286. Early Spring Tillage.—The early stirring of the soil in the spring preparatory to seeding has for its main object the changing of the soil texture so that it will become 1st, warmer, 2d, dryer, 3d, better aerated, 4th, better suited to lessen the rate of evaporation of the deeper soil water, and 5th, to hasten the development of weed seeds so they may be destroyed before the crop is in the way of killing them.

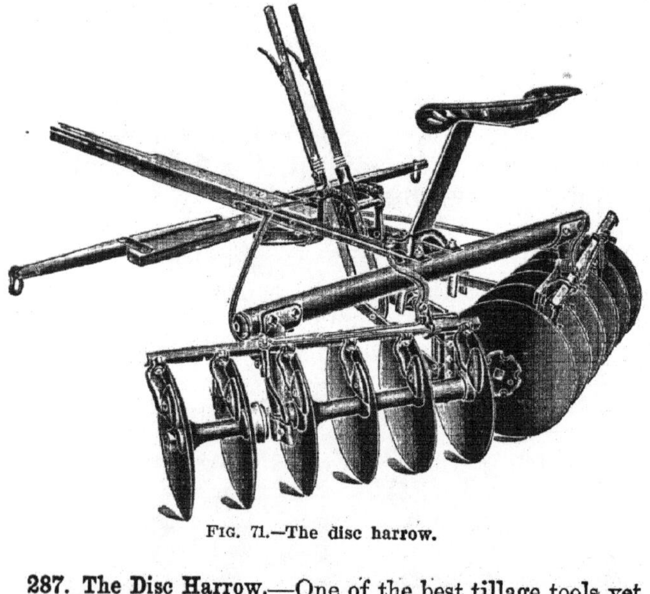

FIG. 71.—The disc harrow.

287. The Disc Harrow.—One of the best tillage tools yet devised is the disc harrow represented in Fig. 71. There is no harrow which so thoroughly pulverizes a soil in the spring after fall plowing as this tool. When set to work deep the draft is heavy but the amount of work it is doing is relatively large. To put a piece of fall plowing in the best shape the harrow should be lapped half and in doing this the furrow between the two sets of discs will be entirely filled and the surface left level.

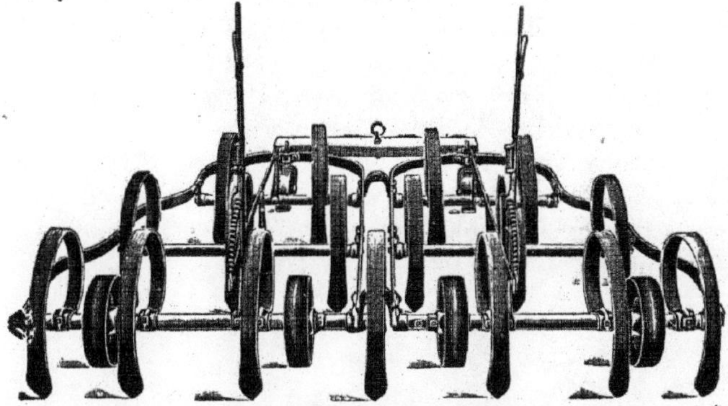

FIG. 72.—Spring-tooth harrow.

Where small grains are to follow corn or potatoes the use of this tool will often make the plow unnecessary.

On the upland prairie soils and others naturally mellow, ground for corn may be plowed in the fall and fitted in the spring with the disc harrow with good results.

288. The Spring Tooth Harrow.—On new land in wooded countries and where the fields are rough and stony the har-

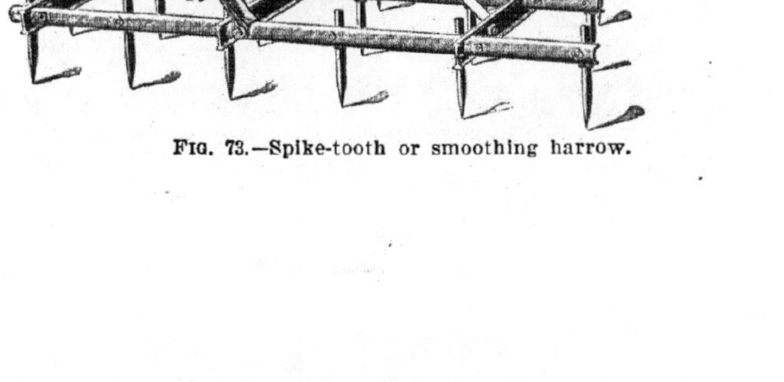

FIG. 73.—Spike-tooth or smoothing harrow.

row represented in Fig. 72 does good work. Its weight forces it into the soil and the elasticity of the teeth prevent them from being broken, but such tools can never do the degree of pulverizing that the disc harrow accomplishes.

289. Smoothing Harrows.—When the soil has been pulverized with the disc or other tool and it is desired to leave the surface more nearly even, or where the soil is naturally very mellow, making less force necessary to change the surface texture, then the heavier weights of tilting harrows, Fig. 73, may be used to great advantage on account of the greater area which may be covered with them in a day and their lighter draft.

FIG. 74.—The planker.

290. The Planker.—It is sometimes desirable to leave the surface particularly smooth without firming it and at the same time to crush lumps. This may be done by means of a planker made of three to five 8- or 10-inch plank bolted together with their edges overlapping as represented in Fig. 74. The tool is best made of oak plank two inches thick and eight to twelve feet long. Such a tool cannot take the place of a roller where it is desired to firm the ground.

291. The Use of the Roller.—The roller is used chiefly when it is desired to firm the surface and to help cover seed, especially when sown broadcast. In other cases it may be used to crush clods or to compress the furrow slices after the sod plow. Again when a green crop like rye or clover has been turned under for manure, or where coarse litter has been plowed under, a roller is needed to compress the soil and establish good capillary connection with the deeper soil water. It is sometimes used to develop a mulch where grain is rolled after it is up.

In all of these cases weight is one of the essential features of the tool. A roller for tillage should have a weight of about 100 lbs. to the running foot and a diameter of about 2 feet.

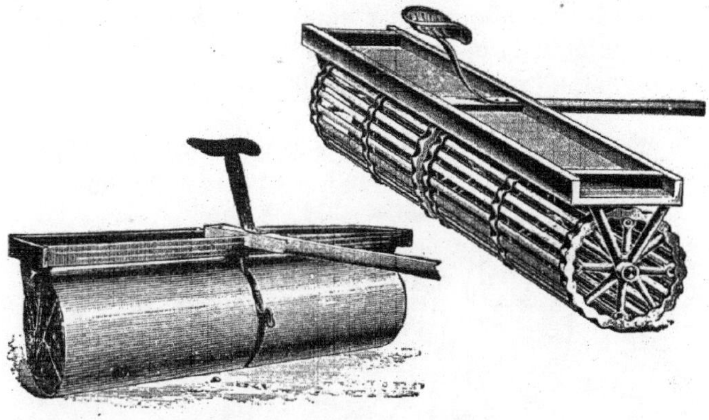

FIG. 75.—Two types of rollers.

Two types of rollers are represented in Fig. 75, the one made of bars being designed to crush clods more completely and to leave the surface ridged so as to be less likely to be influenced by the wind drifting the surface soil.

292. The Harrow Should Follow the Roller.—In most cases when it has been desirable to use the roller to smooth or firm the surface a light harrow should follow it quickly in order to prevent unnecessary loss of soil moisture, because the firming draws the deeper water to the surface, the surface temperature becomes higher in the sunshine and the wind velocity near the smooth surface is greater; each of which favors the rapid loss of water.

293. Danger in the Use of the Roller.—On heavy soils, when they are a little wet, injurious results may follow the use of the roller just after planting or seeding on account of the close packing, excluding the air from the seed, which interferes with quick germination. This danger is greatest where grain has been sown with a drill.

The use of the roller when the soil is a little too wet may also interfere with the formation of nitric acid in the soil by making it too close and too wet. In such a case the immediate use of a light harrow would only retain the moisture and make the rate of nitrification slower.

294. The Plow.—The plow as a tillage tool is used for two distinct purposes, 1st, to alter the texture, forming

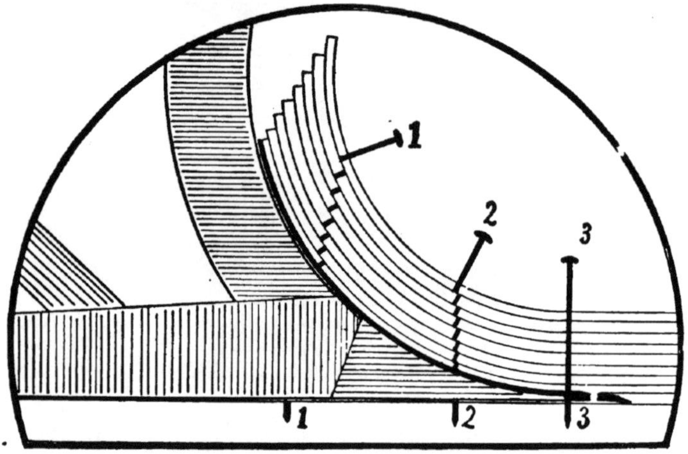

FIG. 76.—Showing the principle of the pulverizing action of the plow.

from a comparatively hard soil a deep and mellow layer of earth; 2d, to bury beneath the surface weeds and other vegetation or manure where it may decay rapidly and be converted into available plant food.

If you will open a book, placing the fingers upon the fly leaf in front and the thumbs under the fly leaf in the back and abruptly bend up the corner it will be seen that every leaf is slipped over its neighbor. What takes place is represented in Fig. 76. Had pins been put through the book before attempting to bend the leaves the bending would have tended to cut the pins into as many pieces as there were leaves, just as seen in Fig. 76.

Now the plow has exactly this kind of effect upon the furrow slice; it tends to make it divide into thin layers which slide over one another just as the leaves of the book did, and it is because of this sort of action that a plow pulverizes a soil as no other tool can.

295. How Plowing May Puddle Soils.—When a soil is too wet its granules are so easily broken that the plow is liable to shear all the coarser ones into two, three, or more slices just as the pin has been sliced in Fig. 76, thus destroying its tilth by puddling it.

296. How Plowing May Correct Texture and Improve Tilth.—If a soil has gotten out of tilth, has become cloddy or has been partly puddled there is a shape of mold board, a stage of soil moisture, and a depth of furrow slice which will help to restore the tilth best and quickest. When such a soil is the least amount too dry to puddle the plow will shear it into the thinnest slices; if still drier the layers will be thicker and will form coarser granules.

When much too dry no shearing can take place at all, and the furrow slice is simply broken into coarse lumps.

If you bend but a few leaves of the book at a time there is but little slipping, but the thicker the pile of leaves the greater is the sliding and the greater is the tendency to shear. So it is in plowing, the deep furrow pulverizes better and puddles worse than the thin slice or shallow furrow.

Again if you bend the leaves gently there is little shearing, but if abruptly the sliding is great. So if you plow with the low mold board of Fig. 77 you disturb the tilth least, puddled the soil least, and leave the texture coarsest; but if the steep mold board of Fig. 78 is used there is the greatest danger of puddling if the soil is too wet and the greatest opportunity to pulverize the soil and improve the tilth if the moisture is right.

297. Forms of Plows.—Plows are made with two fundamentally different shapes depending upon the character of the work which they are expected to do.

If the chief object of the plow is to cut a clean furrow slice and turn it over so as to completely cover whatever may be upon the surface a shape represented in Fig. 77 is used.

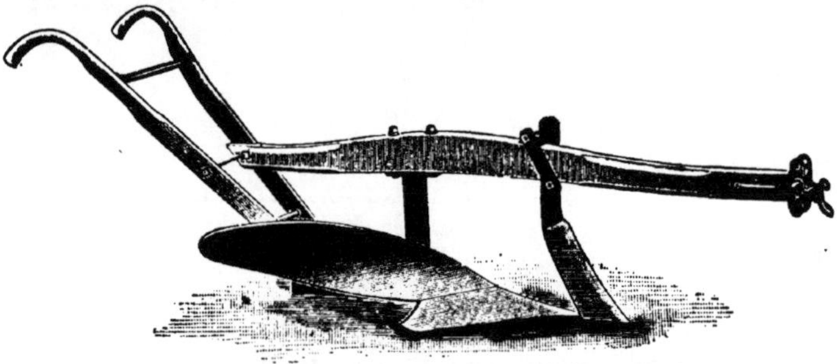

FIG. 77.—Type of sod plow, which pulverizes but little.

If on the other hand the primary object of the plow is to thoroughly pulverize the soil, making it deep and mellow, a form represented in Fig. 78 must be used. Then according as one or the other of these two chief objects vary in importance shapes of plows will be chosen which are intermediate between these two extremes.

298. Kind and Condition of Soil and Shape of Plow.—It must be clear from the mechanical action of the plow that its form should be adapted to the soil. If the soil has a tendency to be too open and porous, and is naturally coarse grained, like the sandy soils, it should be plowed with a steep mold board, a little over wet and as deep as other conditions will permit, so as to break down the granulation and secure the closer texture.

If the soil is generally too close in texture, is heavy and soggy, it needs the less steep mold board used when the soil is a little dry so as to shear into thicker layers and form granules of larger size.

If plowing must be done when the soil is a little too wet

use the less steep mold board and plow as shallow as other conditions will allow.

If a soil has become a little too dry and is not pulverizing fine enough, use the steeper mold board and plow deep for this will split it into thinner layers, make the soil finer, and the tilth better.

299. The Kind of Soil, the Shape of the Mold Board, and the Draft of the Plow.—Since the steepest mold board bends the furrow slice most and pulverizes most, it is clear that the work done is greatest, and hence that the draft will be most.

Since deep plowing pulverizes more than shallow plowing the work done is more than in proportion to the depth.

Since clay soils have more and larger granules which must be sheared in two in plowing than sandy soils do, the labor of plowing must be greater.

Since the granules of the soil are not as strong when the soil is moist as when dry it plows much easier, when in good condition. But if the soil has become too dry and yet must be plowed, it should be plowed deeper rather than shallower. This is necessary to pulverize better, to get more moist soil on the surface for the immediate seed bed, and to quicker moisten and bring into condition the layer which has become too dry.

300. The Sod Plow.—The sod or breaking plow is constructed so as to reduce the draft as much as possible by doing only the work needed to cut and turn over the furrow slice. This is accomplished by making the mold board very long and slanting so that the furrow slice is bent and twisted as little as possible, as shown in Fig. 77; the chief work being to cut it and roll it bottom up.

The extremely oblique edge of the share in the breaking plow reduces the draft in cutting off the roots by allowing the cutting to be done gradually and with a drawing cut, just as it is easier to cut off a limb by letting the blade of the knife slant backward, drawing it across.

The extremely oblique construction of this plow too, makes it easier to hold it steady when passing and cutting off strong roots or other obstruction.

FIG. 78.—Type of pulverizing plow with steep moldboard.

301. The Pulverizing or Stubble Plow.—It will be seen from Fig. 78 that this plow has a much steeper mold board and much less oblique plowshare, the object being to bend the furrow slice as abruptly as possible before it is turned over, for this is what pulverizes the soil, giving it the loose, fine, open texture sought.

302. Mellow Soil Plows.—Soils which are sandy and naturally very mellow may be plowed with a plow having the mold board less steep and more like that of Fig. 79 in shape. With such a form as this the team may cut a wider furrow, and thus cover the ground more rapidly, because the draft is less.

When soils are very heavy and stiff it may also be desirable to use this type of plow, simply because the draft would be too heavy for the team with the type which pulverized the soil more.

Again very loose soils which have an extremely fine texture and tend to clog will often clear better from the less steep mold board because the pressure comes more obliquely against the surface.

303. Draft of Stubble Plows.—The amount of labor involved in plowing a field is so large under the best possible conditions, and it is so easy to make it unnecessarily large, that it is important to understand the principles upon which the draft depends.

Mr. Pusey in England, in 1840, made a series of trials on the draft of plows in soils of different kinds, using 10 different plows. We have combined his results and give them in the table below:

Table showing the draft of plows in tests made in England and in America.

Kind of soil.	No. of plows.	Size of furrow.	Total draft.	Draft per sq. in. of furrow
			Lbs.	Lbs.
Loamy sand	10	5 in. x 9 in.	227	5.04
Sandy loam	10	5 in. x 9 in.	250	5.55
Moor soil	10	5 in. x 9 in.	280	6.22
Strong loam	10	5 in. x 9 in.	440	9.72
Blue clay	10	6 in. x 9 in.	663	12.27
Sandy loam (J. C. Morton)	5	6 in. x 9 in.	566	10.48
Stiff clay loam (N. Y. 1850)	14	7 in. x 10 in.	407	5.81

Prof. J. W. Sanborn made an extended series of trials in 1890 in Missouri and later in Utah and the average of all his trials gives a draft of 5.98 lbs. per sq. inch of the cross section of the furrow slice. Separating these trials historically, omitting those in the blue clay in England, the results stand:

English trials 1840, mean draft 7.41 lbs per sq. inch.
American trials 1850, " " 5.81 " " " "
" " 1890, " " 5.98 " " " "

304. Draft of Sod Plow With and Without Coulter.—A set of trials with a sod plow near the type of Fig. 77, in clover sod 2 years old, when the moisture present was about as high as it is prudent to work the soil, gave results as follows:

	Size of furrow.	Total draft.	Draft per sq. in.
		Lbs.	Lbs.
Sod plow with wheel coulter...... ..	5.575 in. x 15.08 in.	296.25	3 524
Sod plow without coulter...........	5.325 in. x 14.5 in.	343.75	4.453
	Difference.....	47.50	.929

Besides doing the work better the coulter diminished the draft 20.86 per cent.

A later series of observations was made on a clover sod with the same sod plow provided with a wheel coulter, but at a time when the soil was dryer than when the other measurements were made. The results found were:

	Size of furrow.	Total draft.	Draft por sq. in.
		Lbs.	Lbs.
Clover sod without coulter	6.47 x 11.61 in.	714.35	10 80
Clover sod with coulter..	6.413 x 12.47 in.	664.82	8.616
	Difference....	49.53	2.184

In this set of trials the coulter has reduced the draft 25.34 per cent.

305. Draft of Sod Compared With Stubble Plow.—Another set of trials were made at the time of (304 to compare the stubble type of plow, Fig. 78, with that of Fig. 77, and the results are given below:

	Size of furrow.	Total draft.	Draft per sq. inch.
		Lbs.	Lbs.
Stubble plow without coulter........	5.872 x 14.31 in.	452 4	5.384
Sod plow without coulter............	5.325 x 14.5 in.	343.75	4.453
	Difference.....	108.65	.931

In this case the shape of the plow altered the draft 20.9 per cent., and the difference is probably a measure of the difference in the amount of pulverizing done by the two plows.

306. Influence of Difference of Soil Moisture on the Draft of Plows.—By combining the data in the two tables of (304) with reference to the degree of moisture in the soil when the trials were made we have the results given below.

	Sod plow with coulter. Draft per sq. in.	Sod plow without coulter. Draft per sq. in.
Soil rather dry..........	8.616	10.80
Soil in best condition,..	3.524	4.453
Difference.........	5.092	6.347

From this comparison it is clear that the draft of the plow is very much modified by the condition of the soil. The results show the draft more than doubled when the soil was dryer.

FIG. 79.—Type of moldboard suited to mellow soils requiring little pulverizing.

307. The Draft of Sulky Plows.—It is generally claimed that the draft of sulky plows is less than that of the free-swimming types because the friction of the sole and land-side is transferred to the well oiled bearings of the carriage. The few records accessible do not show a material gain, when the influence of the weight of the carriage and driver are not deducted, but where the draft is no greater on the team with the man riding than when walking, and the plow

16

can be handled with equal facility, there is an evident advantage in riding plows such as Fig. 80.

FIG. 80.—Sulky or riding plow.

308. The Line of Draft.—It is very important in the handling of a plow that the *line of draft* be just right and such that a line connecting the center of draft A, Fig. 81, in the mold board with the place of attachment to the plow bridle shall also lie in the plane of the traces, as shown in

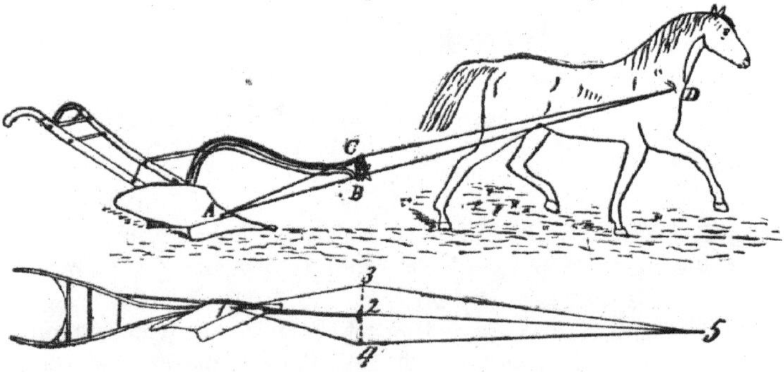

FIG. 81.—Direction of the line of draft for plows.

the cut by the line A, B, D. If for any reason the line of draft becomes a broken one as A, C, D or 1, 3, 5 or 1, 4, 5 instead of 1, 2, 5 the draft of the plow is made heavier.

The greatest care should be exercised to have the length of the traces, or the hitch at the plow bridle such that the plow "swims free," requiring little or no pressure at the handles to guide it. If a steady pressure in any direction is required at the handles something is wrong and the team is doing more work than is necessary as well as the man holding the plow.

309. The Scouring of Plows.—There are certain soils, whose texture is such that the most perfect plow surface fails to shed them completely and in such cases the shapes approaching the sod-plow are more successful. But it is a matter of greatest moment that the mold board possess not only an extremely hard finish, so as not to be scratched by stone or grit in the soil, but it must also possess an extremely close texture so as to be susceptible of a very high polish. If the metal itself is coarse grained there will be inequalities even in the bright surface in which the fine soil particles may lodge and thus clog the plow.

310. Care of the Plow.—Too great pains cannot be taken to maintain a bright clean surface on all polished parts of the plow and the necessary care to do this will always pay; this caution is doubly important where the soils are inclined to clog.

Whenever a plow is laid by, even for a few weeks, its bright surfaces should be thoroughly cleaned, wiped dry and coated with a layer of the thick mineral lubricant used for journal bearings, to prevent rusting. A little rusting may practically ruin a plow for use in a soil which tends to clog and a single winter of rusting may injure a plow more than a full season of heavy service in the field.

311. Keeping the Plow in Form.—A plow cannot render heavy and long continued service without getting out of proper form. The point becomes dull, too short and assumes the form shown in Fig. 82, instead of that in Fig. 83. In this worn condition the inclination of the mold

FIG. 82.—Showing point of plow worn into bad form.

board to the furrow slice is changed, the plow tends to run on its point, is more difficult to hold, the draft becomes heavier and poorer work is done with it.

FIG. 83.—Showing point of plow in good form.

The heel of the share C in Figs. 84 and 85 is especially liable to get into bad form and dull, causing the plow to

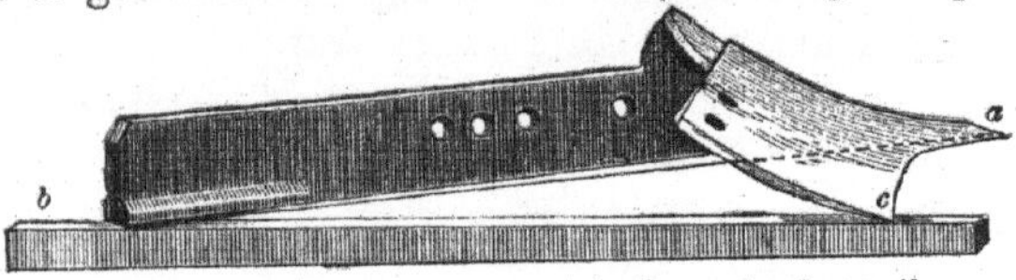

FIG. 84.—Showing heel of plow in form for dry soil.

wing over to the land and draw harder, not only because it is dull but because a steady pressure must be exerted at the handles to prevent the plow from tipping to land.

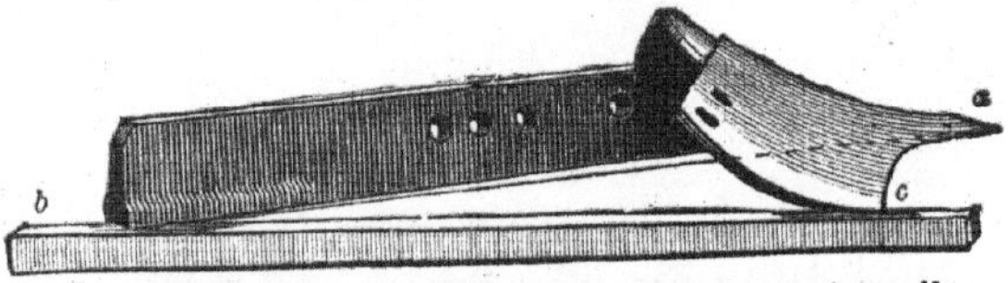

FIG. 85.—Showing heel of plow in form for moist soil.

It is sometimes necessary to change the form of the plow to suit a harder or more mellow condition of the soil. When the soil is dry and hard the heel needs to be set down, as shown at C, Fig. 84, and the point may need to dip even more than in Fig. 83, but when the soil is wet and mellow the shape shown in Fig. 85 is required to prevent it drawing too deeply into the ground.

In taking the share to the shop for sharpening or setting the landside should accompany it in order that the blacksmith may have a guide in giving it the proper shape.

312. The Jointer Attachment.—One of the most useful attachments for a plow is known as a jointer, represented in Fig. 86. This tool is used to great advantage when considerable material needs to be turned under, such as long stubble, coarse manure or in turning under a green crop for manure. When this is used with the drag chain in the furrow very long weeds can be completely laid under the surface, leaving the ground in excellent shape.

FIG. 86.—Plow with jointer.

When sod ground is to be plowed deep and left in shape for immediate pulverizing to fit it for crops this tool will often render excellent service by cutting out a section of the sod, turning it into the bottom of the furrow, where it will be completely covered, at the same time leaving the upper edge of the furrow slice composed only of comparatively loose earth.

313. Subsoil Plow.—One of the most widely used forms of sub-soil plow is represented in Fig. 87. It is intended to be used in the bottom of an ordinary furrow, one plow following the other in doing the work.

Extremely good judgment is required in the use of the subsoil plow to avoid puddling, which is sure to result from using the tool when the subsoil is too wet. In humid climates the dangers are greatest in the spring and least in the fall, and it must be kept in mind that the surface soil may be in good condition to plow when the subsoil is much too wet.

FIG. 87.—Sub-soil plow.

In semi-arid climates the dangers of injuring the soil texture are much less and it is under such conditions that subsoiling is likely to prove most profitable, tending as it does to increase the available moisture for crop production.

OBJECTS, METHODS AND TIMES OF PLOWING.

314. Depth of Plowing.—The best depth to plow at a given time, on a given soil, for a given crop must be decided on the spot after exercising good judgment with a knowledge of the needs and conditions. There can be no "rule of thumb" for plowing.

As a general rule in humid climates the plow never should go deeper than to turn over the surface or dark colored layer of weathered soil. If deeper plowing is done, turning up the unweathered subsoil, the productiveness of the field will be reduced.

It is very desirable to develop and maintain a deep soil; this is clearly proved by the heavier crops which always grow upon "back furrows" and the scanty ones which grow in "dead furrows" as compared with the rest of the field. When a soil is thin and the subsoil is close and heavy it is only safe to deepen it gradually by plowing a little deeper each year or two, turning under as far as possible coarse manure, stubble and green crops to make the soil open and form humus in it.

Fall plowing may usually be as deep as the soil will permit, down to 6, 7 or even 8 inches, but the cases are relatively few where it is important to plow deeper than 6 or 7 inches. Where plowing is for small grains to be sowed at once the depth may usually be shallow, 5 inches or less, as these thrive best in a shallow seedbed.

315. Best Condition of Soil for Plowing.—There is a condition of moisture peculiar to each and every soil at which it will be left with the best texture after plowing, requiring the least amount of finishing work to put it in final condition. If the soil is too wet the crumb structure so essential to a clay soil will be partly destroyed and the soil puddled; if too dry the furrow slice will not shear in thin layers and the soil will not be pulverized fine. The water content should be such that the damp soil squeezed in the hand will hold its form but will easily crumble to pieces and not be at all pasty.

Sod ground can always be plowed a little wetter than corn, potato or stubble ground because the roots lessen the danger of puddling and the shearing effect of the plow is less.

316. Treatment of Ground After Plowing.—Ground plowed late in the fall, to act as a mulch, to allow the moisture to penetrate deeply and to have its texture altered by thawing and freezing, should be left with the natural furrow surface rough and uneven.

If plowed in the spring when the ground is a little over wet and the turned furrow shows large polished surfaces the ground should be gone over with a harrow but not immediately, for if the soil is a little too wet it should be allowed to dry just enough so as to crumble perfectly.

If the soil is already a little too dry and a crop is to be put on at once then the harrow should follow the plow closely, otherwise the soil will become lumpy and the whole furrow slice may become too dry for the best germination.

If the plowing is for corn, potatoes or the garden and is done some time before the ground is to be planted then the surface is better left as it would be for fall plowing, provided the soil is in good condition when plowed, because it will form a better mulch, it will take the rains better, be less likely to become too much compacted by the rains and will harrow down better when planting time comes.

317. Plowing for Corn in the Fall.—On soils which are naturally mellow, where large areas are to be planted and the spring's work is crowded it is often best to plow for corn late in the fall, just before freezing. If such ground is to be manured it can be plowed in then to advantage or if the manure is not too coarse it may be applied as a surface dressing during the winter and disked in the spring. If the soils are very heavy and have a tendency to run together with the spring rains then there is danger that the disc may not be able to bring the field into condition.

318. Plowing Sod.—There are two methods of plowing sod, 1st, skim-plowing, usually in the fall, turning over a thin sod to kill the turf, expecting to cross plow in the spring deep enough to bury the sod and turn up enough soil to work up fine and form the seed bed. 2d. Plowing deep enough at first to provide a sufficient soil to work up with a disc harrow and give the desired depth of seed-bed. The latter method usually requires less time but the draft is heavier. It is usually best in such cases to go over the surface with a heavy roller to press the sod home and lessen the danger of the disc turning them over.

319. Plowing Under Manure.—If manure is coarse or the soil light it is usually better to place it under a deep furrow because it needs more moisture to rot it and in heavy soils it will let the air penetrate more deeply into the soil. In such cases it is better to do the plowing in the fall or as early in the spring as the soil will permit. If the ground is a little too dry when plowed and seeding time is at hand the field should be thoroughly harrowed and firmed, using the heavy roller if necessary in order to establish good capillary connection with the deeper soil. If this is not done the soil above is liable to become too dry.

When the manure is well rotted it may be left nearer the surface to advantage, except in the sandy soils where the air penetrates so deeply as to cause too rapid decomposition of the manure.

320. Plowing Under Green Manure.—Where a crop is turned under for green manure it is usually best to plow deep, to use the jointer and the drag-chain if necessary to get everything well and deeply buried. If a considerable body of material is turned under thorough firming of the soil after plowing will be beneficial.

In green manuring good judgment is always required not to let the crop turned under exhaust the soil moisture too completely, for when this has occurred a new crop starts under very unfavorable conditions, both because of lack of water and immediately available plant food, for the soluble salts are used up with the water by the green manure crop.

320. Early Fall Plowing.—In regions and at times where there is a deficiency of rain, where the soil is light and when the amount of soil leaching is small it is often desirable to plow as early in the fall as the crop has been removed from the ground, in order to save soil moisture and to enable the nitrates and other soluble salts to develop in sufficient quantity for the next season. Where crops hold the soil moisture low it may even become necessary in dry climates to raise one only every other year because the plant food and the crop cannot be produced by the available moisture of a single season. But early fallowing in the fall will often render the full year unnecessary.

GROUND WATER, WELLS AND FARM DRAINAGE.

CHAPTER XII.

MOVEMENTS OF GROUND WATER.

Of the water which falls upon the land one portion finds its way at once, by surface flow, into drainage channels; a second portion is evaporated where it fell, while a third enters the ground. That portion which enters the ground and is not returned by capillarity or root action constitutes the body of ground water which is the source of supply for wells and springs and which requires removal by land drainage when too close to the surface.

322. Amount of Water Stored in the Ground.—In most localities after passing a certain distance below the earth's surface a horizon is reached where the pore space in the soil, sand and rock is filled with water or nearly so. When these pore spaces are large, so that water can flow through them readily, wells sunk beneath the surface fill with water to the level of the ground water surface.

In sands and sandstones lying below drainage outlets the amount of water may be as high as 15 to 38 per cent. of the total volume of the rock so that where a country is underlaid with broad and thick sheets of sandstone, such as the Potsdam and St. Peters in Wisconsin and further south, or the Dakota formation in the west, there is the equivalent of from 15 to 38 feet of water on the level for every 100 feet in thickness of the rock formation, and abundant supplies of water can always be found in such places.

The loose sands and gravels have a pore space of 20 to

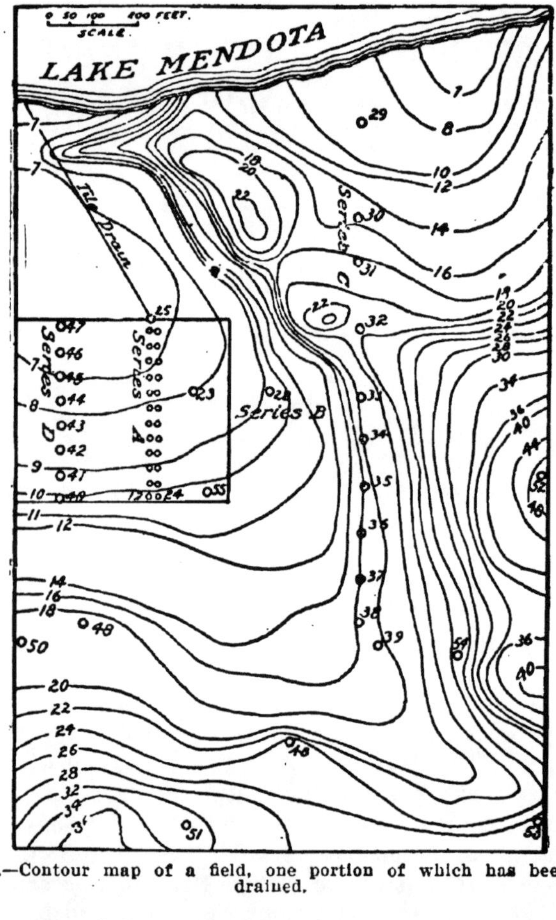

FIG. 88.—Contour map of a field, one portion of which has been tile drained.

38 per cent. of their volume so that where these lie below the ground water surface and their volume is large an abundance of water exists.

In the soils and clays the pore space is even larger than it is in the sands and this too may be filled with water but here the texture is usually so close that a well sunk in such

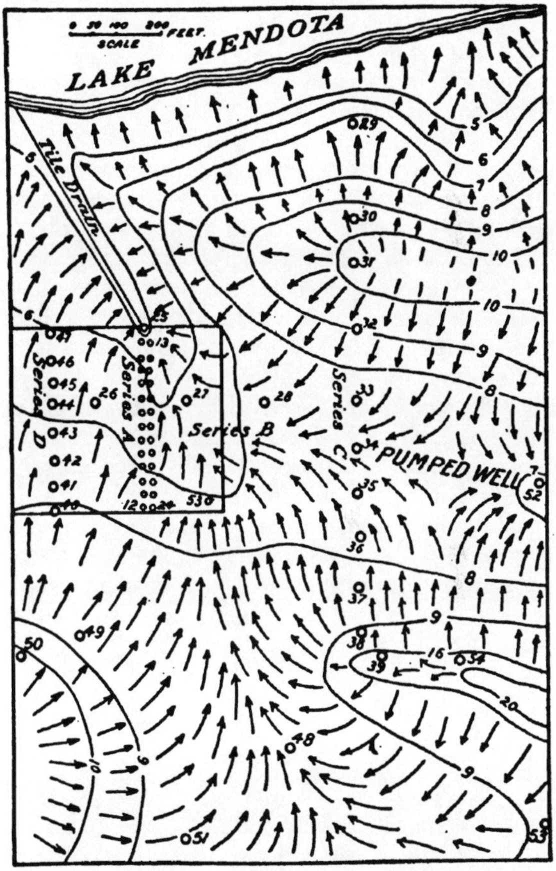

FIG 89.—Contour map of the ground water surface under the field of Fig. 88.

material fills with water so slowly that they cannot serve as sources of water supply.

Even in the hard crystalline rock, like marble and

granite, there may be as much as .4 of a pound of water in each cubic foot, but here again the texture is too close to permit such water to become available in wells.

323. The Ground Water Surface.—As the rains which fall in a given locality percolate beneath the surface they fill the pore spaces between the soil grains and raise the level of the ground water. If none of this water drained away and none of it were lost by evaporation the whole soil would have its pore spaces filled with water and the surface of the ground water would coincide with the surface of the land. As it is, as soon as the surface of the ground water ceases to be level drainage begins and the water under the higher land is lowered until a condition is reached when the rate of drainage laterally exactly equals the rate of accumulation of water from the rains.

In Figs. 88 and 89 are shown the contours of the surface of a section of land and of the ground water beneath, both sets of contours being referred to the same datum plane, Lake Mendota, into which the water is draining. Here, it will be seen, the ground water stands highest where the surface is highest and lowest where the land is lowest. The arrows show the lines of flow and make it clear why the tile drained area needed that treatment.

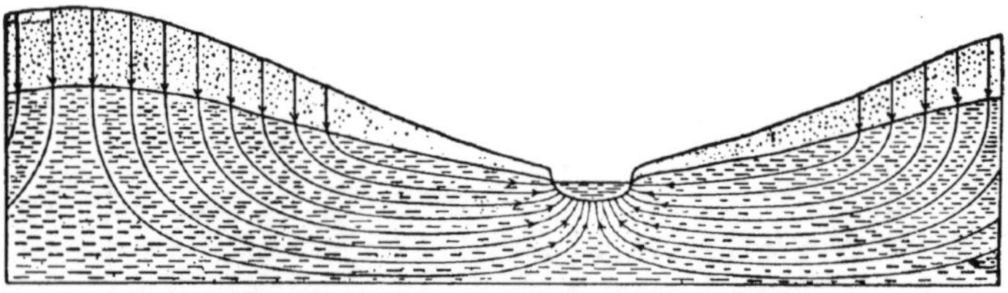

Fig. 90.—Showing lines of flow of ground water during seepage into a stream.

324. Seepage.—Almost everywhere under the land areas there is a slow movement of the ground water from higher to lower levels destined ultimately to reach some drainage outlet. This movement is known as seepage and Fig. 90 is a cross-section showing how the water flows from the adjacent higher lands and enters the channels of streams, the beds of lakes and even the ocean itself.

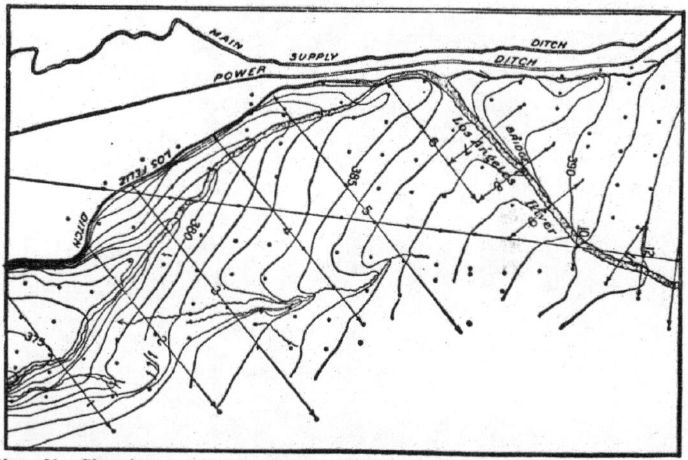

FIG 91.—Showing contours of ground water surface in the vicinity of Los Angeles River, Cal.

325. Growth of Streams.—The water which maintains the low stage flow of streams finds its way into channels all along the banks and bottoms rather than at isolated places in the form of springs, entering in the manner stated in (**324**). In Fig. 91 is represented the ground water surface in the valley of the Los Angeles river, California, where it is seen to rise back from the stream and up the valley. This river must be draining the adjacent higher land and it was found by actual measurement that the growth of this stream in 11 miles was 60 cubic feet of water per second; the water all entering by slow general seepage, there being

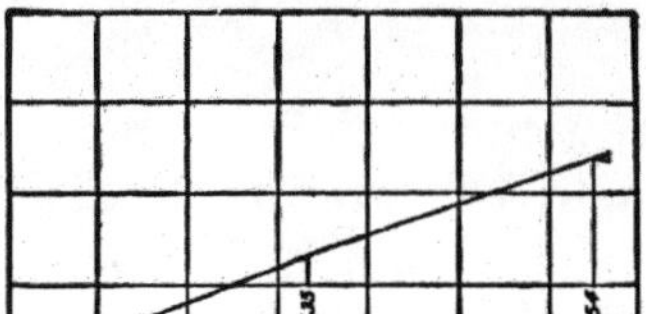

FIG 92.—Profile showing increase of the Los Angeles river by seepage in 25,978 feet.

no visible springs or streams anywhere along the line. Fig. 92 shows the increase in 25,978 feet, determined by gauging.

326. Changes in the Level of the Ground Water.—The level of the ground water in a given section is usually subject to changes, the surface rising and falling with the season and with the rainfall of the place. The change may be as much as 5 or 6 feet in a single season, as represented in Fig. 93, and when a series of dry or of wet years follow in

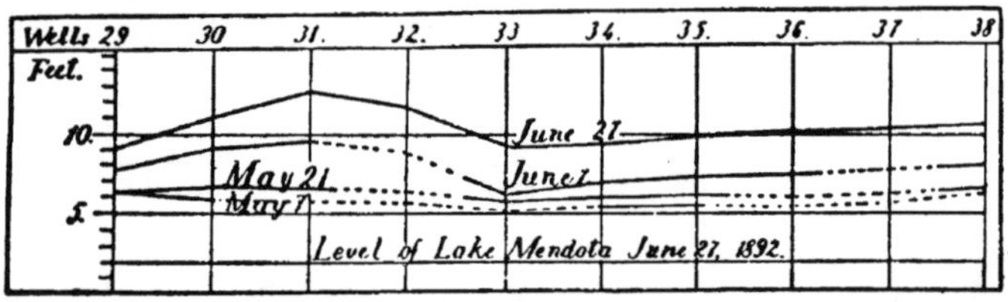

FIG 93.—Showing changes in the level of the ground water surface during the season.

succession the changes may be larger than this. It is clear from these facts that in digging wells whose water comes from near the surface of the ground water the bottom should be carried deep enough into the water bearing beds to leave it below the lowest stages of the ground water.

327. Elevation of the Ground Water through Precipitation and Percolation.—In Fig. 94 is represented the unoccupied space in eight feet of five grades of sand, above standing water, after 2.5 years had been allowed for percolation under conditions where no evaporation could take place from the surface. The unshaded portions of this figure represent the relative amounts of space into which rains may percolate for each grade of sand, as compared with the whole area of the diagram; that is to say, if an inch of rain were to fall upon the whole surface of the diagram and it were occupied with the No. 100 sand the space into which the rain could descend is measured by the unshaded area under 100; so for each of the other sands.

It will be seen from the diagram that up to 12 inches above the ground water surface the space into which water can settle in either sand is very small and hence that a small amount of percolation will produce a relatively large elevation of the ground water surface at first.

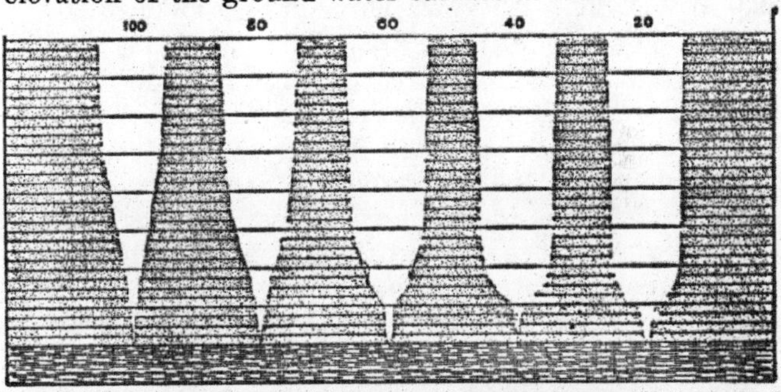

FIG. 94.—Showing the amount of unoccupied space in completely drained sands. Space between long rules, one foot.

In a tank filled with rather coarse sand and provided with glass gauge tubes, as represented in Fig. 112, p. 293, to show the level of the ground water surface, a single pound of water added to the 14 square feet of surface raised the level of the ground water .31 inch. In another trial 16.435 lbs. of water or .226 inch raised the surface 6.7 inches. In still another trial the withdrawal of 33.575 lbs. of water from the tank, or .461 inch, lowered the ground water 9.05 inches.

In the table below are given the amounts of water re-

Table showing amount of rain necessary to raise level of ground water after thorough drainage.

Grade of sand.	1 foot.	2 feet.	3 feet.	4 feet.
	Inches.	*Inches.*	*Inches.*	*Inches.*
No. 20	0.874	4.379	8.550	12.81
No. 40	.433	3.551	7.795	12.19
No. 60	.579	2.701	6.454	10.80
No. 80	.370	1.592	4.080	7.573
No. 100	.242	1.030	2.635	5.131

quired to raise the surface of the ground water 1, 2, 3 and 4 feet in the sands of Fig. 94, after thorough drainage has taken place.

328. Law of Flow of Water Through Sands and Soils.—It has been generally claimed that the velocity of flow of water through sands and soils is directly proportional to the effective pressure and inversely proportional to the length of the column through which the flow is taking place. This means that to double the pressure will double the rate of flow but to double the length through which the water must flow will decrease the rate one half. A law analogous is formulated for the flow of fluids through capillary tubes and under certain conditions of pressure and dimensions the law has been nearly fulfilled, both with sands and capillary tubes.

In practical measurements[1] of flow it is found that the flow through some sands and some capillary tubes increases faster than the pressure while in others it does not increase so rapidly.

The law of flow here referred to has been designated "Darcy's Law" and has been expressed by the formula

$$V = k\frac{P}{h}$$

where

V is the velocity,
P is the difference in pressure at the ends of the column.
h is the length of the column.
k is a constant depending upon the size of the soil grains, the amount of pore space and the viscosity of the fluid.

329. To Compute Flow of Water Through a Column of Sand, Soil or Rock.—Under the conditions where Darcy's law may be fulfilled the amount of discharge may be computed by means of the formula derived by Slichter[2] and given below:

---

[1]Nineteenth Annual Report, U. S. Geol. Survey, Part II., p. 202.
[2]Nineteenth Annual Report, U. S. Geol. Survey, Part II., pp. 301-322.

$$q = 10.22 \frac{pd^2 s}{\mu hk} \text{ c. c. per second} \qquad (1)$$

where

p is the pressure in c. m. of water at 4° C.

d is the diameter of the soil grains in millimeters.

s is the area of the cross-section in sq. c. m.

μ is the coefficient of viscosity.

h is the length of the column.

k is a constant whose log. is taken from the table, p. 123.

and 10.22 is a constant whose log. is [1.0094.]

If the pressure is measured in feet of water at 4° C., the length in feet, the area of cross section in square feet, the time in minutes and the diameter of the soil grains in millimeters the formula is

$$q = .2012 \frac{pd^2 s}{\mu h k} \text{ cubic feet per minute.} \qquad (2)$$

If the flow of water occurs under a temperature of 10° C. or 50° F. the formula may be written

$$q = 15.30 \frac{pd^2 s}{h k} \text{ cubic feet per minute.} \qquad (3)$$

Problem.—A cylinder 4 feet long, having a cross section of 2 sq. ft., is filled with sand whose grains have an effective diameter of .15 mm. What will be the flow of water through it under an effective pressure of 12 feet, when the temperature is 50° F. and the pore space is 35 per cent.?

Substituting these values in equation (3) we get, taking the value of k from the table, page 123.

$$15.3 \frac{12 \times (.15)^2 \times 2}{4 \times 31.62} = .06532 \text{ cu. ft. per minute.}$$

Problem.—What would be the flow in cubic feet per minute under the same conditions except at a temperature of 68° instead of 50° F.? In this case use formula (2) and the results are, taking the coefficient of viscosity at 68° F. at .0101 from the table below:

$$.2012 \frac{12 \times (.15)^2 \times 2}{.0101 \times 4 \times 31.62} = .06741 \text{ cu. ft. per minute.}$$

TABLE III.— *Coefficients of viscosity for water for various temperatures centigrade.*

θ=temperature centigrade.	μ=coefficient of viscosity.	θ=temperature centigrade.	μ=coefficient of viscosity.
0	0.0178	10	0.0131
1	0.0172	11	0.0128
2	0.0166	12	0.0124
3	0.0161	13	0.0120
4	0.0156	14	0.0117
5	0.0152	15	0.0114
6	0.0147	16	0.0111
7	0.0143	17	0.0109
8	0.0138	18	0.0106
9	0.0135	19	0.0103
10	0.0131	20	0.0101

330. Observed and Computed Flows Compared.—When sands have been sorted into grades of nearly uniform size and the effective diameter determined by the method of (**143**) and then the flow of water through them measured in such an apparatus as is represented in Fig. 95 the observed and computed flows are related as given in the next table.

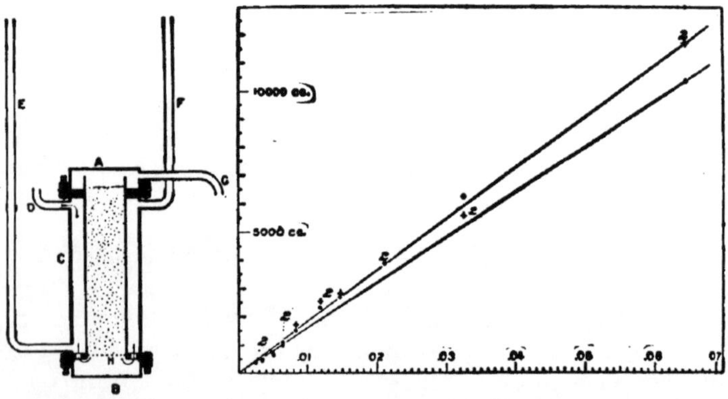

FIG. 95.—Showing apparatus for measuring the flow of water through sands and the relations of flow to the diameters of the sand grains. Lines show theoretical flow; dots, observed flow.

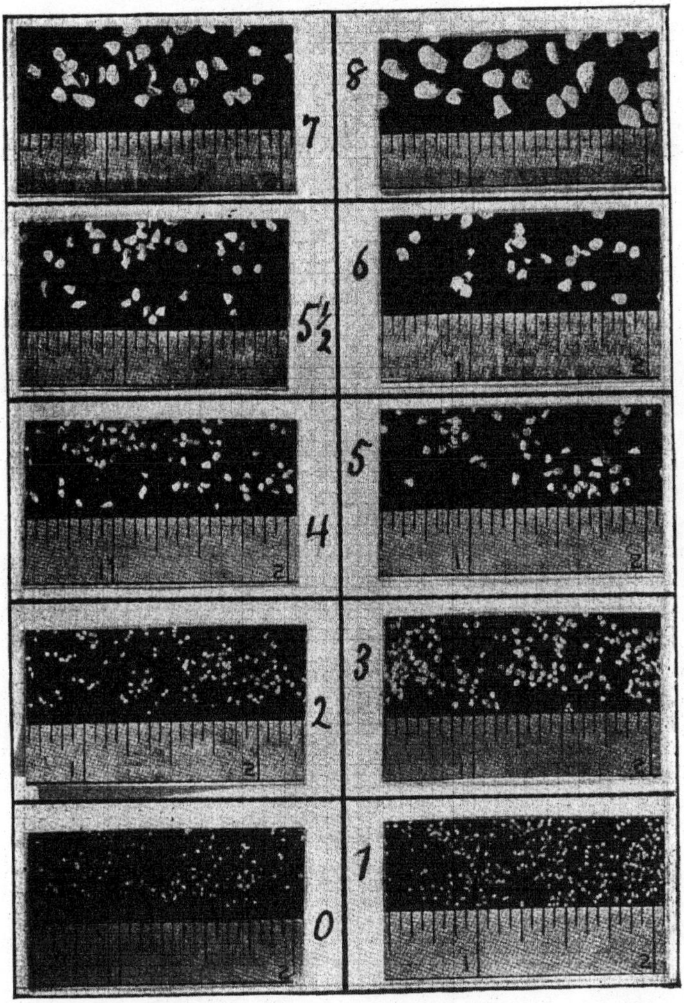

FIG. 96.—Showing the sand grains referred to in table on p. 266. Natural size.

Table showing observed and computed flow of water through simple sands of different diameters under a pressure of 1 c. m. of water.

Grade of sand.	Diameter of grains.	Observed flow.	Computed flow.
	m. m.	gms.	gms.
8	2.54	2,296	2,277
7	1.808	1,080	1,132
6	1.451	756	757
5½	1.217	542	522
5	1.095	504.6	453.2
4	.9149	329.2	297.5
3	.7988	210.0	193
2	.7146	138.6	122
1	.6006	94.8	80.6
0	.5169	72.3	66.8

The agreement between the observed and computed flows is not as close as could be wished but when it is observed that the flow of air, from which the diameters were computed, was not measured through the same sample as the one through which the flow of water was measured, that the pieces of apparatus were not the same and that the flow varies, theoretically, as the squares of the diameters of the soil grains, it must be conceded that there is much more than a chance agreement.

The samples of sand used in these trials are represented full size in Fig. 96.

331. Relation of Observed Flow to Diameter of Soil Grains.—If the squares of the diameters of the sand grains represented in Fig. 96 are plotted as abscissas and the observed and computed flows as ordinates their relations will be as shown in Fig. 95, where it is clear that the rates are such as to agree reasonably well with the squares of the diameters of the grains.

332. Relation of Pressure to Flow Through Sands.—Most experimenters along this line have found that while there is a general tendency for the flow to increase directly as the pressure there are nevertheless conditions which prevent these relations being realized in experiment, in some cases the flow being systematically too fast and in others too slow.

A series of observations by Welitschkowsky and Wollny

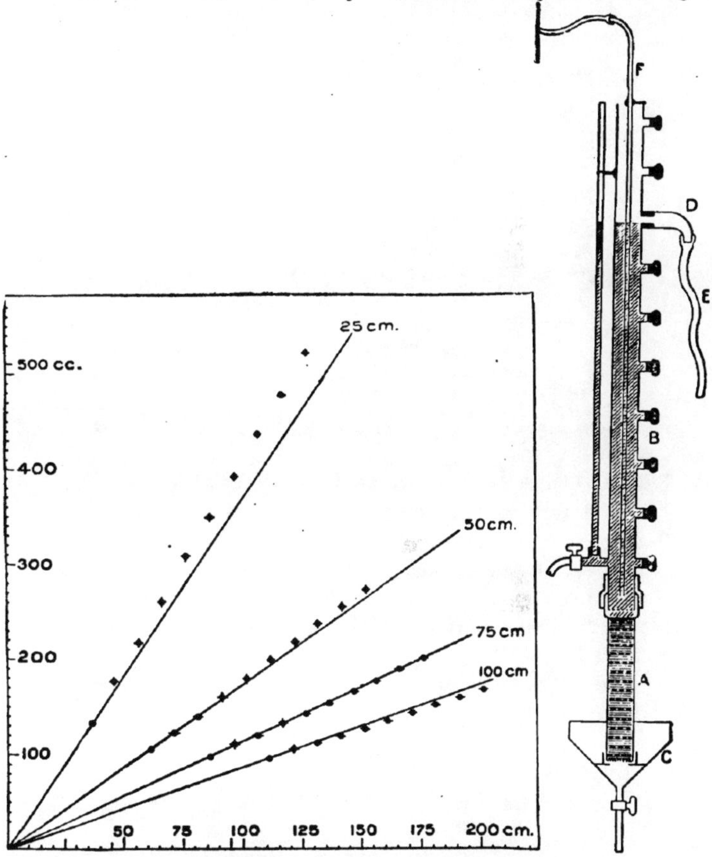

FIG. 97.—Showing apparatus of Welitschkowsky and the relation of pressure to flow of water observed by him.

and the apparatus with which they were secured are represented in Fig. 97. It will be observed that where the columns of sand used by Welitschkowsky were 25 c. m. and

50 c. m. long the flow increased faster than the pressure; but when the column was 75 c. m. long the flow increased directly as the pressure, while when it was made 100 c. m. long then the flow did not increase as rapidly as the pressure.

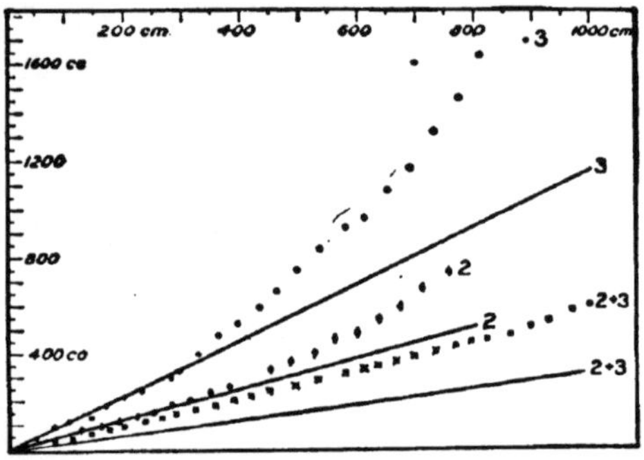

FIG. 98.—Showing the observed relation of pressure to flow of water through sandstone, as measured in the apparatus of Fig. 99.

333. Relation of Pressure to Flow Through Sandstone.—When the flow of water is measured through sandstones such as constitute most water-bearing beds it is often found that here, as in the sands, the flow may increase in a much higher ratio than the pressure. Three series of such observations are plotted in Fig. 98, and the apparatus used is shown in Fig. 99.

Where the flow does not increase as rapidly as the pressure the departure from the theoretical flow has been explained by assuming that the currents become turbulent and thus reduce the discharge; but no satisfactory reason has yet been assigned to the cases where the flow increases faster than the pressure.

334. Observed Rates of Flow of Water Through Sands and Sandstones.—The observed rates of flow of water through the series of sands represented in Fig. 96, when expressed in cubic feet per minute per square foot of section and per foot of length, under a gradient of 1 in 10, is given below:

	No. 8	7	6	5½	5	4	3	2	1	0
Cu. ft. per min.	5.23	3.65	1.85	1.36	1.22	.82	.51	.33	.23	.18

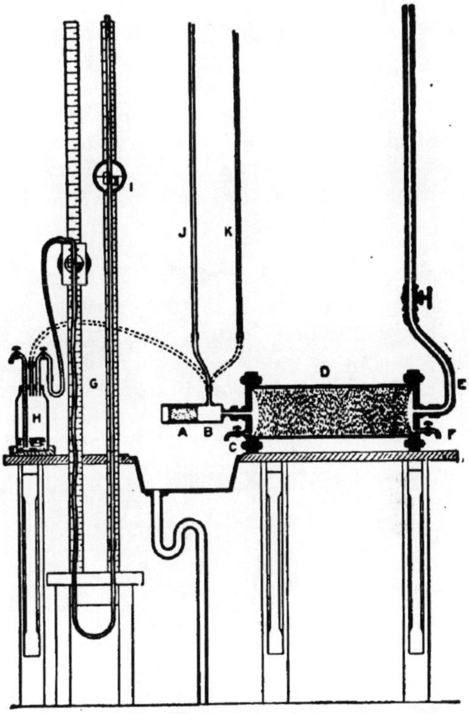

FIG. 99.—Apparatus for measuring the flow of water through sandstones, under different known pressures.

According to Darcy's law, if these sand columns had their lengths increased 10, 100 and 1,000 times the discharges observed would be only $\frac{1}{10}$, $\frac{1}{100}$ and $\frac{1}{1000}$ of those given. In the case of four sandstones the rates of flow were so slow that 10 days were required for .29, .34, 2.45 and .14 cubic feet of water to be discharged under the conditions for the sand.

335. General Movement of Ground Water Across Wide Areas.—The waters which supply artesian wells and many springs, where the discharges take place through openings in overlying impervious beds, are often obliged to travel long distances, even 100 or more miles, before reaching their outlets. But this cannot occur with such low rates of flow as those observed in (**334**) and it is clear that nearly the whole movement across long distances must take place through rock fissures and along bedding planes, the water seeping out of the rock into these as it does into river channels and lines of tile drains.

336. Fluctuations in the Rate of Flow of Ground Water.—When arrangements are made to automatically record the rate of discharge of water from springs, artesian wells or lines of tile drains it is seen that the flow is not uniform, varying not only with the season, but often daily and even hourly.

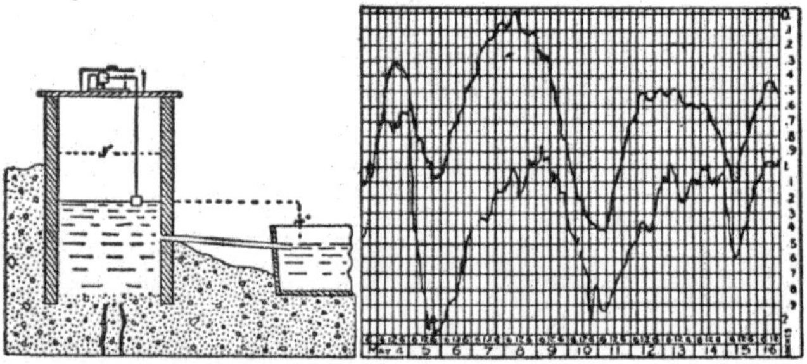

FIG. 100.—Showing observed barometric changes in the rate of flow of water from a spring, and the apparatus for recording it. Lower curve, record of spring.

In Fig. 100 is shown an autographic record of the discharge of water from a spring during 13 days, together with the changes in barometric pressure as recorded by a barograph 45 miles to the west of the spring. The method of recording the changes is also represented in the same figure. The changes in the rate of discharge from the spring, which are associated with changes in the pressure of the atmosphere, amount to as much as 8 per cent. of the total normal flow.

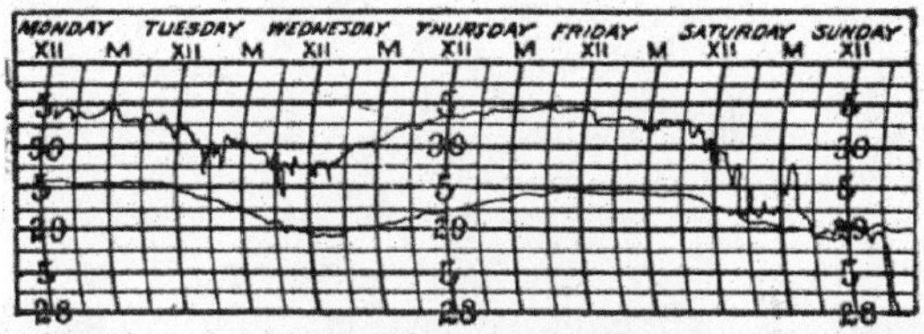

FIG. 101.—Showing the barometric changes in the rate of seepage into tile drains. Lower curve, drain.

337. Barometric Changes in the Discharge of Water from Tile Drains.—Using the same means for recording the rate of discharge of water from tile drains it was shown that changes occur here which are entirely analogous to those recorded from the spring, and Fig. 101 shows a week's record of the changes both in atmospheric pressure and in the rate of discharge from a system of tile drains. In this system changes in the rate of flow as great as 15 per cent. of the mean have been recorded, entirely independent of rainfall and apparently due solely to changes in atmospheric pressure.

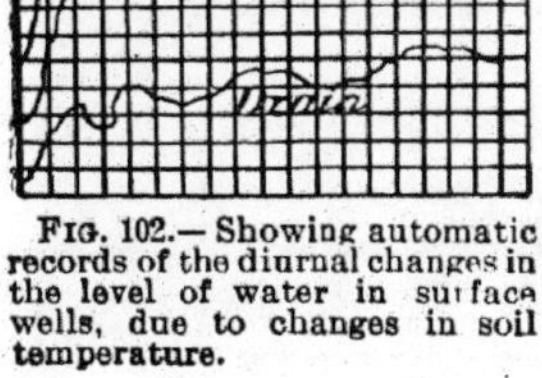

FIG. 102.—Showing automatic records of the diurnal changes in the level of water in surface wells, due to changes in soil temperature.

338. Diurnal Changes in the Rate of Discharge from Tile Drains.—Besides the changes associated with changes of barometric pressure referred to in (337) there may also be diurnal changes in the rate of discharge which are due to the diurnal changes which take place in the soil air above the ground water. As the air expands under the heat absorbed it presses downward upon the water, causing it to drain away faster, which makes it much as if a rain had occurred and percolation had increased the hight of the ground water itself. Fig. 102 shows the changes which did occur in the level of the water in surface wells near the system of tile drains in question. The curves were produced at the same time by self-recording instruments. Fig. 103 shows another series of diurnal fluctuations where the changes in level were measured twice daily, in the morning and at night, and Fig. 104 shows the conditions under which these changes occurred. The lower curve represents the changes in the inner well while the upper curve shows those in the outer well where the water percolated from above the stratum of clay under the influence of the air pressure caused by the diurnal changes in temperature.

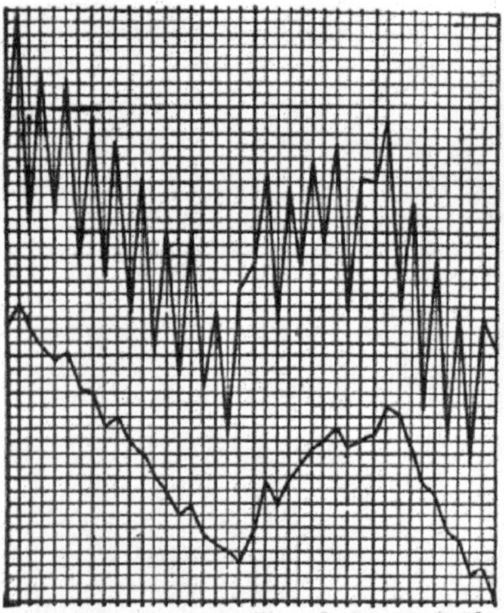

FIG. 103.—Showing diurnal changes in the level of the ground water measured twice daily in surface wells.

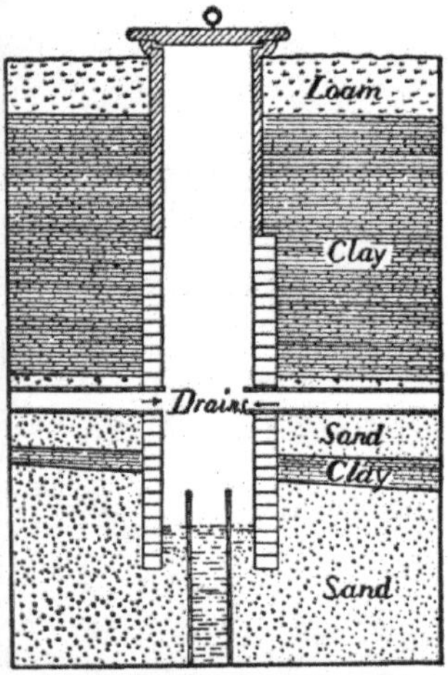

FIG. 104.—Showing the soil conditions under which the changes of Fig. 103 took place.

339. Fluctuations in the Level of Water in Wells.—In all ordinary wells, whether they are deep or shallow, the water is seldom at rest, the surface continually either rising or falling through varying distances, and Fig. 105 is a record of one such series of changes which it will be seen are nearly

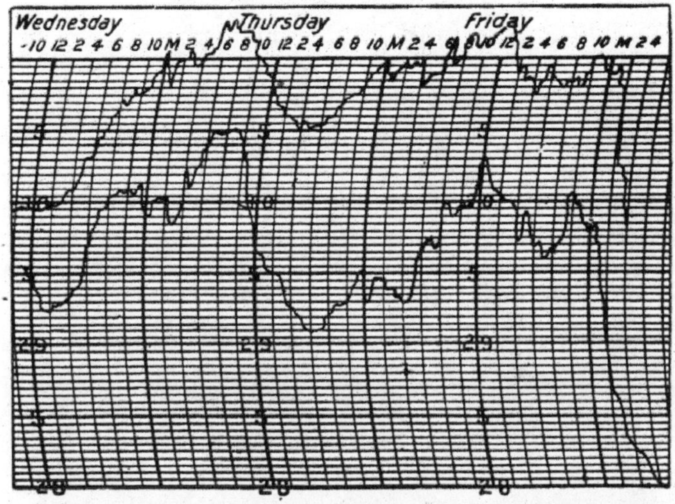

FIG. 105.—Showing fluctuations in the level of water in a well and simultaneous fluctuations in the rate of discharge from a spring half a mile distant.

FIG. 106.—Showing sudden and large fluctuations in the level of water in a well, during times of thunder showers, due to sudden changes in pressure.

coincident in phase with those which occurred in the discharge of water from a spring. Changes much more violent than these and of shorter duration are shown in Fig. 106. Fluctuations like these occur at times of violent thunder storms and are due to changes in air pressure and not to rainfall. In this case the changes occurred in a drilled well 60 feet deep with 6 inch steel casing to rock and the changes in the level of the water were so great that the instrument had to be set over three times to keep the pen on the record sheet.

CHAPTER XIII.

FARM WELLS.

340. Essential Features of a Good Well.—The essential features of a good well are: (1) Ample capacity to supply pure, clear, cold water. (2) A location which renders it not likely to be contaminated by seepage from surface impurities. (3) A casing or curbing which is vermin proof at the top and if possible water-proof in its upper 10 to 20 feet.

341. The Capacity of a Well.—The capacity of a well should always, if possible, be much greater than the probable demands which will be put upon it, and it should not be possible in a few hours to pump it dry with an ordinary pump.

In working the ordinary domestic pump about 20 strokes are made per minute and these will fill a pail with 20 to 24 pounds; this is at the rate of about a cubic foot or 7.5 gallons in 3 minutes and a good well should be able to supply water at this rate for several hours without failing.

The domestic animals on the farm will need water at the rate of more rather than less than a cubic foot per each 1,000 lbs. of weight per day. A cow giving a heavy flow of milk often takes nearly 2 cubic feet of water in 24 hours.

Five cows, during 120 days in winter, averaged 85.4 lbs. per head when the water was warm and 77.3 lbs. when it was cold. At this rate the equivalent of 40 adult cows would need 3,416 lbs. of water or 54.7 cubic feet and this would require, at the rate assumed above for pumping, 2 hours and 45 minutes to supply them.

342. Geological Conditions Which Give the Best Wells.— The largest and best supplies of well water are usually found in the extensive sandstone formations and wherever these are within easy reach the well should be sunk into them deep enough to have 20 or more feet of percolating sandstone surface. Next to the sandstone formations as sources of water supply stand the fissured limestones which either overlie sandstones or are so related to the surface soil that water from them can percolate down into the fissures and through them reach the well when sunk so as to connect with a system of these fissures.

Again beds of sand between beds of clay often give large supplies of pure cold water.

In many localities artesian or flowing wells can be secured and some of the conditions under which these originate are represented in Fig. 107.

343. Conditions which Influence the Capacity of a Well.— The rate at which water can enter a well depends upon five prime factors: (1) The size of the grains of the water-bearing beds and the pore space. (2) The depth of the well in the water-bearing bed. (3) The amount the water is lowered in the well when pumping. (4) The diameter of the well. (5) Whether the well is in or near a system of fissures.

344. Influence of Size of Grains and Pore Space on the Capacity of the Well.—From the fact that the flow of water through sands is nearly proportional to the squares of the diameters of the soil grains, and is greater the larger the pore space, it is clear that these are very important factors in determining the capacity of wells. It has been computed that when all other factors are the same the capacities of two wells, in sands having the diameter of grains of .15 mm. and .25 mm. and pore spaces of 30 per cent. and 32 per cent., are to each other as 5.234 to 18.01 or one is over three times the other. It is therefore clear that when the sand grains and pore space are small the other well factors must be made enough larger to compensate.

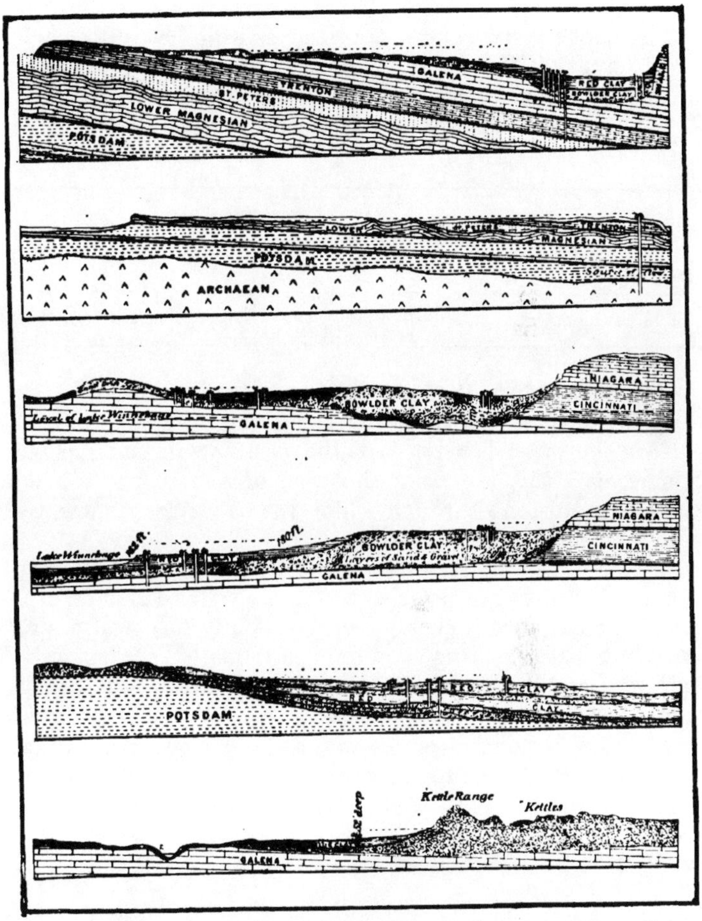

Fig. 107.—Showing geological conditions under which artesian wells are formed.

The capacity of a 6-inch well sunk 100 feet into sandstone having different sizes of sand grains but with uniform pore space of 32 per cent. and a temperature of 50° F. give computed flows under a pressure of four feet as follows:

	Size of grains in m. m.							
	.02	.04	.06	.08	.1	.2	.4	.6
Cu. ft per min................	.047	.189	.580	.757	1.083	4.73	18.93	57.96

345. Influence of Depth on the Capacity of a Well.—When other conditions are the same the greater the depth of a well in the water-bearing beds the greater will be its capacity because this increases the area of the section of the sand or sandstone through which the water may enter the well.

If a 6-inch well is sunk just to the surface of a water-bearing bed the area through which the water can enter it is only 28.27 square inches. So, too, if a 6-inch well casing ends in a water-bearing sand only so much water can enter this well as can flow through a circle of sand 6 inches in diameter.

If the well penetrates the water-bearing bed one foot so that water can enter the sides as freely as it enters the bottom then the percolation surface will be increased to

$$28.27 + 226.2 = 254.47 \text{ sq. in.}$$

making the section of flow nine times as great. Leaving the bottom of the well out of consideration it is clear that doubling the depth of the well in the water-bearing beds doubles the area for water to enter and hence it is a matter of the greatest importance to secure a sufficiently large percolating surface in the water-bearing beds. This capacity increases in a somewhat slower ratio than the depth, as indicated in the table below.

Table showing the flow in a 6-inch well sunk different depths into 200 feet of water-bearing sandstone where the pore space is 32 per cent. and the diameter of the grains .25 m. m.

	Depth of well in feet.								
	4.	8.	12.	16.	20.	40.	80.	100.	200.
Flow in cubic feet per minute..............	1.003	1.818	2.544	3.265	4.08	7.68	14.88	18.49	36.02

346. Influence of Pressure on the Capacity of a Well.—Since the flow of water through sands and sandstone is somewhat nearly proportional to the effective pressure it is clear that the depth of water in the well at low water stage should be great enough to permit its surface to be lowered until the needed pressure to force the water into the well is developed.

If, in pumping, the water in a well is lowered 4 feet the pressure developed will be about that of four feet of water and to lower it 8, 12, 16 or 20 feet will increase the pressure 2, 3, 4 and 5-fold. This relation being true it is clear that not only should there be an ample depth of water in the well but the cylinder of the pump should be so placed as to enable the full depth to be utilized.

In the case of a 6-inch well sunk 100 feet into water-bearing sandstone 200 feet thick having a pore space of 32 per cent. and diameter of grains of .25 mm. the capacity of the well under different pressures is computed to be as follows:

Amount the water is lowered in the well in pumping.

	1	2	4	8	12	16	20
Cu. ft. per minute....	1.8483	3.6966	7.3932	14.7864	22.1796	29.5728	36.966

347. Influence of the Diameter of the Well on its Capacity.—The capacity of wells when they extend any considerable depth into the water-bearing beds does not increase as rapidly with increase of diameter as might be expected, and Slichter computes that three wells 2 inches, 6 inches and 12 inches in diameter respectively, if sunk 100 feet into a bed of sandstone having sand grains .25 mm. in diameter and a pore space of 32 per cent. will have capacities in cubic feet per minute as follows, when the water is lowered 20 feet:

	Diameter. 2 inch.	Diameter. 6 inch.	Diameter. 12 inch.
Cubic ft. per minute....	31.90	36.94	44.45

These amounts are on the assumption that the walls of the well or casing offer no resistance to the discharge, which of course is not true, and the 2-inch well could not discharge the amount indicated under the pressure of 20 feet although that amount could enter the well if it were removed fast enough.

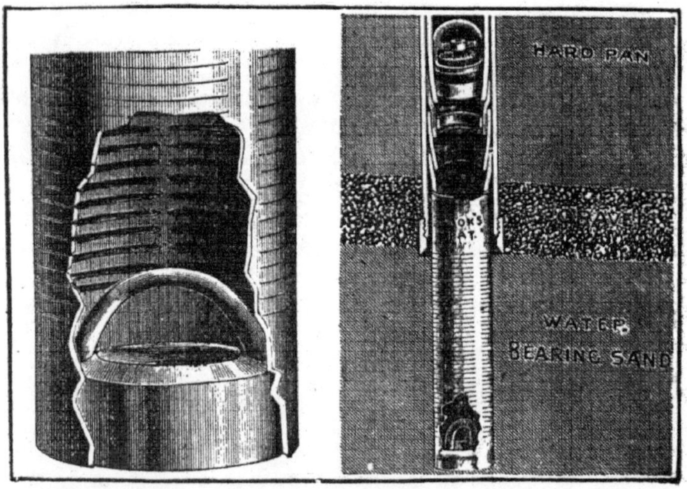

FIG. 108.—Shows a good form of sand strainer made by sawing slots in brass tubing.

It is clear from these results that for most wells there is little gained in making them larger in diameter than is needed to provide accommodation for the pump.

348. The Use of Sand Strainers.—Where water must be procured in loose sand, especially if it is fine, some form of sand strainer should be used unless the well is an open one and even then a suitable point will often greatly increase the capacity.

The difficulty in getting water rapidly from loose sand grows out of its tendency to move with the water, filling up the well or the suction pipe or cutting out the valves. Since the specific gravity of sand is only about 2.65 just as soon as a pressure greater than 3 feet is developed to force the water out of the sand the sand must move with it unless there is something to prevent it.

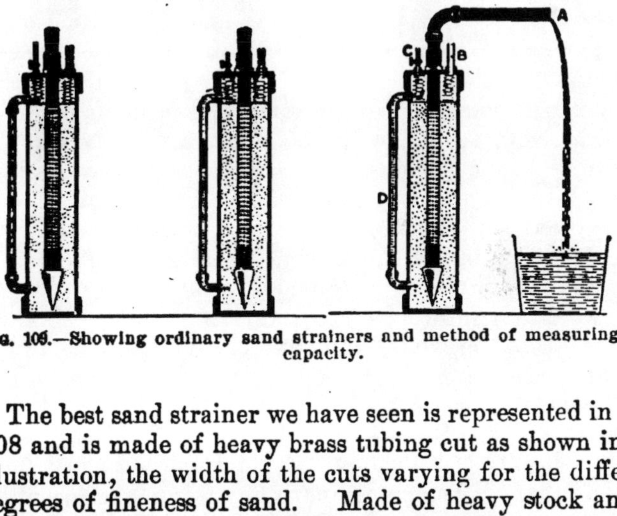

FIG. 109.—Showing ordinary sand strainers and method of measuring their capacity.

The best sand strainer we have seen is represented in Fig. 108 and is made of heavy brass tubing cut as shown in the illustration, the width of the cuts varying for the different degrees of fineness of sand. Made of heavy stock and of one kind of metal it is not liable to corrode and clog as with the common form represented in Fig. 109.

349. Capacity of Sand Strainers.—The capacity of sand strainers varies essentially in the same way as wells of similar dimensions would, made in the same kind of material. The longer the strainer, the coarser the sand and the greater the pressure the larger will be the capacity.

In Fig. 109 is represented a method used in measuring the capacity of three Gould Sand Strainers, Nos. 50, 80 and 90, each 18 inches long, and the table below gives the results secured.

Table showing the rate of flow through three drive well points.

Pressure feet.	No. 50. Lbs. per min.	No. 80. Lbs. per min.	No. 90 Lbs. per min.
2	6.6	2.2	.57
4	13.2	4.4	1.14
6	19.8	6.6	1.72
8	26.4	8.8	2.29
10	33.0	11 0	2.86
12	39.6	13.2	3.43
14	46.2	15.4	4.00
16	52.8	17.6	4.58
18	59.4	19.8	5.15
20	66.0	22.0	5.72

The sand about the No. 50 strainer had a diameter of .294 mm., that about the No. 80 .172 mm., and about the No. 90 .085 mm. The table shows under these conditions about 2 minutes of steady flow, under a pressure of 12 feet, are required for the No. 50 strainer to supply sufficient water for a single cow one day; 6 minutes for the No. 80 and more than 20 minutes for the No. 90 strainer.

It would therefore be necessary to use a strainer 54 inches long in the No. 80 sand and one 17 feet long in the No. 90 sand to supply the water obtained through the No. 50 strainer.

350. Capacity of a Pump on a Sand Point and on an Open Suction Pipe.—When an ordinary pump is connected up in the manner represented in Fig. 110, so as to draw water through the sand point or through the open suction, the capacity of the pump under the two conditions may be very different. In the case of a two and a half inch cylinder working on an 18 inch No. 50 sand strainer, or on the open suction pipe as represented in the illustration, when 20 strokes would fill the pail from the open suction it required

35, made at the same rate, to raise the same amount of water, and the energy required to do the work was much greater. The increased labor was due to the fact that the water came in too slowly through the sand point to fill the space behind the piston as rapidly as it was raised and a vacuum was formed; into this the piston fell when the pressure was released and the water for only about half a stroke could be secured.

Sand strainers give a fair well in very coarse material where one of sufficient size can be placed in a water-bearing bed of sufficient thickness, but generally they can be depended upon for only small amounts of water. For wind-mill service they are less satisfactory because of the greater power required to work the pump.

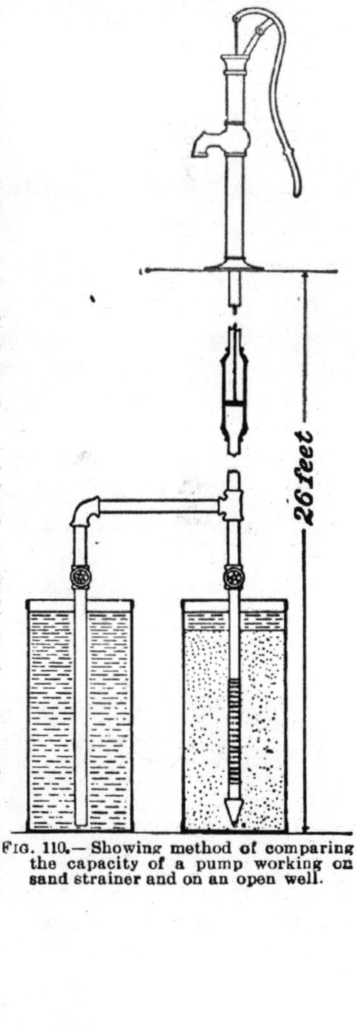

FIG. 110.—Showing method of comparing the capacity of a pump working on sand strainer and on an open well.

351. Depth of the Well.—An important feature of every well, where the water is intended for domestic or stock use, is a sufficient depth to prevent the quick entrance of water from the surface and to maintain a constant low temperature. This depth should usually exceed 20 feet and even where water is found nearer the surface than this it is better, if the water-bearing beds will permit of it, to go 30 or more feet and then place the pump so as to draw the water from the bottom where it is coolest and freshest.

Both depth of soil, to act as a filter, and time to bring about changes in surface waters, to free them from organic matter, are required in order to render the water falling upon the ground pure and suitable to drink.

352. Temperature of Well Water.—The zone of lowest ground temperature is generally from 20 to 70 feet below the surface and in this zone the coldest waters are procured. Above 20 feet the waters will be colder in winter and warmer in summer and below 70 to 75 feet the water generally becomes warmer from the internal heat of the earth.

The normal temperature of the coldest well water in a locality is usually from 2 to 4 degrees higher than the mean annual air temperature of the place, and in Wisconsin this ranges from 43° in the northern portion to about 50° in the southern portion.

353. Well Casing or Curbing.—Everything considered there is probably nothing better for a curbing or casing for a well than the 6 inch lap-weld steam pipe. The same pipe galvanized is better because it will not rust out so quickly. The great advantage of this kind of casing is that it is so completely water tight and at the top can be so securely closed as to prevent insects and vermin falling in.

Next to the steel casing is one made of cement tile or glazed sewer tile with their joints set in cement. Where a well is to have a brick or stone curbing the upper 10 feet should be laid in cement and plastered with the same on the back to exclude surface water and vermin.

354. Top of the Well.—In finishing a well the casing should be carried 12 to 18 inches above the surrounding surface and then earth be graded up to it so as to secure perfect and quick removal of all surface water.

Where a steel casing is used the well platform is best made by screwing a wide flange on the top and then bolting the pumphead directly to this, having first drilled holes through both to receive the bolts. This arrangement secures a very solid and perfectly tight platform. Around this plank may be laid, or better still, a block of cement.

CHAPTER XIV.

PRINCIPLES OF FARM DRAINAGE.

Both irrigation and drainage are usually looked upon as arts whose application to agriculture are required only in special cases; but a broader and more helpful conception is that all fertile fields must be both well irrigated and thoroughly drained.

It is true that over much the larger portion of the earth's surface the water required for the growth of crops is supplied by the natural rainfall, and when this is timely and sufficient it is the best and ideal irrigation, done by nature's hand.

It is again fortunately true that most land areas have acquired such surface features that the excess of rainfall is opportunely removed by percolation and seepage or surface flow; and this is nature's method of land drainage.

The fundamental fact is that all lands must be irrigated or watered and drained and in special cases nature's efforts need to be supplemented.

355. Necessity for Drainage.—There are several imperative demands for the drainage of farm lands:

1. The removal of the more soluble salts formed by the decay of rock and organic matters, because when the soil water becomes too strong in soluble salts it either poisons the plant or renders the root hairs inactive by causing them to shrivel. If these soluble salts which plants cannot use are not removed the soil comes into the condition known as alkali lands, upon which little vegetation can grow.

2. The water in the soil needs to be frequently changed or replaced by a fresh supply containing an abundance of

atmospheric oxygen because the roots of plants and microscopic life tend to exhaust this supply. If the soil is not drained the water in it becomes stagnant in a sense, the rains which fall simply running off the surface, leaving the soil water the same as was there before the rain.

3. Farm lands must be drained in order to render them sufficiently firm to permit the farm operations.

4. Soils must be drained in order to provide room for soil air. **(238.)** **(251.)**

5. The excess of water must be removed to permit the soil to become warm enough for plant growth. **(268.)** **(271.)**

356. Conditions which Require Drainage.—The cases in which it becomes desirable to supplement natural drainage fall into five classes:

1. Comparatively flat lands or basins upon which the water from the surrounding higher lands collect.

2. Areas adjacent to higher lands where the structure is such as to permit the water which sinks into the high land to flow or seep under and up through the low ground, making them wet.

3. Lands inundated regularly by the rise of tides or frequently by the overflow of rivers.

4. Extremely flat lands in wide areas which are underlaid near the surface by a thick, close, nearly impervious stratum of clay, such as were formerly old lake bottoms.

5. Lands like rice-fields, water-meadows and cranberry marshes where water is applied in excessive quantities at stated times and must be removed again quickly.

357. Deep Drainage Increases Root Room.—No plant can utilize the resources of the soil to the best advantage unless there is provided for it an abundance of root room. In all well drained soils the roots of most cultivated crops spread themselves widely and to a depth of 2.5 to 4 or more feet. When conditions are such as to permit crops to do this the best growth and largest yields result.

Proper drainage so lowers the ground water surface that roots are able to penetrate to their normal depth, and Fig. 111 shows how the roots of corn have been massed together near the surface because of too much water in the soil below, and Fig. 45, p. 147, shows the apparatus with the corn growing in it.

358. Drainage Increases the Available Moisture.—When the roots of a crop are forced to develop so close to the surface as shown in **(357)** the first effect is to exhaust the soil of its moisture so much as to leave it too dry and so lessen the capillary rise that, although there is an abundance of water in the soil below, it cannot be brought to the roots and the soil below is too wet to permit the roots to go to the moisture.

On the other hand if the ground water is lowered the roots are permitted to advance deeper, making it unnecessary for the water to move up as high and leaving the soil more moist, and so capillary action stronger and capable of lifting water higher and faster. **(198.)** **(199.)**

359. Soil Made Warmer by Drainage.—Whenever soils are kept continuously wet, so that large amounts of water evaporate from their surfaces, the temperature is low. Two thermometers having their bulbs side by side, one left naked and the other covered with a close fitting layer of wet muslin, will often show temperatures as much as 20° different, the wet one colder, made so by the evaporation of water. The teakettle on the stove has the temperature of its bottom held constantly near 212° by the evaporation of the boiling water, showing the cooling power of water when evaporating.

During early spring differences in soil temperature at the surface, due to differences in drainage, may often be as great as 12°.

The differences in the amount of moisture in clayey and sandy soil often cause a difference of 7° F., in the surface

FIG. 111.—Showing how the roots of corn are forced to develop near the surface when the soil is not drained. See apparatus, Fig. 45, p. 147.

foot, when both are well drained, and as much as 5° in the second and third feet.

360. Soil Better Ventilated by Drainage.—The change of air in wet soils after they have been well drained is very much more thorough and this is perhaps the greatest benefit due to drainage.

There are several ways in which thorough drainage leads to a more rapid exchange of air in the soil:

1. Lowering the ground water enables both the roots of plants, and animals like earthworms and ants, to penetrate the soil more deeply, leaving passageways larger and freer than existed before.

2. When the deeper clays come to dry after being drained shrinkage checks are formed in great numbers and through these the air moves more freely.

3. With the deeper penetration of soil air nitrates are more freely formed, and with the larger amounts of soluble salts the clay is flocculated, making a more granular texture, which again admits the air more freely.

4. When lines of tile are laid under a field 50 to 100 feet apart they furnish an opportunity, with every change in atmospheric pressure and of soil temperature, to force air into and out of the soil, and so a line of tile laid in the soil becomes a system for air circulation.

5. With every heavy rain which causes percolation, where the water can flow away, a volume of fresh air is drawn into the soil after it, completely changing the air.

361. Kinds of Drains.—There are two types of drains: (1) closed and beneath the surface after the manner of underground water channels; and (2) open, such as ditches, which are in function like natural river channels.

The closed forms are usually most effective, least in the way, require less expense in maintenance and are most durable and should generally be adopted, but there are cases where surface ditches must be used.

In the earlier history of underdraining closed drains were

made by laying bundles of twigs in the bottom of the ditch and covering them, expecting the water to trickle through the passageways left. In other cases two or three round poles were covered in the bottom of the ditch or two slabs were laid edge to edge with their round sides down. Two boards were sometimes set on edge V-shaped, with opening down.

More permanent closed drains were made by filling the bottom of the ditch with cobblestone, by setting flat stone on edge V-shape, by setting two lines of stone on edge and covering with flat stone and even by using four stone for top, bottom and sides. In other cases brick were used in place of stone and some even made tile out of blocks of peat, cutting semi-cylindrical cavities in the faces of square blocks of peat, then laying these together to form the waterway. Most of these devices, however, must be looked upon as makeshifts rather than as permanent improvements, and have largely gone out of use.

The modern tile, made of hard burned clay, is cylindrical in form and usually in 1-foot lengths with diameters ranging from 2 to 12 or more inches.

3C2. Essential Features of Drain Tile.—A good drain tile should be hard burned, giving a clear ring when struck. It is much more important to have them hard burned and strong than it is to have them open and porous. Soft burned tile which give little or no ring when struck are much more liable to crumble down under the action of frost. We have visited one field drained with soft burned tile laid 2.5 to 3.5 feet deep and, in less than five years after laying, holes appeared in the field in many places. On digging in these places it was found that the tile had crumbled into small chips, caused by freezing.

Tile are sometimes made from clay containing pebbles of limestone which when burned are converted into lime. These lumps of lime bedded in the tile slack as soon as water enough reaches them and by their expansion the tile are broken. It will often happen that such tile may be laid in place and covered before the slacking occurs.

Besides being hard burned, strong, giving a clear ring when struck and free from lime the tile should be smooth and straight, with square cut ends and true circular outline so that they may be laid with close joints which will exclude silt.

363. How Water Enters Tile.—The texture of a tile is like that of common brick and will allow water to flow readily through the walls, but even were the walls water tight the water could still find access to the tile through the joints formed by the abutting sections as rapidly as it can be brought by ordinary soils requiring drainage.

Measurements made of the rate of percolation through 2-inch Jefferson, Wisconsin, tile showed a flow of 8.1 cubic feet per 100 feet of length in 24 hours, under a pressure of 23.5 inches, when surrounded by clear water only. When the same tile were bedded in a fine clay loam, so that the water had to percolate through the soil, the discharge was reduced to 1.62 cubic feet per 24 hours and per 100 feet.

364. The Use of Collars.—It has sometimes been the custom to use collars to slip over the joints formed by the meeting of the sections of the tile, with the idea of better excluding the silt and of holding a better alignment. The collars are short sections of a size of the tile large enough to slip over the joints readily.

The use of collars is not advisable, first, on account of the greater cost, and second, because when good tile are properly laid they are not needed.

365. Depth at which Drains Should be Laid.—It is seldom necessary to lower the ground water more than four feet below the surface and except in very springy places a depth of 3 feet will answer most purposes.

Since the level of the ground water changes with the season and since many lands which are benefited by drainage are only too wet during the spring it may be best to lay the drains only so deep as is needful to bring the field into condition for working in due season, and in such cases tile placed 2.5 to 3 feet, rather than 3.5 to 4 feet, will usually be found sufficient for general farm crops.

When tile are placed needlessly deep not only is the cost greater but, in all of those cases where there is an underflow of water from the higher land, the level of the ground water is drawn down earlier in the season to such a depth that the crop will get less advantage by the subirrigation resulting from the capillary rise of the underflowing water into the root zone.

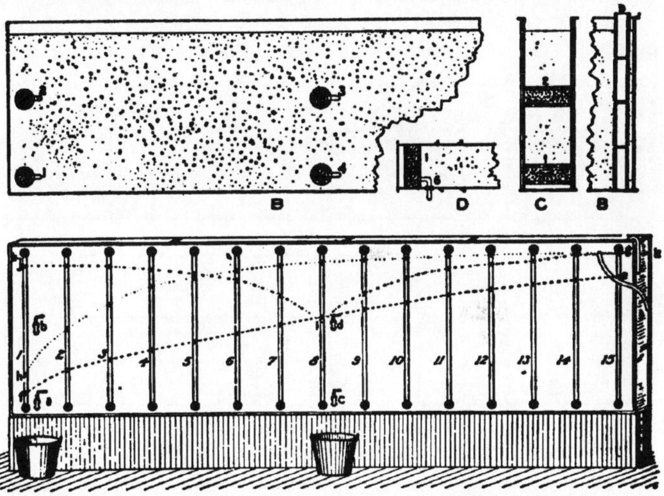

FIG. 112.—Representing an apparatus for demonstrating the slope of the ground water surface back from a tile drain and the changes in pressure when discharge is taking place. A. front elevation of tank, with a, b, c, d, faucets from drain tile, and 1, 2, 3, ...15, pressure gauges; B, B, vertical sections lengthwise, with 1, 2, 3, 4, tile and faucets, and 5, supply tile at end; C, cross-section, with 1, 2, tile; D, section at 1 in B, showing connection of faucet with tile.

366. Rise of Ground Water Away from Drainage Outlet.—If reference is made to the contour map of the ground water surface, Fig. 89, p. 257, it will be easy to compute the gradient of the ground water surface as it rises back from the lake. In well 29, 150 feet from the lake, the water stood on a certain date 7.214 feet above the level of the water in the lake, thus showing a mean rise or gradient of 1 foot in 20.79 feet. In the same locality, but outside the area represented by the map, a well stands 1,250 feet back from the lake and in this the water has a level 52 feet above the lake or drainage outlet, which gives a mean gradient or rise of 1 foot in 24.4.

In Fig. 112 is represented an apparatus for demonstrating the position of the surface of the ground water and the difference of pressure at different distances away from and above a drain tile, and Fig. 113 shows the observed differences of pressure under two sets of conditions.

In Fig. 114 is also represented the general slope of the ground water surface and the modification of it by a line

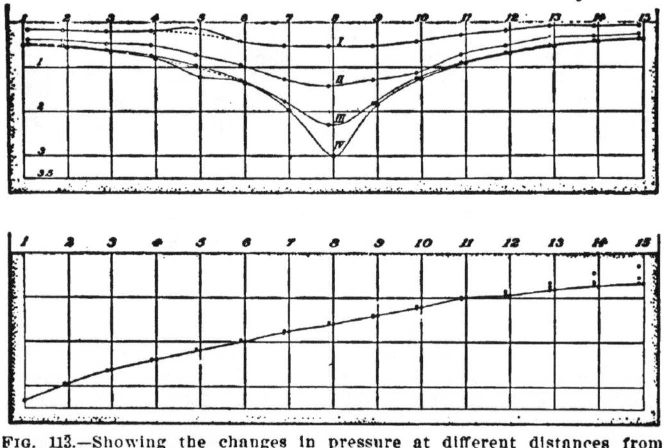

FIG. 113.—Showing the changes in pressure at different distances from the tile drain when the water is flowing. The lower curve shows the pressure when the flow is from the stopcock a, Fig. 112, and the upper set of curves represent changes which occurred during a period of flow from the stopcock c, Fig. 112.

of infiltration pipes, which is in effect a tile drain. The rate of rise of the ground water back from a tile drain is one of the chief factors in determining the distance apart the drains should be placed in the field.

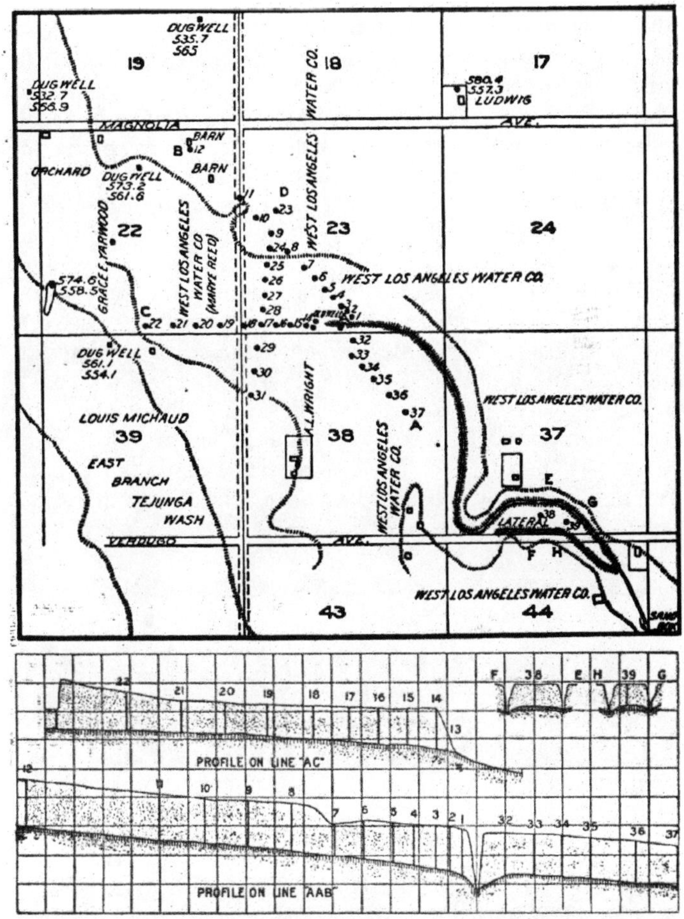

FIG. 114.—The upper portion is a diagram of flume of West Los Angeles Water Company and vicinity. Numbered dots show where level of ground water was measured in wells sunk for the purpose, and correspond with numbers on lower part. Lower part, profiles of the surface of the ground water in the vicinity of the West Los Angeles Water Company. The heavily shaded line is the ground water surface. Each square represents 100 by 10 feet.

367. Distance Between Tile Drains.—There are three prime factors which determine the distance between tile drains. 1. The effective size of soil grains and pore space of the subsoil through which the water must move to reach the drain. If the subsoil has a close fine texture the resistance to the flow will be great, and hence the water surface will rise faster back from the drain, bringing it near the surface sooner and making it necessary to place the lines closer together.

2. The depth at which the drains are placed. It is clear, that when it is desired to hold the water midway between a line of tile a certain distance below the surface, that the deeper the tile are placed the further they may be apart, and Fig. 115 illustrates both this point and the first.

3. The interval between rainfalls sufficiently heavy to produce percolation. In regions where the rainfall is both heavy and frequent tiles need to be placed nearer together than where the reverse conditions exist.

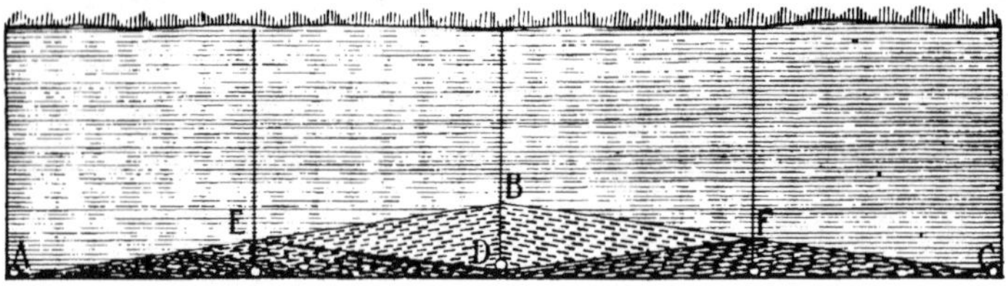

FIG. 115.—Showing the influence of distance between tile drains on the relation of the ground water to the surface of the ground.

In general practice for field crops it is usually sufficient to place the lines of tile from 50 to 100 feet apart. In favorable cases they may be placed even further apart than this and in special cases they may be required as close as 30 feet.

368. Observed Ground Water Surface in a Tile Drained Field.—In Fig. 116 is represented the observed ground water surface in a tile drained field where the lines are 33 feet apart, 3 to 4 feet below the surface and where the subsoil at 3 to 4 feet and below is sand. The slope of the surface was obtained by boring wells with a 4-inch auger between the lines of tile and the measurements were made 48

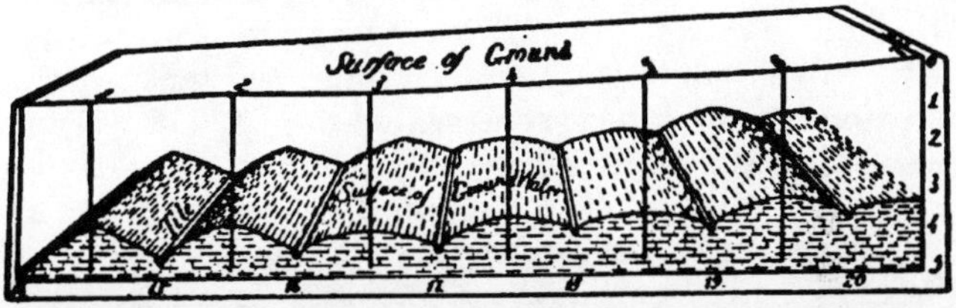

FIG. 116.—Showing the observed conformation of the ground water surface in a tile drained field 48 hours after a rainfall of .87 inch.

hours after a rainfall of .87 inch May 13, when the soil was already well saturated. On this date the highest level above the top of 3 inch tile between any two lines was 1 foot and the lowest .3 foot.

369. Rate of Change in the Contour of the Ground Water Surface Between Lines of Tile.—At the time the data for the last section were taken observations were also made to determine the rate of change in the level of the ground water after the rain, and Fig. 117 represents the differences in the level of the water at and between the tile drains on three different dates. It will be seen that the water fell fastest under the highest ground and on the 16th was below the tile in the upper part of the field.

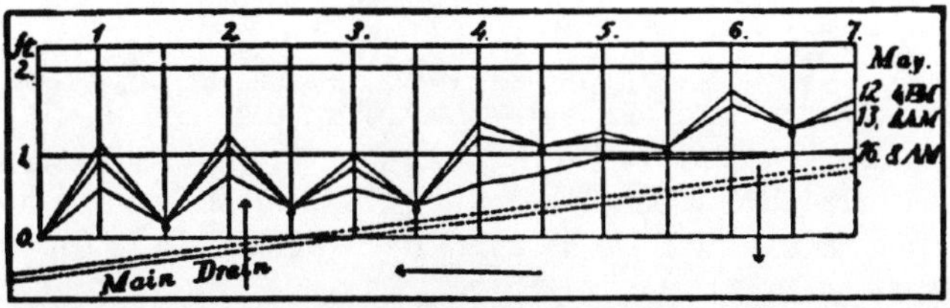

FIG. 117.—Showing changes in the level of the ground water surface in a tile drained field.

This illustration makes it clear also how the tile in the lower portion of the field make their influence felt in the upper portion, the water moving as indicated by the long arrows.

370. Movement of Water where Heavy Clay Soils are Underlaid with Sand.—When a heavy, close soil is underlaid with sand or gravel the movement of water toward the tile drains will be almost entirely through the sand when the conditions are like those represented in Fig. 118. In such cases the rains percolate vertically down into the sand and then move laterally to the tile drains, where it rises to enter them, as shown by the arrows.

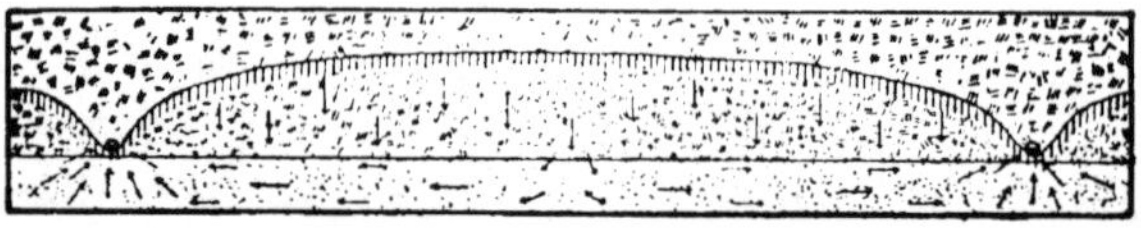

FIG. 118.—Showing how the main flow of water to lines of tile may be through a subsoil of sand when this is present and near the surface.

It is clear that under conditions like these the heavy clay soil above does not determine the distance apart drains should be placed but rather the sand stratum below.

371. Fall or Gradient for Drains.—Generally drains should be given as much fall as the conditions will permit and the gradient should not be less than 2 inches in 100 feet if this can be secured. Cases will occur where less must be accepted and then careful leveling must be done to secure the largest fall available.

It will often happen that the line of lowest ground is quite tortuous, making the distance long, and on this account making the fall small. Frequently in such cases cuts across bends can be made by digging deeper, in this way increasing the fall, as is sometimes done in straightening streams.

372. Uniform Fall Desirable.—Effort should be made to secure throughout the course of a main or lateral drain a uniform fall, and never, where it can well be avoided.

change from a steeper to a less steep grade, because if this is done there is danger that sediment may lodge where the fall is less and close up the drain. The case is different where a change can be made from a small fall to one which is greater, for then whatever sediment is carried by the water along the flatter slope will be carried down the steeper one.

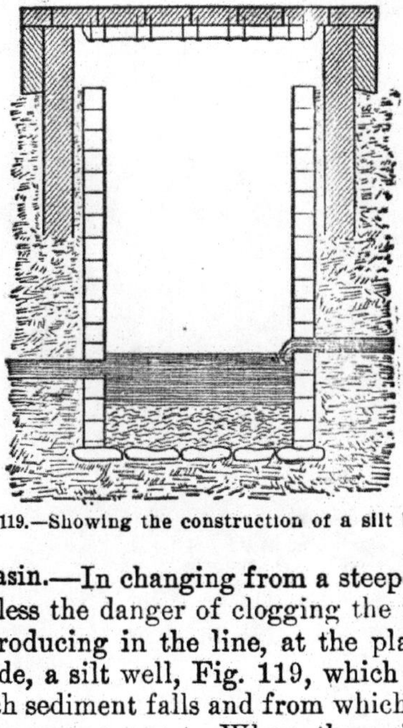

FIG. 119.—Showing the construction of a silt basin.

373. Silt Basin.—In changing from a steeper gradient to one which is less the danger of clogging the tile can be reduced by introducing in the line, at the place where the change is made, a silt well, Fig. 119, which provides still water in which sediment falls and from which it may be removed as often as necessary. Where these silt basins may be small, glazed sewer tile of suitable size may be used for the portion above the ground.

374. Size of Tile.—The proper size of tile can only be definitely stated when the detailed conditions under which the drain is to work are known. They should be large enough to remove in 24 to 48 hours the excess water of the heaviest rains likely to occur.

1. Where single drains are laid here and there in irregular order to drain low places larger tile are required than where a whole area is systematically treated, because in the former cases a larger per cent. of surface water from surrounding higher lands will flow upon the low areas under which the drains are laid.

2. The greater the fall the smaller the tile may be,—doubling the grade increasing the carrying capacity nearly one-third.

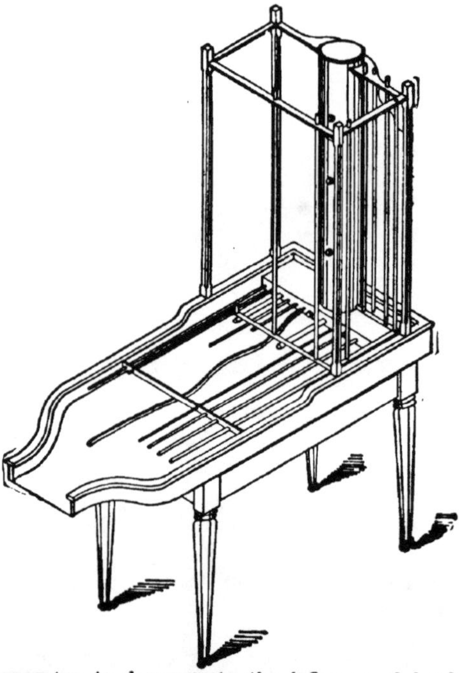

FIG. 119a.—Apparatus to demonstrate the influence of head, diameter, length, and bends on the rate of discharge of water through lines of tile and water pipe.

3. The areas of cross section of tile increase with the squares of their diameters:

If their diameters are in the ratio of 2, 3, 4, 5, 6, 7, their areas will be in the ratio of 4, 9, 16, 25, 36, 49, but as the friction on the walls of small tile and the disturbance due to eddies set up at the joints are greater in proportion to the amount of water carried the capacities of tile, running full, increase faster than the squares of their inside diameters.

4. It is seldom advisable to use tile smaller than 3 inches in diameter because so little variation above or below a true grade will fill them with sediment.

5. The size of mains must vary with the area they are to drain, with their fall and their length. C. G. Elliott states that where drains are laid 3 feet or more deep, and on a grade not less than 3 inches in 100 feet, a 2-inch main not more than 500 feet long will drain 2 acres.

A three inch tile will drain	....	5 acres.
A four " " " "	....	12 "
A five " " " "	....	20 "
A six " " " "	....	40 "
A seven " " " "	....	60 "

He specifies further that a 2 inch main should not be laid longer than 500 feet and a 3 inch not longer than 1,000 feet.

375. A Practical Illustration of Sizes and Distances Apart of Drains.—The sizes of mains and sub-mains, the sizes of laterals, the lengths of each size used and the distance between drains may be most clearly and briefly stated by citing a practical example. The case selected is an 80 acre field laid out under the direction of C. G. Elliott where the soil is a rich black loam approaching muck in its lowest places and at 2.5 feet underlaid with a yellow clay subsoil. The fall of the main is not less than 2 inches in 100 feet, the laterals being more rather than less. This area is represented in Fig. 120.

The main begins with 1,000 feet of 7 inch tile carrying the water from 80 acres of flat land surrounded by level fields. Next follow 1,200 feet of 6 inch, then 600 feet of 5 inch and closing with 157 feet of 4 inch tile into which no laterals lead. Nothing smaller than 3 inch tile are used for laterals and the least distance between them is about 150 feet.

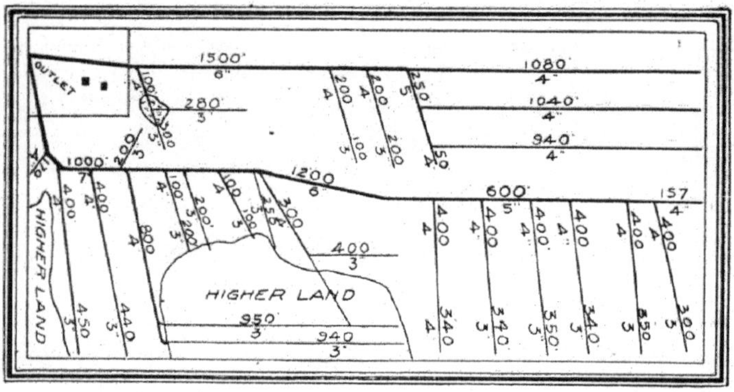

FIG. 120.—Drainage system of 80 acres. Double lines, mains; single lines, laterals. Numbers give length and diameter of tile. (After C. G. Elliott.

376. Outlet of Drains.—Much care should be exercised in selecting the location for, and in placing, the outlet. It should if possible have a free outfall as shown at A, Fig. 121, rather than to end beneath water as at B.

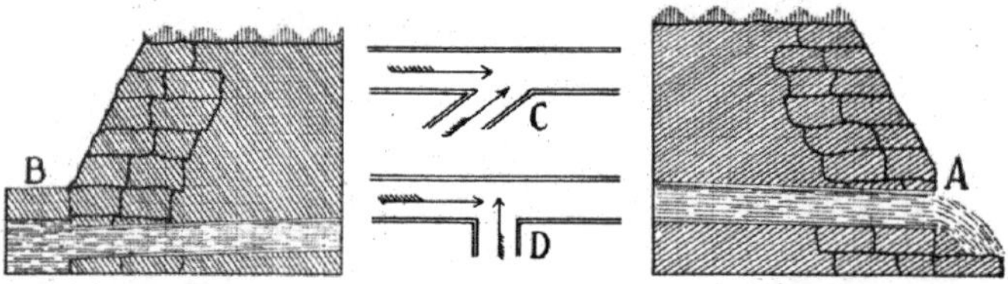

FIG. 121.—A, proper outlet for drain; B, improper outlet; C, proper junction of lateral with main; D, improper junction.

To avoid injury from freezing in cold climates the last 10 to 16 feet of the main should end in glazed sewer tile or in the glazed drain tile; and the outlet should be guarded with masonry and covered with a grating to keep out animals.

377. Connecting Sub-main with Main.—Where a sub-main joins a main the connection should be made at an acute angle as represented at C, Fig. 121, rather than at right angles as at D. If this is not done silt will collect on account of the reduced velocity caused by the meeting of the two streams. It is best in such cases to use the manufactured junction tile.

378. Joining Laterals with Main.—The junction of a lateral should if possible be made above the axis of the main, cutting a hole through the main with a tile pick; this is to avoid the clogging of the lateral. Where the fall is great enough to admit of doing so one of the best unions with a main is represented in Fig. 122, the end of the lateral being thoroughly plugged with a stone bedded in clay, or better with 3 or 4 inches of cement.

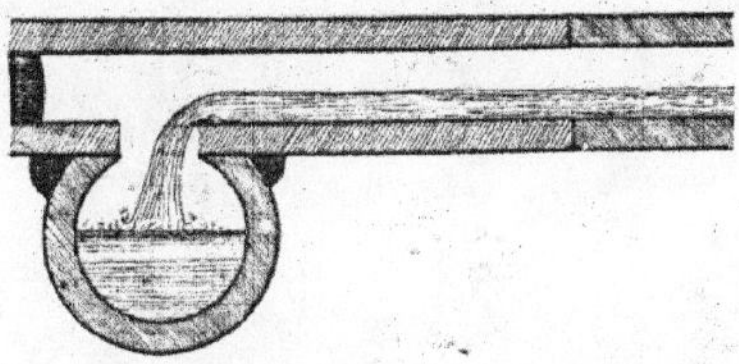

FIG. 122.—Method of connecting lateral with main drain. (After Jul. Kuhn.)

Where, on account of small fall, the lateral must approach the main low down it should be connected in the oblique manner represented in Fig. 121 at C.

379. Obstructions to Drains.—The demand for water by trees is so great that they must not be permitted to grow within 3 or 4 rods of a line of tile which has water running in it during any considerable portion of the growing season. Fig. 123 represents two bunches of European larch roots taken from 6 inch tile which they had completely closed. A small rootlet entered at the joint, where it grew, branched and expanded until its fibrils collected so much silt as to completely close the drain. The willow, poplar, elm, larch and soft maple are among the trees most likely to make trouble in this way.

FIG. 123.—Roots of European larch removed from a 6-inch tile drain, which they had effectually clogged.

380. Laying out Drains.—Careful study should be given to the best manner of laying out a system of drains; the aim being to secure the greatest fall, the least amount of digging, the least outlay for tile and the most perfect drainage. To secure these results drains must be laid so that no two lines are taking the water from the same territory, the outlets must be as few as possible and only as large tile used as are needed to do the work.

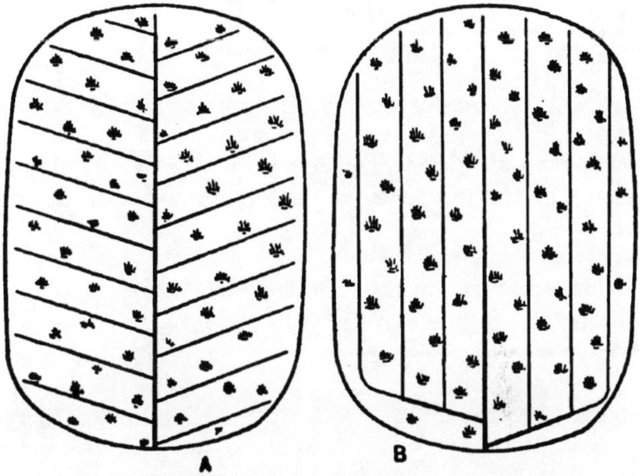

FIG. 124.—Two systems for laying out drains.

In Fig. 124 drains are laid out by two systems for the same area of 14 acres with the lines 100 feet apart. By the system A 625 feet of 4 inch main and 3,020 feet of 3 inch laterals are required; while by the system B only 550 feet of 4 inch and 2,830 feet of 3 inch tile are required to cover the ground so as to secure equal drainage. It will be seen that in the system A the ends of all the laterals traverse 50 feet of territory drained by the main.

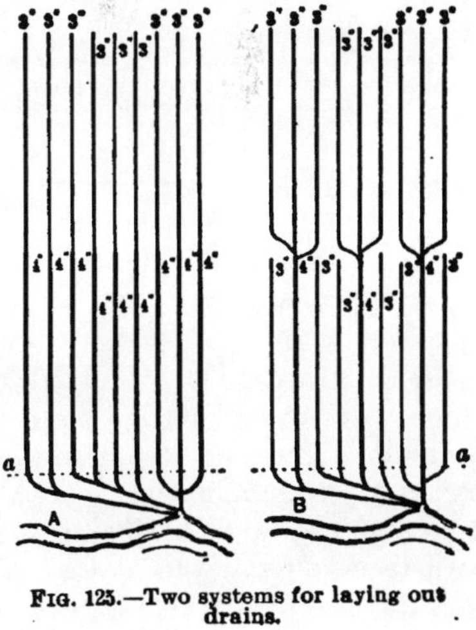

FIG. 125.—Two systems for laying out drains.

When long lines of tile must be laid, requiring more than one size, three systems have been used: 1st, that represented at A, Fig. 124; 2d, that at A, 125 and 3rd, that at B, 125. In the second case, covering an area 2,000 feet by 900 feet, above the line a a, 9,000 feet of 4 inch and

9,000 feet of 3 inch tile are laid 100 feet apart; but following the third system only 3,000 feet of 4 inch and 15,300 feet of 3 inch render the same service with a saving of about $33.00 for tile.

Usually no single system can be followed but the slope and shape of the land will require a combination of two or more.

381. Intercepting Surface Drainage.—In very many cases where drainage is required the necessity is caused by the collection of surface waters from the surrounding higher lands. It may often be possible in such cases to avoid a large part of the expense of underdrainage by intercepting and controlling the surface waters, collecting them into surface drains and leading them away as represented in Fig. 126. In this case the water is collected into a surface ditch before it reaches the low area and is carried around on the higher ground. It is specially important to use this method in cases where low areas are surrounded on all sides by a rim of land high enough to prevent the construction of underdrains.

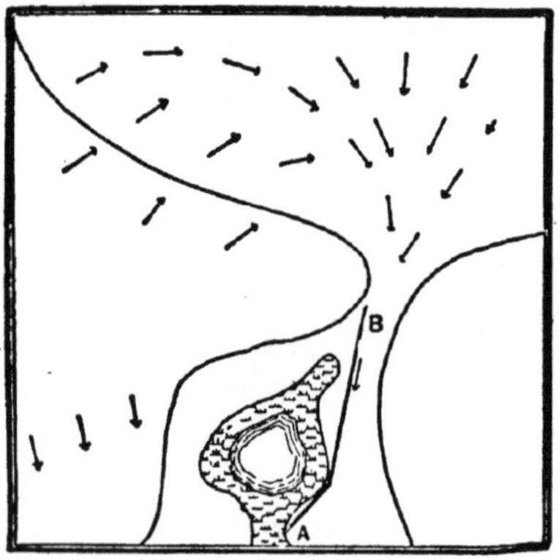

FIG. 126.—Method of intercepting surface drainage. A, B, surface ditch. (From Irrigation and Drainage.)

382. Construction of Surface Drains.—Where surface waters are to be handled as in (381) it can usually best be done by constructing broad and comparatively shallow runways, which can be kept in permanent grass, the width and slope of the ditch being such that a wagon and mower can readily be driven along and across it. Such waterways should usually be 1 to 2 feet deep and 10 to 15 feet wide with sides sloping gently to a flat bottom which can carry a considerable volume of water slowly without being eroded.

383. Intercepting the Underflow from Higher Lands.—In a very large number of cases lands require drainage because of the underflow of water from the adjacent higher land in the manner indicated in Fig. 127. In such cases,

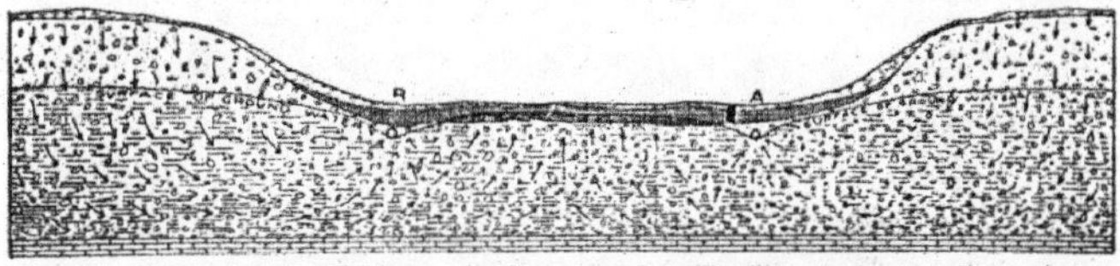

FIG. 127.—Showing how lines of tile may be placed at A and B to intercept the underflow from the higher land.

when drains are laid along the foot of the hill below the ground water surface, as represented at A and B, much of the seepage water will rise into the drain and be conveyed away rather than flow on under the flat land beyond. When such corrections as these are made it may even be unnecessary to underdrain the flat land or when the drains at the foot of the hill do not fully correct the evil the cost is made relatively less.

384. Draining Basins Without Outlets.—There frequently occur sinks or ponds entirely surrounded by rims too high to permit drainage outlets to be constructed across them. Such cases must be met in special ways. 1. Occasionally such basins are underlaid with gravel or sand which is well drained and the water is retained on the surface only by a comparatively thin stratum of clay subsoil. When this is true, one or more wells may be sunk through the clay into the sand or gravel, as represented in Fig. 128, and filled with cobblestone and gravel. Into this underdrains may be led from various directions to collect the water and bring it to the subterranean outlet thus provided.

2. Where several acres must be drained the above method would hardly be practicable even if the underdrainage conditions were favorable. It is possible, how-

ever, to arrange in such a manner that a good windmill will drain a considerable body of land, where only the underflow must be dealt with and the lift is less than 20 feet. One method of draining by wind power is illustrated in Fig. 129 where A is one of a number of closed drains

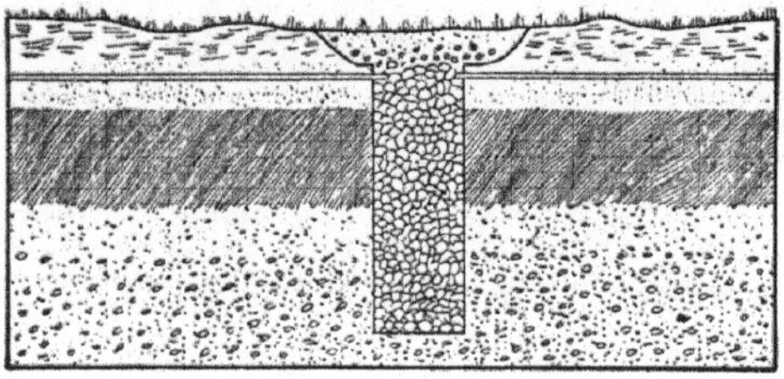

FIG. 128.—Method of draining sinks.

leading to a collecting basin, D, which is connected with the well from which the water is discharged through the pump into the drain C. If the area is small or the capacity of the pump large the water may discharge directly into the well, which may be provided with a float to throw the

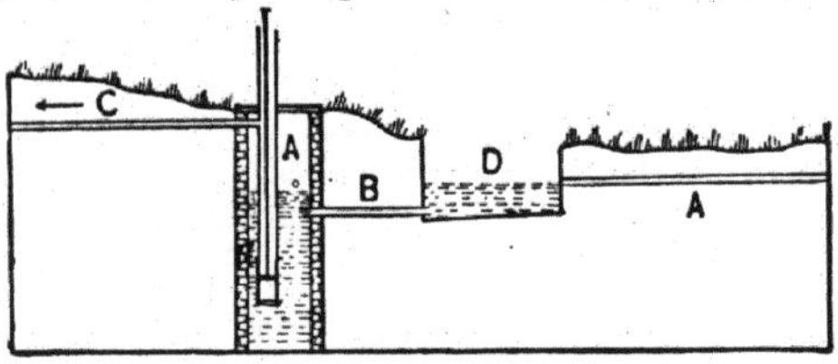

FIG. 129.—Method of draining sinks by wind power. (From Irrigation and Drainage.)

mill out of gear when the water is getting too low for the pump. The object of the well is to permit the mill to work during the winter.

3. In still other cases it may be practicable to lay the sink off into lands separated by broad, open and rather deep ditches, into which the water from the lands could drain and where evaporation would be much more rapid than from the soil. To increase the rate of evaporation of water from the ditches lines of water loving trees, like the willow, could be planted, but these would interfere with

cropping. The better plan would be to utilize the ground with a crop which would endure the shallow drainage.

385. Lands Requiring Surface Drainage.—There are many wide stretches of very flat land which can only be drained through surface channels. Such are the districts which in recent geologic times were lake bottoms, over which a heavy sheet of close textured clay was deposited. Soils like these have subsoils so close that were there plenty of fall and good opportunity to find outlets for drains the rains could not reach the drains freely enough to meet the needs of crops.

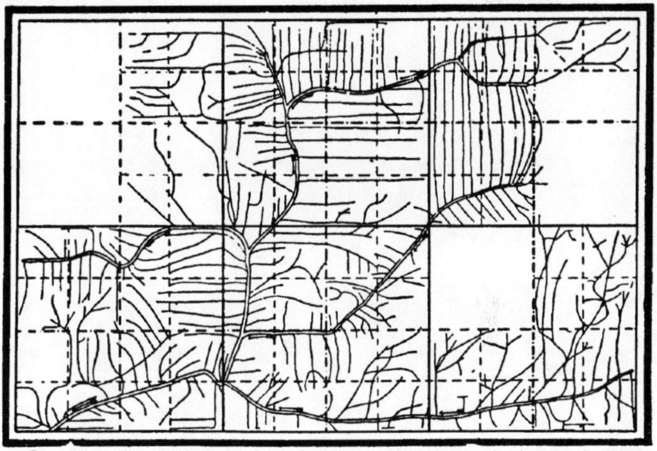

FIG. 130.—Plan for drainage of lands of the Illinois Agricultural Company, Rontoul, Illinois. (After J. O. Baker.) The smallest squares are 40 acres; double lines show open ditches; single lines are tile drains.

Such fields must be plowed in narrow lands with the dead furrows in the direction of greatest fall in order to provide a quick removal of the surplus rains.

Other districts are so flat that the rains have not yet been able to cut sufficiently deep river channels to drain the fields enough for agricultural purposes. The soil may be porous enough, even a coarse sand, and yet for lack of natural drainage channels remain too wet to till.

In such cases deep open ditches must be provided to convey the water out of the country, serving as outlets for underdrains laid in the adjoining fields. A district of this type of land drainage is represented in Fig. 130, covering nearly six square miles. The double lines represent deep open ditches and the single lines underdrains.

Another drainage system of this sort in Mason and Tazwell counties, Ill., has 17.5 miles of main ditch 30 to 60 feet wide at the top and 8 to 11 feet deep. Leading into these mains there are five laterals 30 feet wide and 7 to 9 feet deep, the whole system embracing 70 miles of open ditch for the purpose of providing outlets for underdrains.

CHAPTER XV.

PRACTICE OF UNDERDRAINAGE.

The best work in underdraining can only be done by the man who has a thorough grasp of the principles of the art and who has had enough practical experience to make him perfectly familiar with the essential details as they vary with soil, topography, climate and crop conditions.

There are many cases of local drainage where the area and expense involved are small, where the farmer having a fair knowledge of the principles of drainage can supervise or do his own work, but when large areas are to be underdrained, where the fall is small and the surface conditions complex, it will be safest to entrust the leveling and staking out of the mains and laterals ready for the ditcher to a competent and thoroughly reliable drainage engineer.

Indeed it will generally be best and more economical to let the whole job if it is large and difficult to a man of experience who has established a reputation for reliable work. Even in the matter of digging the ditch, and particularly in giving it its finish, as well as in placing the tile, drainage engineers find it difficult to find men who have the patience, the feeling of responsibility and the practical skill to do it well. A man who has the right frame of mind and the skill to do this finishing and most important work well is much more to be trusted than the farmer himself who has so many duties to distract his attention and tempt him to rush the job.

But while the general farmer should not be encouraged to attempt the draining of large and difficult areas on his own place it is quite important for him to have a clear conception of the general principles of drainage and of what constitutes thoroughly good detail practice.

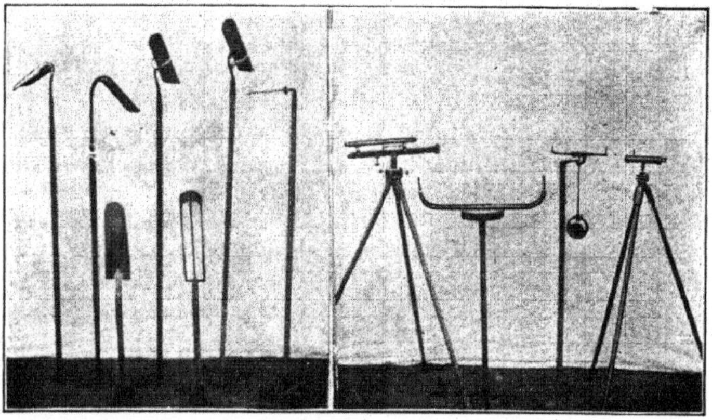

FIG. 131.—Showing forms of drainage tools.

383. Means for Determining Levels. —As a general rule the laying out of a system of drains should only be attempted with good instruments, two of which are represented in Fig. 131. Where a good drainage level cannot be had the best substitute is the water level, one form of which is represented in Fig. 131 and another in Fig. 132; which consists of a piece of gas pipe about 3 feet long mounted on a standard and provided with two elbows into which are cemented two pieces of water gauge glass. When the instrument is filled with water the surfaces in the two tubes stand on a level and can be used to sight across. To move the instrument close the ends of the tubes with corks.

As a substitute for the gas pipe a piece of rubber tubing may be used or a piece of garden hose.

A less reliable level can be improvised by arranging an arm upon a standard upon which a carpenter's level may be set. Or a still more crude level may be made from a carpenter's square mounted on a horizontal arm on which a plumb bob is suspended, with which to set the square with its long arm level.

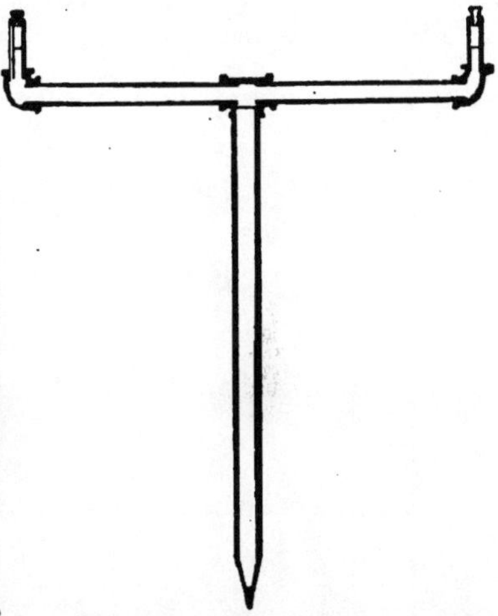

FIG. 132.—Showing one form of water level.

387. Leveling a Field.—In determining the differences of level, in different parts of a field it is desired to drain, the simplest method for the inexperienced person is to lay out the field into squares of 100 or more feet, driving short stakes at the corners:

Set the instrument at a, Fig. 133, midway between the stations I-1 and I-2 and record the reading of the target placed upon the stake at I-1 in the table in the column headed "back-sight" which is assumed for illustration to be 4 feet. Next turn the instrument upon stake I-2, when its distance below the level is found to be 3.8 feet and is entered in the column headed "fore-sight." This shows that the ground at I-2 is

$$4 \text{ ft.} - 3.8 \text{ ft.} = .2 \text{ ft.}$$

higher than station I-1.

In the column headed "Elevation" the first station is given arbitrarily a hight of 10 feet above an assumed datum plane to avoid minus signs. The level is now transferred to b and the distance of I-2 below the instrument found to be 4.2 feet which is entered in the column "back-sight" as before. Turning now upon I-3, its reading is found to be 4 feet and this is entered in the column "fore-sight."

The difference in level between the back sight and fore sight shows the difference in level between the two stations and is placed in the column headed "difference." The first difference added to the datum, 10, gives 10.2, the hight of station I-2 above the datum plane. The second differ-

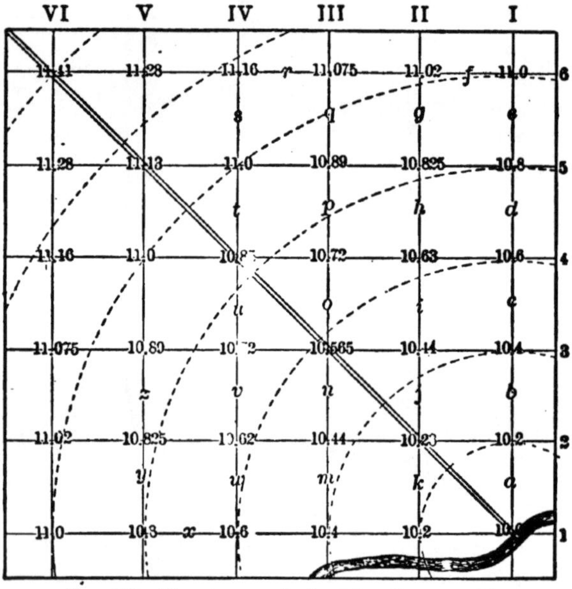

FIG. 133.—Showing method of leveling a field.

ence, .2, added to the elevation of station I-2 gives 10.4, the elevation of station I-3 above datum. In this manner the level is moved from station to station until e is reached when it is transferred to f and back sights and fore sights taken as before, and entered in the table to connect the first line of observations with the new one just begun.

Proceeding as before the level is moved from f to g and then through h, i, j, k and l to m and so on until the field is all completed. When proceeding from higher to lower levels the differences must be subtracted rather than added to obtain the elevation of the lower station. Fig. 134 shows the relation of the level to the target rod along a single line of stations shown in profile.

Table giving data obtained in leveling field of Fig. 133.

Station.	Back-sight.	Fore-sight.	Difference.	Elevation
I-1	4			10
I-2	4.2	3.8	.2	10.2
I-3	3.8	4	.2	10.4
I-4	4	3.6	.2	10.6
I-5	3.9	3.8	.2	10.8
I-6	4	3.7	.2	11
II-6	3.8	3.98	.02	11.02
II-5	3.9	3.995	.195	10.825
II-4	4	4.095	.195	10.63
II-3	4.1	4.19	.19	10.44
II-2	3.9	4.26	.16	10.28
II-1	3.8	3.98	.08	10.2
III-1	4	3.6	.2	10.4
III-2	3.9	3.96	.04	10.44
III-3	4.2	3.775	.125	10.565
III-4	4.1	4.045	.155	10.72
III-5	3.8	3.93	.17	10.89
III-6	4.1	3.625	.185	11.075
IV-6	4	4.185	.085	11.16
IV-5		3.84	.16	11

388. Contour Map of Field.—When the field has been laid out as represented in Fig. 133, and the elevations of the several stations transferred to the map, the figures show at

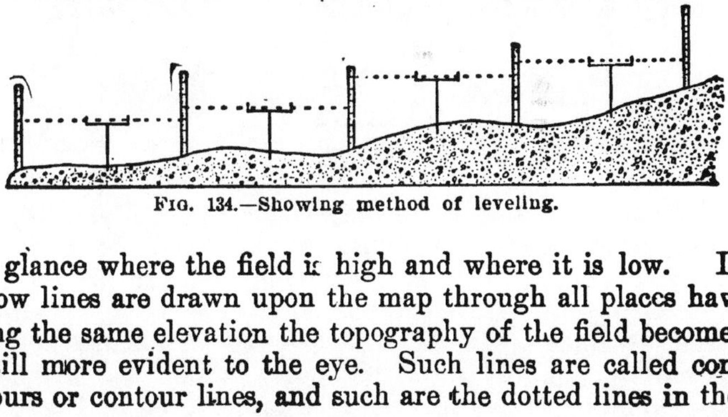

FIG. 134.—Showing method of leveling.

a glance where the field is high and where it is low. If now lines are drawn upon the map through all places having the same elevation the topography of the field becomes still more evident to the eye. Such lines are called contours or contour lines, and such are the dotted lines in the map.

389. Location of Mains and Laterals.—It is clear from the contour map that the highest station in the field is VI—6 and the lowest I-1. If then we are seeking the steepest fall or gradient for the main it will be found along a straight line connecting these two stations. Of course no field will be found with so regular a slope as this but the principle is no less true for being so simply stated.

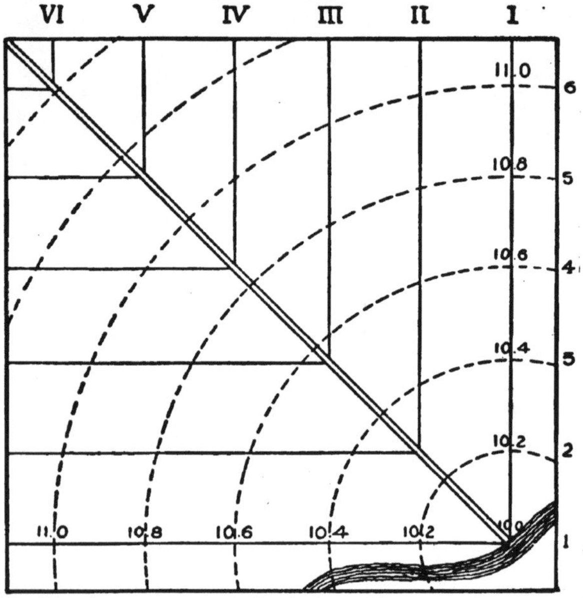

FIG. 135.—Showing a system of tile drains laid out on the leveled field of Fig. 133. (From Irrigation and Drainage.)

If such a field is to be drained by placing laterals 100 feet apart about the maximum fall for them, and the minimum amount of tile and ditching, will be secured by placing the laterals along the lines of leveling, in which case the lines I, II, III, IV, V, VI will constitute the laterals on one side of the main and the lines 1, 2, 3, 4, 5, 6 the laterals on the other side, as represented in Fig. 135. Since the lines I and 1 are both radii of the same circle and have the same elevation at their outer extremities the fall or gradient will be the same or .2 of a foot per 100 feet, as shown on the contour map, but along the lines V and 5 the gradient will be .15 feet per 100 feet or 1.8 inches instead of 2.4 inches per 100 feet along the lines I and 1. The fall is therefore not uniform for all the laterals nor can it be when they are placed along parallel lines.

If the field required drains every 50 feet then a greater mean fall could be secured and less tile would be required if a system like that of Fig. 136 were adopted.

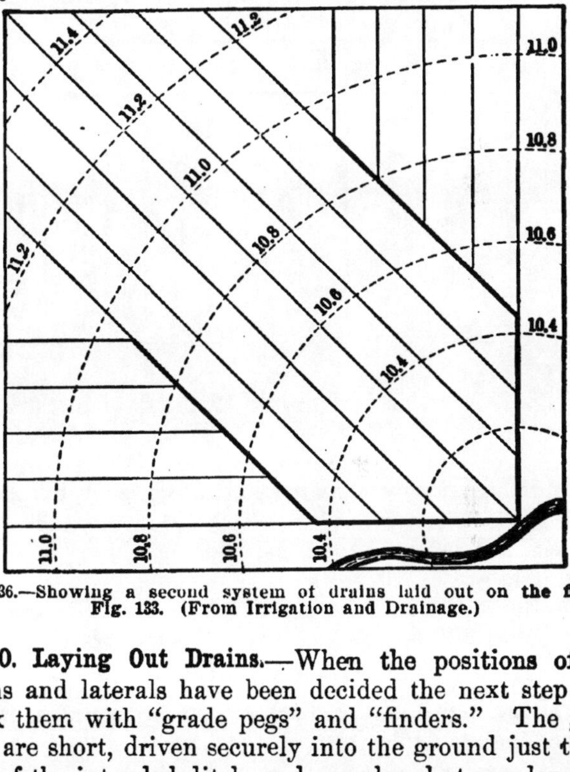

FIG. 136.—Showing a second system of drains laid out on the field of Fig. 133. (From Irrigation and Drainage.)

390. Laying Out Drains.—When the positions of the mains and laterals have been decided the next step is to mark them with "grade pegs" and "finders." The grade pegs are short, driven securely into the ground just to one side of the intended ditch, and are placed at regular intervals apart. To one side of the grade pegs are placed longer ones called "finders" upon which is to be recorded the depth below the grade peg the ditch is to be dug.

391. Determining the Grade and Depth of the Ditch.—In doing this work the leveling begins at the outlet and the steps are the same as those already described for the field leveling, the results being recorded in a table calling for two more columns when worked out than were needed in the field work. These are indicated in the table below:

Table showing Field Notes for determining depth of ditch and grade of drain.

Station	Back-sight	Fore-sight.	Difference.	Elevations	Grade line	Depth of ditch.
Outlet	7	—	—	7	7	0
0	4	—	3	10	7	3
50	3.97	3.87	.13	10.13	7.12	3.01
100	4.2	3.83	.14	10.27	7.24	3.03
150	4.1	4.08	.12	10.39	7.36	3,03
200	3.95	3.99	.11	10.5	7.48	3.02
250	3.87	3.82	.13	10 63	7.6	3.02
300	4	3.69	.18	10.81	7.72	3.09
350	4.25	3.83	.17	10.98	7.84	3.14
400	4.08	4.1	.15	11.13	7.96	3.17
450	4.05	3.96	.12	11.25	8.08	3.17
500	3.97	3 95	.1	11.35	8.2	3.15
550	3.75	3 97	—	11.35	8.32	3.03
600	—	3.74	0.1	11.36	8.14	2.93

In Fig. 137, which is a profile of the data in the table showing the outlet of the drain at A, the first stake at O and the second at 50, etc., up to 600, both the lines of grade and the datum plane are shown. On each numbered stake is given the depth of the ditch below the top of the grade peg, and below the peg has been set the hight of the bottom of the ditch above the datum plane.

Since the outlet in this case is 7 feet above datum and the surface at 600 feet is 11.36 feet the total fall is

$$11.36 \text{ feet} - 7 \text{ feet} = 4.36.$$

But if the depth of the ditch at the upper end is made 2.92 feet the available fall will then be

$$4.36 \text{ feet} - 2.92 \text{ feet} = 1.44.$$

Since the ditch is 12 times 50 feet long the fall will be

$$\frac{1.44}{12} = .12 \text{ feet per 50 feet.}$$

or .24 feet per 100 feet. At each 50 foot station then the bottom of the ditch above datum plane will be found by adding .12 foot, to 7 feet, which is the height of the outlet, for that of the second station; then .12 feet added to this gives the third station and so on, thus:

7, 7.12, 7.24, 7.36, 7.48, 7.60, 7.72, 7.84, 7.96, 8.08, 8.20, 8.32, 8.44.

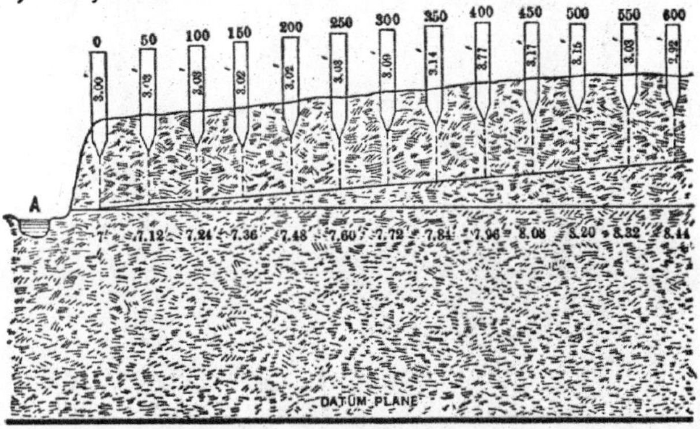

FIG. 137.—Profile of ditch staked ready for digging, with depths for the ditch at the several stations.

If these numbers are subtracted from the hights of the surface of the ground at the respective places the difference will be the depth the ditch must be dug at those places, and the figures which are placed upon the finders for the instruction of the men in digging. These figures are given in the table in the column "depth of ditch."

The experienced drainage engineer with accurate telescope level makes the details of leveling, establishing the grade and marking the grade pegs simpler than here given but it is not safe for a farmer with a cheap level to follow his methods.

392. Changing from One Grade to Another.—It may happen in laying out the ditch that it is impracticable to follow a single grade on account of having to dig too deep in some places or of leaving the tile too close to the surface in others. Suppose in the last profile (391) the ditch was to be 500 feet longer and that in this 500 feet there had been

FIG. 138.—Showing tile distributed in the field ready for use.

a rise of but 6 inches. It is clear that to hold a single grade, making the upper end of the ditch 2.92 feet deep, would require a greater depth in other portions than necessary. But if the grade is changed at the 600 foot station so as to give a fall of

$$\frac{.5 \text{ ft.}}{5} = .1 \text{ ft. per } 100 \text{ ft.}$$

a sufficient depth will be secured and labor in digging saved.

FIG. 139.—Showing the ditching line and the commencement of digging.

393. Ditching Tools.—In digging a ditch it is a matter of first importance to have suitable tools; and whatever else is chosen the men should be provided with first class spades, kept sharp and free from rust. The spade which gives the best satisfaction has a long, thin, narrow and curved blade. The curvature is of first importance in giving greater stiffness and allowing the blade to be made thinner and lighter. The spade should be narrow and thin to enable the user to force it full length into the soil with the pressure of the foot and so as to be able to leave the bottom of the ditch narrow, removing as little earth as possible.

In Fig. 131 are shown two forms of spades, four tile hoes, which are used in finishing the bottom of the ditch and removing the loose earth, and a tile hook, used in placing the tile. The series of half tones shows these different tools in use.

394. Making the Ditch Narrow and Straight.—To make the ditch straight a strong light line is stretched taut near the surface and 4 inches back from the edge. If the ditch is to be only 2.5 to 3 feet deep it need be no wider at the top than one foot, as shown by the length of tile in Fig. 139. Where the ditch must be 4.5 to 5 feet and receive a 6 inch tile, as shown in Fig. 141, it must have a width at the top of 15 to 18 inches.

The ditcher is trained to cut the walls straight with an even slope to the bottom so as to leave a straight line along the bottom to receive the tile. In Fig. 140 it will be seen that four men are working in line to complete the depth of the ditch which is 4.5 feet at the place.

395. Shaping the Bottom and Bringing It to Grade.—In Fig. 141 the man in the foreground is using the tile hoe to clean out the last loose earth and to bring the bottom to grade and proper shape to receive the tile. The grade is secured by stretching the ditcher's line tight, and on the slant the bottom of the ditch is to be given, and at a known hight above it. It is then only necessary for the experienced man to use a measuring rod to secure the depth and grade desired.

FIG. 140.—Showing four men in line digging a main ditch.

When the requisite skill and judgment have not been acquired for this work the man is provided with a measuring stick with a sliding arm which extends at right angles to the rod and long enough to reach the grade line. It is then only necessary to hold the rod or "ditcher's square" plumb to know whether the ditch has the depth desired.

396. Placing the Tile.—When the ditch has been finished the tile are laid with the tile hook, as represented in Fig. 142. With the aid of this tool they are placed rapidly and accurately without getting into the ditch. Great care should always be taken to turn and shift the tile until a perfectly close joint is made all around. It does not do to simply have them meet on the upper edge, they should fit squarely and closely through the entire circumference and if necessary tile too much warped to permit of this must be discarded.

Some prefer to place the tile with the hand, standing in the ditch upon them, covering them as rapidly as laid with 4 to 6 inches of earth, taking care to get it thoroughly packed and not to get the tile out of alignment.

The greatest care should be exercised to pack the earth thoroughly about the joints so as to avoid large open cavities through which the water may rush during heavy rains, washing dirt into the tile.

Tile laying should begin at the outlet of the main, proceeding upward to the first lateral, where the junction should be made and tile enough laid in the lateral to permit the main to be partly filled. The main may then be carried on until the next lateral is reached, when this should be commenced as before. Care should be exercised not to leave the upper end of an unfinished line of tile open for heavy rains to wash mud into it. If the line cannot be finished before the rain the end may be guarded by closing it with a board, brick or bunch of grass.

FIG. 141.—Showing method of cleaning the ditch and finishing the bottom, with a tile hoe, ready for the tile.

Fig. 142.—Laying the tile from the top of the ditch, using the tile hook.

Fig. 143.—Showing method of filling the ditch, after the tile have been covered.

397. Filling the Ditch.—After the tile have been placed and covered with the first layer of earth the balance may be put in by any convenient method. A common and expeditious way is represented in Fig. 143 where a plow is drawn by a team attached to a long evener. For the finishing the ordinary road grader makes an efficient tool.

Still another method is to use a light board scraper provided with handles to be held against the bank of earth, which is drawn into the ditch by a team on the opposite side drawing from a rope and backing when the scraper is emptied.

PRINCIPLES OF RURAL ARCHITECTURE.

CHAPTER XVI.

STRENGTH OF MATERIALS.

A knowledge of the principles governing the strength of materials is helpful along many lines of farm practice and particularly in the construction of farm buildings.

398. A Stress.—When a post is placed upon a foundation and a load of two thousand pounds set upon it the post is undergoing or opposing a stress of two thousand pounds. When a rope is supporting a load of one thousand pounds in a condition of rest it is subject to a *stress* of one thousand pounds. The joists under a mow of hay are subjected to a *stress* measured by the tons of hay which they carry.

399. Kinds of Stress.—Solid bodies may be subjected to three kinds of stress which tend to break them and will do so if the stress is great enough. These are:

1. A crushing stress, where the load tends to crowd the molecules closer together, as when kernels of corn are crushed between the teeth of an animal.
2. A stretching stress, as where a cord is broken by a load hung upon it.
3. A twisting stress, as where a screw is broken by trying to force it into hard wood with a screw-driver.

400. Strength of Moderately Seasoned White and Yellow Pine Pillars.—Mr. Chas. Shaler Smith has deduced, from experiments conducted by himself, the following rule for strength of moderately seasoned white and yellow pine pillars:

Rule.—*Divide the square of the length in inches by the square of the least thickness in inches; multiply the quotient by .004 and to this product add 1; then divide 5,000 by this sum and the result is the strength in pounds per square inch of area of the end of the post. Multiply this result by the area of the end of the post in inches, and the answer is the strength of the post in pounds.*

In applying this rule in the construction of farm buildings the timbers should not be trusted with more than one-fourth to one-sixth of the theoretical load they are computed to carry, because the theoretical results are based upon averages, and there is a wide variation in the strength of individual pieces.

Table of breaking load in tons, of rectangular pillars of half seasoned white or yellow pine firmly fixed and equally loaded, computed from C. S. Smith's formula.

Length in feet.	Dimensions of rectangular pine pillars in inches.													
	4x4	4x6	4x8	4x10	4x12	6x6	6x8	6x10	6x12	8x8	8x10	8x12	10x10	10x12
	tons	tons	tons	tons	tons	tons	tons	tons	tons	tons	tons	tons	tons	tons
8....	12 1	18.1	24.2	30.2	36 3	44.5	59.3	74.1	88.9	101.7	126 9	152.3	182.7	219.2
10....	8.7	13.0	17.4	21.7	26.1	34.6	46.2	57.7	69 2	84.2	105.3	126.3	158.6	190.3
12....	6.5	9.7	12.9	16.1	19 4	27.2	36.3	45.4	54 4	69.7	87.1	104.5	136.7	164.0
14....	5.0	7.4	9.9	12.4	14.9	21.7	29.0	36 2	43.5	57.9	72 3	86 8	117 4	140.9
16....	3.9	5.9	7.8	9.8	11.7	17 7	23.5	29.4	35.3	48.4	60.6	72.7	101.0	121.2
18....	.....	.....	.....	.....	.....	14.6	19.4	24.3	29.1	40.8	51.0	61.2	87.2	102 6
20....	.....	.....	.....	.....	.....	12 2	16.2	20 3	24.3	34.8	43.4	52.1	75.7	90.8
22....	.....	.....	.....	.....	.....	10.3	13.7	17 2	20 6	29.9	37.4	44.8	65.8	79.0
24....	.....	.....	.....	.....	.....	8.8	11.7	14 7	17.6	25.9	32.3	38.8	57.9	69.4

In the application of the rule for the crushing load for posts in barn building the length referred to is the greatest distance between any supports which prevent the post from bending.

401. Bearings for Posts.—In order that a post may carry its maximum load it is important that it rests squarely upon its support and that the load carried presses squarely upon the post. If the ends of the post are not square or if the bearing is out of true so that the strain comes upon one edge the carrying power is greatly lessened.

402. Tensile or Stretching Strength of Timber.—The tensile strength of materials is measured by the least weight which will break a vertical rod one inch square firmly and squarely fixed at its upper end, the load hanging from the lower end. Below are given the results of experiments with different varieties of wood, but the strengths vary greatly with the age of the trees, with the part of the tree from which the piece comes, the degree of seasoning, etc.

Elm	6,000 lbs. per square inch.
American hickory	11,000 lbs. per square inch.
Maple	10,000 lbs. per square inch.
Oak, white and red	10,000 lbs. per square inch.
Poplar	7,000 lbs. per square inch.
White pine	10,000 lbs. per square inch.

403. Tensile or Cohesive Strength of Other Materials.—

American cast iron	16,000 to 28,000 lbs. per sq. inch.
Wrought iron wire, annealed	30,000 to 60,000 lbs. per sq. inch.
Wrought iron wire, hard	50,000 to 100,000 lbs. per sq. inch.
Wrought iron wire ropes, per sq. in. of rope	38,000 lbs. per sq. inch.
Leather belts, 1,500 to 5,000, good	3,000 lbs. per sq. inch.
Rope, manila, best	12,000 lbs. per sq. inch.
Rope, hemp, best	15,000 lbs. per sq. inch.

404. Transverse Strength of Materials.—When a board is placed upon edge and fixed at one end as represented at A, Fig. 144, a load acting at W puts the upper edge under a stretching stress.

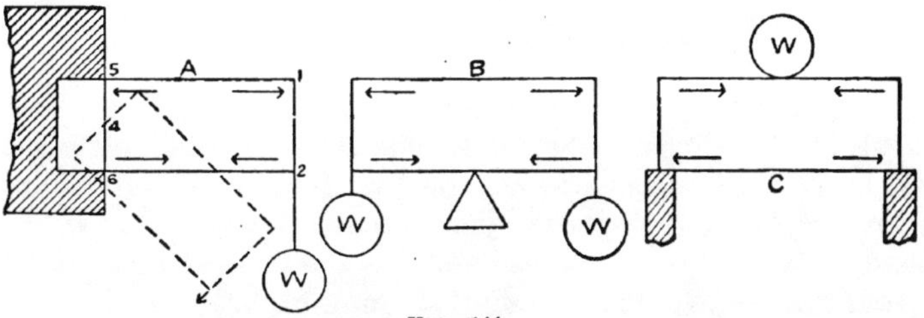

FIG. 144.

We know from experience that in case the board breaks under its load when so situated the fracture will occur somewhere near 5–6. Now in order that this may take place there must be, with white pine, according to **(402)** a tensile stress at the upper edge of ten thousand pounds to the square inch, and if the board is one inch thick the upper inch should resist a stress of 10,000 pounds at any point from 5 to 1; but we know that no such load will be carried at W. The reason for this, and also for its breaking at 5 rather than at any other point, is found in the fact that the load acts upon a lever arm whose length is the distance from the point of attachment of the load to the breaking point, wherever that may be, and this being true the greatest stress comes necessarily at 5.

If the board in question is 48 inches long and 6 inches wide, it will, in breaking, tend to revolve about the center of the line, 5–6, and the upper three inches will be put under the longitudinal strain but, according to **(402)**, is capable of withstanding

$$3 \times 10{,}000 \text{ lbs.} = 30{,}000 \text{ lbs.}$$

without breaking; but in carrying the load at the end as shown, this cohesive power is acting at the short end of a bent lever whose mean length of power arm is one-half of 4–5 or 1.5 inches, while the weight arm is forty-eight inches in length. It should therefore only be able to hold at W 937.5 pounds, for

$$\text{as } P \times PA = W \times WA,$$
$$\text{we have } 30{,}000 \times 1.5 = W \times 48.$$
$$\text{whence } W = \frac{45{,}000}{48} = 937.5 \text{ lbs.}$$

When a board, in every respect like the one in A, Fig. 144, is placed under the conditions represented in either B or C, Fig. 144, it should require just four times the load to break it, because the board is practically converted into two levers whose power-arms remain the same, but whose weight-arms are only one-half as long each.

405. The Transverse Strength of Timbers Proportional to the Squares of their Vertical Thicknesses.—Common experience demonstrates that a joist resting on edge is able to carry a much greater load than when lying flat-wise. If we place a 2x4 and a 2x8, which differ only in thickness, on edge their relative strengths are to each other as the squares of 4 and 8, or as 16 to 64. That is the 2x8, containing only twice the amount of lumber as the 2x4 will, under the conditions named, sustain four times the load. The reason for this is as follows: In Fig. 145 let A represent a 2x4 and B a 2x8. In each of these cases the load draws lengthwise upon the upper half of the joist, acting through a weight-

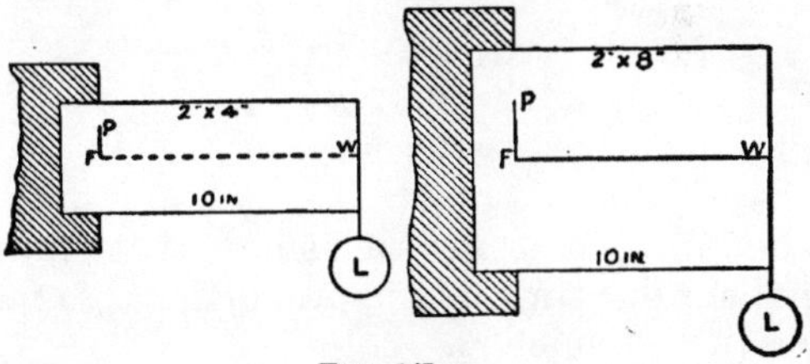

FIG. 145.

arm F, W, ten inches in length, to overcome the force of cohesion at the fixed ends, whose strength, according to **(402)** is ten thousand pounds per square inch, or a total of

$$2 \times 2 \times 10,000 \text{ lbs.} = 40,000 \text{ lbs. in the } 2 \times 4 \text{ joist,}$$
$$\text{and of } 2 \times 4 \times 10,000 \text{ lbs.} = 80,000 \text{ lbs. in the } 2 \times 8 \text{ joist.}$$

These two total strengths become powers acting through their respective power-arms F, P, whose mean lengths are, in the 2x4 joist, one inch, and in the 2x8 joist, two inches.

Now we have (**531**)

$$P \times P\,A = W \times W\,A,$$

and substituting the numerical values, in the 2x4 joist, we get

$$4 \times 10,000 \times 1 = W \times 10$$
$$\text{or } 4 \times 10,000 = 10\,W,$$
$$\text{and } W = 4,000.$$

Similarly, by substituting numerical values in the case of the 2x8 joist we get

$$8 \times 10{,}000 \times 2 = W \times 10,$$
$$\text{or } 16 \times 10{,}000 = 10\,W,$$
$$\text{and } W = 16{,}000.$$

It thus appears that the loads the two joists will carry are to each other as 4,000 is to 16,000, or as 1 is to 4; but squaring the vertical thickness of the two joists in question we get, for the 2x4 joist

$$4 \times 4 = 16,$$
and for the 2×8 joist
$$8 \times 8 = 64;$$

but 16 is to 64 as 1 is to 4, which shows that the transverse strengths of similar timbers are proportional to the squares of their vertical diameters.

406. The Transverse Strength of Materials Diminishes Directly as the Length Increases.—It will be readily seen from an inspection of Fig. 145, that lengthening the pieces of joists, while the other dimensions remain the same, lengthens the long arm of the lever, while the short arm remains unchanged; and since the force of cohesion remains unaltered, the load necessary to overcome it must be less in proportion as the lever arm upon which it acts is increased. Thus, if the 2x8 in Fig. 145 is made 20 inches long, we shall have,

$$P \times PA = W \times WA$$

and by substituting the numerical values we get

$$80{,}000 \times 2 = W \times 20$$
$$W = 8{,}000$$

instead of 16,000, as found in **(405)**.

407. The Constants of the Transverse Breaking Strength of Wood.—Since the laws given in **404, 405,** and **406** apply to all kinds of materials, it follows that the actual breaking strength of different kinds of materials will depend upon the cohesive power of the molecules as well as upon the form and dimensions of the body which they constitute. The breaking strength of a beam of any material is always in proportion to its breadth, multiplied by the square of its depth, divided by its length, or

$$\frac{\text{Breadth} \times \text{the square of the depth}}{\text{length}}$$

and if the breadth of a piece of white pine in inches is 4, its depth in inches 10, and its length in feet 10, we shall have, taking the length in feet,

$$\frac{4 \times 10 \times 10}{10} = 40.$$

Now if we find by actual trial, by gradually adding weights to the center of such a beam, that it breaks at 18,000 pounds, including half its own weight, the ratio between this and forty will be

$$\frac{18,000}{40} = 450,$$

and as this ratio is always found for white pine, when the breadth and depth are taken in inches and the length in feet, no matter what the dimensions of the timbers may be, 450 is called its breaking constant for a center load.

For other materials this constant is different, and has been determined by experiment and given in tables in various works relating to such subjects. The following are taken from Trautwine.

408. Breaking Constants of Transverse Strength of Different Materials.—

WOODS.

American White Ash	650 lbs.
Black Ash	600 lbs.
American Yellow Birch	850 lbs.
American Hickory and Bitter nut	800 lbs.
Larch and Tamarack	400 lbs.
Soft Maple	750 lbs.
American White Pine	450 lbs.
American Yellow Pine	500 lbs.
Poplar	550 lbs.
American White Oak	600 lbs.
American Red Oak	810? lbs.

METALS.

Cast iron	1,500 to 2,700 lbs.
Wrought iron, bends at	1,900 to 2,600 lbs.
Brass	850 lbs.

409. To Find the Quiescent Center Breaking Load of Materials having Rectangular Cross-sections, when Placed Horizontally and Supported at Both Ends.—In placing joists and beams in barns it is important to know the breaking load of the timbers used. This may be determined with the aid of the following rule and the table of constants given in **(408)**:

Rule.—*Multiply the square of the depth in inches by the breadth in inches and this by the breaking constant given in* (**408**); *divide the result by the clear length in feet and the result is the load in pounds.*

But in the case of long heavy timbers and iron beams one-half of the clear weight of the beam must be deducted because they must always carry their own weight.

$$\text{Breaking load} = \frac{\left.\begin{matrix}\text{Square of}\\ \text{depth}\\ \text{in inches}\end{matrix}\right\} \times \text{breadth in inches} \times \text{Constant}}{\text{Length in feet.}}$$

What is the center breaking load of a white pine 2x12 joist 12 feet long?

$$\text{Breaking load} = \frac{12 \times 12 \times 2 \times 450}{12} = 10,800 \text{ lbs.}$$

What is the breaking load for the same 10 feet long? 14 feet long? 16 feet long? 18 feet long? Solve the same problems for other woods.

410. General Statements regarding the Quiescent Breaking loads of Uniform Horizontal Beams.—If the center quiescent breaking load be taken as 1, then, when all dimensions are the same, to find the breaking load:

(1) When the beam is fixed at both ends and evenly loaded throughout its whole length, multiply the result found by **(409)** by two.

(2) When fixed at only one end and loaded at the other, divide the result obtained by **(409)** by four.

(3) When fixed only at one end and the load evenly distributed divide the result obtained by **(409)** by two.

(4) To find the breaking load of a cylindrical beam, first find the breaking load of a square beam having a thickness equal to the diameter of the log and multiply the result by the decimal .589.

411. Breaking Load of Rafters.—In finding the breaking load of timbers placed in an oblique position, as shown in Fig. 146, take the length of the rafter equal to the horizontal span A, C, and proceed as in **(409)** and **(410)**.

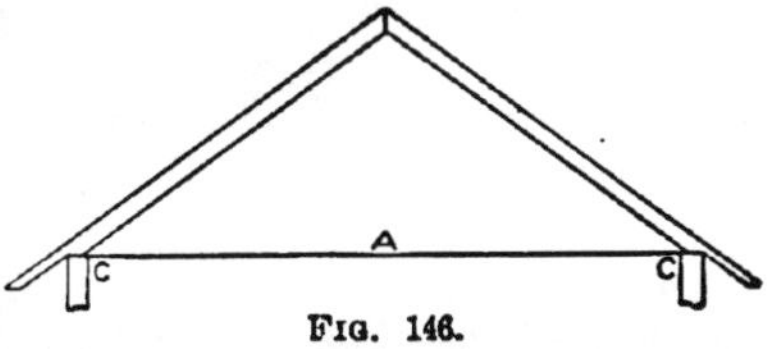

FIG. 146.

412. Table of Safe Quiescent Center Loads for Horizontal Beams of White Pine Supported at Both Ends.—In this table the safe load is taken at one-sixth of the theoretical breaking load. This large reduction is made necessary on account of the cross-grain of timbers and joists and the large knots which weaken very materially the pieces. Where a judicious selection is made in placing the joists, laying the inherently weak pieces in places where little strain can come upon them, much saving of lumber may be made.

Depth in inches.	Span 10 feet. Breadth.			Span 12 feet. Breadth.			Span 14 feet. Breadth.			Span 16 feet. Breadth.		
	2 in.	4 in.	6 in.	2 in.	4 in.	6 in.	2 in.	4 in.	6 in.	2 in.	4 in.	6 in.
	lbs.	lbs.	lbs.	lbs.	lbs.	lbs.	lbs.	lbs.	lbs.	lbs.	lbs.	lbs.
4....	240	4[illegible]0	720	200	400	600	17	344	516	150	300	450
6....	540	1,0[illegible]0	1,620	450	900	1,350	386	772	1,158	3[illegible]6	672	1,008
8....	960	1,920	2,880	800	1,600	2,400	6[illegible]0	1,372	2,058	600	1,200	1,800
10...	1,500	3,000	4,500	1,250	2,50[illegible]	3,750	1,072	2,144	3,216	9[illegible]6	1,872	2,808
12....	2,160	4,3[illegible]0	6,480	1,800	3,600	5,400	1,544	3,088	4,63[illegible]	1,350	2,700	4,0[illegible]0

	Breadth.			Breadth.			Breadth.			Breadth.		
	4 in.	10 in.	12 in.	8 in.	10 in.	12 in.	8 in.	10 in.	12 in.	8 in.	10 in.	12 in.
	lbs.	lbs.	lbs.	lbs.	lbs.	lbs.	lbs.	lbs	lbs.	lbs.	lbs.	lbs.
4....	960	1,200	1,440	800	1,000	1,200	688	860	1,032	600	750	900
6....	2,160	2,700	3,240	1,800	2,250	2,700	1,544	1,930	2,316	1,344	1,680	2,000
8....	3,840	4,800	5,760	3,200	4,000	4,800	2,744	3,430	4,116	2,400	3,000	3,610
10....	6,0[illegible]0	7,500	9,000	5,000	6,250	7,500	4,288	5,360	6,432	3,714	4,680	5,601
12....	8,64[illegible]	10,800	12,960	7,200	9,000	10,800	6,176	7,720	9,264	5,400	6,750	8,166

413. Selection of Lumber to Increase Carrying Capacity.—It is possible to greatly increase the carrying capacity of a lot of joists or of a set of beams by giving attention to the lumber used, selecting the evidently strongest pieces for use where it is known the heaviest strains will come. Sometimes a joist should be reversed or turned the other side up in order to enable the piece to render its highest service. In the arrangement of joists under a hay bay or granary, where heavy loads are to be carried, the cross-grained pieces and those with exceptionally large knots should be well distributed among the stronger ones, making the evidently weak come between those evidently above the average in strength.

414. Braces.—There are two principles underlying the use of braces to give greater strength to lumber. (1) That of equalizing the load, making it fall more heavily upon the stronger members. (2) That of shortening the free span.

The first case is illustrated in the rows of bridging used between the joists in a floor. In these cases when a weak member is bridged between two stronger ones a portion of its load, because it yields soonest, is thrown by the bridging upon the stronger, and stiffer floors are thus secured and the breaking of individual pieces prevented.

Braces in nearly all cases are, in principle, either posts or else they are suspension rods which allow the strength of the material to be utilized unaffected by the principle of leverage, the stress being a direct pull or a push, bringing into play the full tensile or crushing strength of the material.

To shorten the free span of an 18-foot joist or timber two feet at each end by means of suitable braces is increasing its carrying power 28.5 per cent.

It is much more important to pay strict attention to these matters of strength at the present time than in former years both because lumber is higher and often of much inferior quality.

415. Constructing Timbers from Two-inch Lumber.—It is often not only cheaper but better to construct 8x10 or 8x12 beams by putting together 2x10 or 2x12 plank, the timber thus constructed often being stronger than a solid one would be because weak places are more likely to be distributed so as to give a greater mean strength. It is of course not true that a 10x10 so made would be stronger than a solid timber of the same dimensions if both were of highest grade lumber.

416. Form of Barn Frame.—During pioneer days, when saw mills were none or few, it was much easier to secure the needed stability for a barn by hewing a few heavy timbers of suitable length and putting them together with braces than it was to use the 2 inch lumber now so common in the frames of dwelling houses.

Since the old type of barn frame was depended upon to

give the needed stability, little or no support coming from the siding or sheeting, it was necessary to use large timbers

FIG. 147.

and to frame them together and brace them very securely making a structure costly both in material and labor.

417. Plank Frame.—The high price of lumber has led to an effort to imitate the construction of the old hewn timber frame barn in the construction of essentially the same type of frame but using plank spiked together instead of timbers. This type of frame is represented in Fig. 147.

The frame so made is strong and not as expensive as one of heavy timbers at the present prices but it is neither as simple in construction nor as cheap as a frame for most barns can be made. Now that the conditions which made the heavy timber frame a necessity have disappeared there is no need of imitating it by splicing lumber.

418. Balloon or House Frame.—The reason for not adhering to the old type of barn frame is because it permits of no advantage being taken of the inherent strength of the siding and sheeting to give the barn its needed ability to withstand wind pressure.

When the two inch lumber used in the plank frame is treated as studding and the siding and sheeting are put on horizontally, and securely nailed, the whole covering of the barn then braces it from all sides and does double duty by largely dispensing with braces. To distribute the plank, using them as studding rather than building them into timbers forming bents, does not give them less power to withstand pressure from within or without and much less lumber, less nails and less labor are required.

Where the building is long and broad so as to require the sides to be tied, bents may be used and made in the ordinary way except that less lumber need be used at the walls.

419. The Round Barn Frame.—The strongest possible structure for a barn, with the least amount of lumber in its frame and the least special attention to bracing, is secured

FIG. 148.—Showing frame and general plan for a cylindrical barn. A, barn floor extending around the silo; B, hay bay; C, granary; and T, tool room.

when the barn is made cylindrical in form and the studding set upon the circumference of a circle as represented in Figs 148 and 149. In this type of barn not only is the smallest number of studding required to form the outer

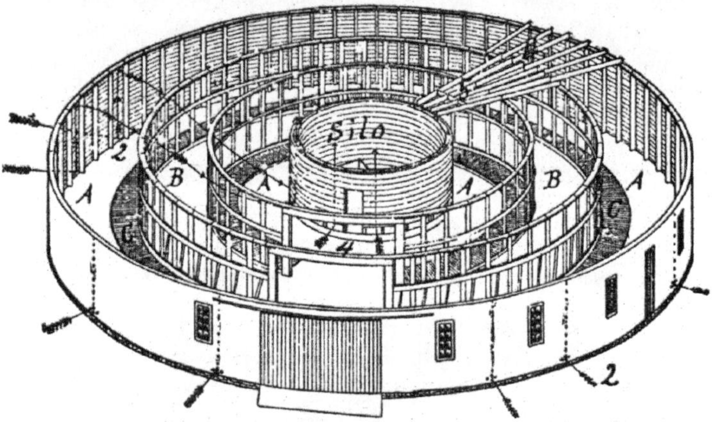

FIG. 149.—Showing frame and general plan of a cylindrical barn. A, driveways behind cattle; B, feed alley; C, platforms for cattle.

part of the frame but smaller sizes can be used, for the reason that every board in the siding is a portion of a hoop which makes spreading impossible, while at the same time they are arched against the wind and take its pressure with a crushing stress.

With barns of this type 2x4 studding set 2 feet apart have ample strength for all diameters up to 40 feet and 2x6 studding is large enough for barns 40 to 100 feet in diameter.

CHAPTER XVII.

WARMTH, LIGHT AND VENTILATION.

CONTROL OF TEMPERATURE.

The life activities manifested in the animal body involve the continuous maintenance of a train of chemical changes which give rise to or maintain them. These chemical changes, like all others, can only begin at a certain temperature; below this they cease; within a certain range they go forward at normal rates; above this temperature reactions occur which interfere with the life activities, making them abnormal or causing them to cease.

420. Automatic Control of Temperature.—The animal body is so constituted that within certain limits the normal temperature of the body may be maintained automatically, if only sufficient food is supplied. If outside conditions are such as to lower the temperature of the body the nervous system reacts, setting in operation a train of changes which evolve heat fast enough to meet the greater loss. If on the other hand the surrounding temperatures are too high and the body is becoming too warm the heat producing reactions are inhibited or perspiration is stimulated to reduce the too high temperature by bringing the blood to the skin, where the temperature may be lowered by the evaporation of water in the same manner that the wet bulb of a thermometer is cooled by the loss of heat which does the work of evaporation.

421. Normal Animal Temperatures.—The normal temperatures which must be maintained within the animal body vary with different species of animals but among the warm blooded forms the range is not wide, as indicated in the table below.

Horse	100.4°F.	to 100.8°F.
Cattle	101.8	to 102
Sheep	101.3	to 105.8 probably 103.6 to 104.4
Swine	100.9	to 105.4
Dog	99.5	to 101.7

Any marked departure from these temperatures in the animal body, either up or down, results in physiological disturbances which injure the health of the animal.

422. Best Stable Temperature.—The data for a rational practice with reference to this point have yet to be determined experimentally. At present rules can be formulated only from general considerations.

Since most of the bodily functions result in the generation of more or less heat and since the temperature must be kept below 100° to 105° it is clear that no active animal should be surrounded by temperatures as high as the normal temperature of the body. One of the main objects of the circulation of the blood through the skin is to lower its temperature before it returns to the interior, so that those parts may be cooled. In our case we become uncomfortable in a surrounding temperature much above 72° and the same is true of our domestic animals.

Stables should then as a rule have a temperature lower than 72° F. but how much must depend upon circumstances. The right surrounding temperature is that which will permit the necessary loss of heat from the body with only the normal rate of perspiration.

Reasoning from general principles it is to be anticipated that animals which are being fed heavily, like fattening swine, steers or sheep, as well as milch cows, will do better in somewhat cooler quarters because (1) the larger activity necessary to produce the extra assimilation desired would develop more heat which must be removed from the body, and (2) because the aim is to induce such animals to eat as much as they can convert economically into flesh and milk and warm quarters must make the demand for food less.

It has been found with man that when fasting and at rest under a temperature of 90° F. he consumed 1,465 cubic inches of oxygen per hour, but under the same conditions except a temperature of 59° F. the amount of oxygen was 13 per cent. greater and the amount of carbon dioxide given off also 13 per cent. greater, showing that a higher rate of consumption of food in the body was maintained and hence that the man would be required to eat more.

It is with the cow and fattening animals as it is with a threshing machine, it requires a higher rate of waste of energy to run the machine rapidly than it does to run it slower, but the saving in time of all employed to manage the machine more than pays for the greater waste. So the cow may require an extra amount of food for temperature maintenance to overcome the cooler quarters but she is likely to eat enough more food to enable her to make more milk and a higher profit when all items of expense are taken into account.

With animals on simply a maintenance ration the aim is to carry them with the least amount of food and hence in as warm quarters as will be healthful.

It seems likely that the best temperature surroundings for animals being crowded will be found between 40° and 50° F. and for animals upon maintenance rations from 50° to 65° or even 70° F.

423. Heat-Proof Construction Impossible.—No enclosure or building can be so constructed that all the heat it contains will be prevented from escaping. If it is kept above freezing through cold winters there must be within the enclosure a source of heat. So, too, no enclosure or building can be so thoroughly made as to exclude all heat and hence it is impossible to build a "cool room" which will not get warmer during the summer unless it contains some means of removing the heat which enters.

The out-door root cellar which does not freeze during

the winter is prevented from doing so by the heat which enters it through the bottom. The same cellar during the summer grows gradually warmer as the season advances and is only relatively cool because part of the heat entering above is conveyed through the bottom into the earth, to restore that which kept the cellar from freezing during the winter. The warm stable which does not freeze is kept so by the heat of the animals sheltered, and the warmly constructed stable only makes less animal heat needed to maintain the temperature; the walls in themselves are not warm. So, too, no garment however made is in itself warm. We call it warm when the loss of heat through it is slow.

424. Means of Controlling Temperature.—When it is desired to construct a room which will be warm in winter or one which will be cool in summer the same principles must be employed in each. In the first case it is desired to retain the heat produced in the room; in the second case to prevent heat coming through the same walls, but from the opposite direction.

To secure either of these ends two essentials of construction must be observed. The walls must be as nearly air tight and as poor conductors of heat as possible. In the construction of a warm house, a warm stable, a cool ice house or a cool curing room for cheese the greatest attention should be paid to securing air tight walls because, no matter how poor conductors are put into the walls, if there are cracks about doors and windows or open joints in the wall, the effect of wind pressure and wind suction will be to change the air in the room so rapidly that it will be difficult to keep it either warm or cold.

425. Solid Masonry Walls.—Stone basements with solid walls are sufficiently warm for stables but they are too good conductors of heat to be suitable for dwelling houses in cold climates where the inside temperature must be maintained at 72° F. Hollow brick walls, when plastered with a close textured mortar, through which air cannot pass readily, are better than solid masonry but are not as warm as those well constructed of all wood and good building paper.

An unplastered brick wall, or a brick wall plastered with coarse lime mortar only, is one of the poorest which can be used either to retain or exclude heat. Its pores are so open that the smallest wind pressure or wind suction causes a ready flow of air through every portion of the wall, changing the air of the room quickly.

For cheese curing rooms, where the temperature is to be held down by means of cold air ducts, masonry walls, even when made air tight, are not suitable because they are such good conductors of heat and so massive that they tend to maintain a uniform temperature in summer somewhat higher than the mean of the air outside.

426. Hollow Masonry Walls.—When stone or brick walls are made hollow they become much warmer in winter and cooler in summer than when built solid because the air is a much poorer conductor of heat. The thickness of the air space is not important and one-half an inch thick is practically as serviceable as one of 6 inches.

Where basement or semi-basement curing rooms for cheese are constructed the upper four feet of the wall should be made with a dead air space to prevent the heat of the warm soil as readily reaching the interior. So, too, in the case of dwelling houses in cold climates, whether they have cellars under them or not, it is important to make the upper 3 or 4 feet of the wall hollow for the reason that the cellar will be warmer and hence the floors under the living rooms above.

427. Brick Veneered Walls.—Where brick are cheap and lumber high, walls made of 2x4 studding sheeted inside and outside with matched fencing and then veneered with brick make a very durable and warm building. The brick will not decay and the expense of nails and frequent painting are avoided.

It does not do to depend upon the brick for warmth, however; they simply take the place of the siding and paint. Where the house is simply sheeted outside with common boards and veneered with brick, and then lathed and plastered inside, the building will be very cold because the wind will go easily through the brick and the cracks in the sheeting.

428. All Wood Walls.—For the construction of dwelling houses, cheese curing rooms above ground and ice houses there is no type of wall so effective and so cheap in first cost as the all wood wall where good building paper is used with the lumber. For a dwelling house a reasonably warm wall is secured when the studding are sheeted outside and in with one layer of tongued and grooved fencing, covered outside with 2-ply acid and waterproof paper and lathed and plastered inside. The inside sheeting is warmer than back plastering and better because it gives a more solid wall, and lath may be used on it for furring.

LIGHTING FARM BUILDINGS.

The lighting of farm buildings is required to secure three important objects: (1) facility in doing work; (2) needs of the animals housed, and (3) healthful conditions.

In the dwelling house much care should be exercised to secure an ample amount of light in the kitchen, in the dining room and especially in the main living rooms. An abundance of light is needed in the kitchen not only to facilitate the work but to make the best intentions and efforts toward cleanliness more certain. It requires an effort to be gloomy and feel ugly in the face of a hearty laugh, and a bright cheerful room has much the same effect upon those who occupy it.

429. Efficiency of Windows.—There are many conditions which affect the efficiency of windows in lighting a building. Trees or buildings near by, which cover a considerable portion of the sky, may reduce the light entering a window very much. Much more light comes from the sky high above the horizon than from low down and hence a porch over a window cuts out a very large share of the light which might enter it.

Buildings which have thick walls require larger windows to admit the same amount of light as would enter through windows in thin walls. Basement stables with heavy stone walls require larger windows because the walls are thick, and so with a brick or stone house.

Windows long up and down admit much more light than windows of the same dimensions with their long axis horizontal because much more light comes from the upper portion of the sky. So, too, windows extending from near the ceiling toward the floor light the room better than when extending from near the floor up.

430. Position of Windows.—Living rooms and stables should if possible be arranged so that the body of light may come from the south side where the direct sunshine may enter the windows. In a dwelling house in the winter this is very important because then the amount of light is smallest at best and the family must be more closely confined and therefore need the direct sun then most. For poultry and for swine south windows are specially desirable. Large windows at the south are not as objectionable for heat in summer as might at first be thought because the sun is so high that a large portion of the direct sunshine is reflected from the glass and prevented from entering the house; but during the winter, when the sun is low, the advantage which comes from its heating effect as well as the light is very considerable.

VENTILATION OF FARM BUILDINGS.

In the physiological sense air is as indispensable to the cow and horse as is water, grain, hay or grass; so, too, is it as essential to the development of power in the steam engine as is the water and the fuel. It is so abundant about us and we procure it usually so unconsciously that its necessity does not occur to us. But when large numbers of animals are housed together in close stables ample provision must be made for the ingress and egress of air.

431. Necessity for Ventilation.—The need of ventilating dwellings and stables grows out of several conditions: (1) The consumption of the oxygen which is the essential ingredient; (2) the exhalation from the lungs of carbon dioxide, moisture, ammonia, marsh gas ($C H_4$) and organic matter; (3) the accumulation in the air of occupied stables and dwellings of bacteria and other micro-organisms as well as solid dust particles.

432. Carbon Dioxide in the Air.—This gas is given off from the lungs with each respiration in nearly the same ratio that the oxygen is removed, hence air once breathed is not only deprived of a portion of its oxygen but it is diluted with an equal volume of carbon dioxide and is therefore rendered doubly unfit for use again.

That air once breathed from the lungs is not suited to further use can be clearly and forcibly proved by filling a quart Mason jar with air from the lungs, by blowing through a rubber tube, and then quickly lowering a lighted taper into it, which is quickly extinguished, showing that the air has lost so much oxygen and gained so much carbon dioxide that the taper cannot burn in it.

433. Moisture from the Lungs and Skin.—The moisture taken with the food and as drink must be again removed from the body and a large portion of it leaves through the lungs and skin in the form of invisible vapor. If the air of a stable or dwelling is not changed with sufficient frequency it becomes so damp as to interfere with the proper action of the lungs and skin in this respect, and it is important that the ventilation should be strong enough to prevent the air becoming too damp.

One of the surest indications of an improperly ventilated stable is the condensation of moisture on the walls, ceiling and floors. It is sometimes remarked that cement floors, and stone basements are objectionable because they "draw moisture," making the air damp. The truth is the stables are insufficiently ventilated and the moisture from the animals condenses upon the cement floor and stone walls simply because these happen to be colder. Instead of "drawing" moisture and making the air damp they have exerted exactly the opposite effect by condensing the moisture from the air, leaving it dryer than if the condensation had not occurred.

434. Ammonia and Organic Matter Removed from the Lungs.—When one passes from the fresh air into an occupied stable or room where the air has been rendered impure from imperfect ventilation a depressed feeling and offensive odor are recognized and sometimes this effect may be so strong as to produce nausea. When these odors and the odor of ammonia can be detected it is positive proof that the air needs changing more rapidly.

Some of the organic matter given off from the lungs is strictly poisonous and so much so as to produce death in a few moments. If a live mouse is kept in a sealed pint fruit jar until it is nearly suffocated, as shown by its action, another mouse introduced into this jar will die at once, while the one which vitiated the air may be removed and it will apparently recover. It appears as if the organic principle eliminated from one animal is more poisonous when breathed by another, even of the same kind.

So poisonous is the organic principle removed from the lungs that Brown-Sequard in 1887 condensed the vapor of expired air and injected 15 cc. of it into a rabbit which died from the effects. Brown-Sequard considered the substance a volatile alkaloid secreted by the lungs.

Water standing over night in a poorly ventilated room or stable comes to have a very disagreeable taste from the absorption of impurities from the air and this is one of the most serious objections to keeping water standing in the stable for cows or other animals.

435. Micro-organisms and Dust in the Air.—It has long been recognized that the air of old and poorly ventilated houses, especially if they are not kept clean, contains many more dust particles, spores and micro-organisms than newer and better ventilated houses do. The same must be true also of stables but in a higher degree. The amount of dust and of organisms as well is almost always more abundant in occupied rooms than in the open air. This would be expected both because of the slowing down of air movements after entering the house, which acts exactly like a silt basin in a line of tile, and because of their production there from various causes.

Strong ventilation tends to remove these organisms and dust particles with the air from the compartments and this is the rational basis for airing a bedroom or any other after sweeping. The air has been filled with both sets of impurities and opening the windows or using any other means of producing a strong current will help to clear the room.

436. Bad Ventilation Predisposes to Disease.—The most helpful health rule which man can adopt for himself or for his domestic animals is to avoid whatever tends to weaken the system and to take advantage of whatever tends to greater vigor.

It sheuld be clcaily recognized that the germs of diphtheria, of tuberculosis, hog cholera and other contagious diseases are liable to be met with almost any day and in any place and that wherever a proper breeding place may be found the disease is liable to start and from it spread by force of greater numbers of germs.

While therefore the micro-organisms usually found in greatest numbers in dusty houses and stables poorly ventilated and cared for are not in themselves a source of danger, the run-down, weakened condition which poor ventilation is sure to engender will certainly tend to start a case of contagious disease and then, with greater numbers of germs in the air to be introduced into the system, animals of greater vigor must succumb to these invisible foes because of their vast numbers.

Ample ventilation then should always be secured, first, as an indispensible condition for maintaining the power to resist disease, and second, in case of disease, to both clear the air and to give the animals an opportunity to defend themselves against this type of foe.

437. Amount of Air Respired.—The amount of air ordinarily taken into and put out of the lungs by man with each respiration is given by different observers as follows:

Herbst	20 — 30 cubic inches
Valentin	14 — 92 cubic inches
Vierordt	10 — 42 cubic inches
Coathupe	16 cubic inches
Hutchinson	16 — 20 cubic inches
Average	15.2 — 46 cubic inches

or an average of about 30 cubic inches.

The amount of pure air which must be breathed in order to supply the oxygen needed by different animals, deduced from Colin's table, is given below:

Animal.	Air Breathed in 24 Hours.		Oxygen Consumed in 24 Hours.	
	Per 1,000 lbs. of weight.	Per head.	Per 1,000 lbs. of weight.	Per head.
	cu. ft.	cu. ft.	lbs.	lbs.
Man	2,833	425	12.207	1.831
Horse	3,401	3,401	13.272	13.272
Cow	2,804	2,804	11.04	11.04
Swine	7,353	1,103	29.698	4.456
Sheep	7,259	726	29.314	2.931
Hen	8,278	24.84	24.84	.075

438. Amount of Air Used Compared with Feed and Water.—A 1,000-pound cow requires daily the equivalent of about 30 lbs. of hay and grain and 70 lbs. of water or, in round numbers, 100 lbs. per head and per day of solid and liquid food.

A cubic foot of air weighs about .08 lbs. hence, from the table in **(437)**, we have

$$2804 \times .08 \text{ lbs.} = 224.32 \text{ lbs.}$$

which shows that a cow needs to be supplied with twice the weight of pure air that she does of food and water combined.

439. Degree of Impurity of Air Permissible.—We are yet without sufficiently exact data to permit this problem to be concisely stated for stables used for domestic animals. In absence of exact data and in view of the unavoidable leakage of air through the walls and about windows and doors we have arbitrarily assumed that if the air is changed in the stable at such a rate that it at all times contains no more than 3.3 per cent. of air once breathed fairly good ventilation would be provided.

440. Rate of Supply of Air to Stables.—On the basis of (439) the number of cubic feet of air per head and per

hour, using the data in the table of (**437**), would be **as stated** below:

For horses............	4,296	cu. ft. per hour per head.
For cows..............	8,542	cu. ft. per hour per head.
For swine.......... ..	1,392	cu. ft. per hour per head.
For sheep.............	917	cu. ft. per hour per head.
For hens..............	81.4	cu. ft. per hour per head.

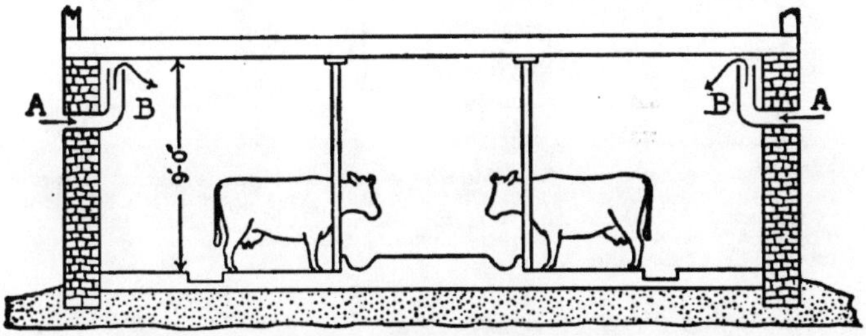

FIG. 150.—Simplest method of taking air into stone or basement stable. A B and A B show where the air enters. These flues may be made out of ordinary 5 or 6 inch stove pipe with elbow, or galvanized iron conductor pipe, or the pipe through wall may be ordinary 5 inch drain tile, with stove pipe and elbow on inside, or the flue may be made of 6 inch fencing.

The weights here assumed are 1,000 lbs. for the horse and cow, 150 lbs. for the hog, 100 lbs. for sheep and 3 lbs. for the hen. With different weights the amounts would change somewhat in proportion to the size of the animals.

441. Capacity of Ventilating Flues.—With the data in the last section, and the number of animals to be provided with air, the capacity of ventilating flues should be such as to ensure an air movement equal the rate given in the table of (**440**). It is practicable to construct ventilating flues through which the air from stables will travel at the rate of 200 to 500 feet per minute without mechanical forcing or the aid of heat, other than that derived from the animals in the stable.

With a ventilating flue 2x2 feet inside measure 20 cows would be supplied when the current in the flue was at the rate of 295 feet per minute. At this rate 40 cows would need two flues 2x2 feet inside measure; 60 cows three; 80 cows four and 100 cows five.

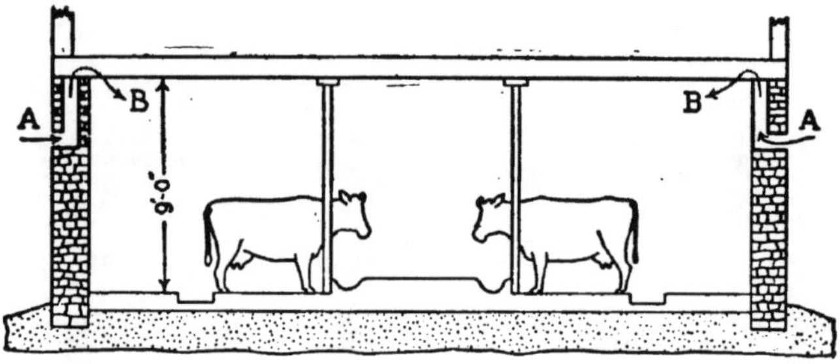

FIG. 151.—Modification of Fig. 150 where on the right a notch is left in the wall when building, so that the flue rises flush with the inside of the wall. While on the left side the flue is shown built in the wall. This may be done by building around 5-inch drain tile or around a box made of fencing.

442. Cubic Feet of Space in Stable per Animal.—It has been customary with sanitary engineers in planning hospitals, prisons, school rooms, etc., to allow so many cubic feet of space per occupant, but the number chosen has not

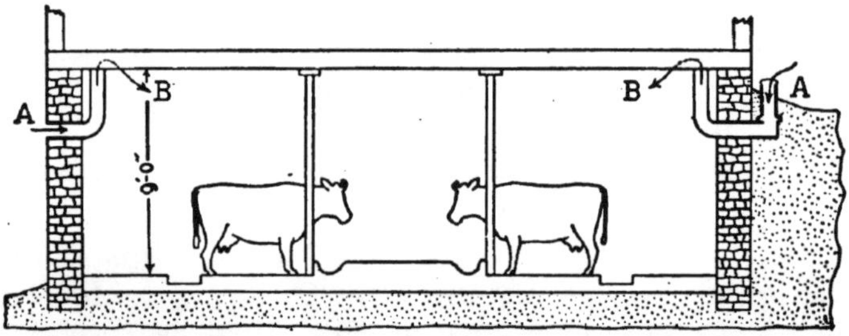

FIG. 152.—Method of taking air into a bank barn on the up-hill or bank side. The air flue is made in the same way as described in Figs. 150 and 151, but on the outside has its end covered as represented at A on the right with a length of 6 or 8 inch sewer tile with its top covered with a cap of coarse wire screen. Drain tile would not answer for the outside exposure at the surface of the ground as frost would cause it to crumble. Wood could be used and replaced after rotting has occurred.

been to supply the proper amount of air but rather to avoid drafts too strong for health and comfort.

It should be distinctly stated that in matters of ventilation it is cubic feet of air rather than cubic feet of space which should be provided, and in the construction of stables the amount of space need be only so much as is required to permit ample room and freedom to care for the animals.

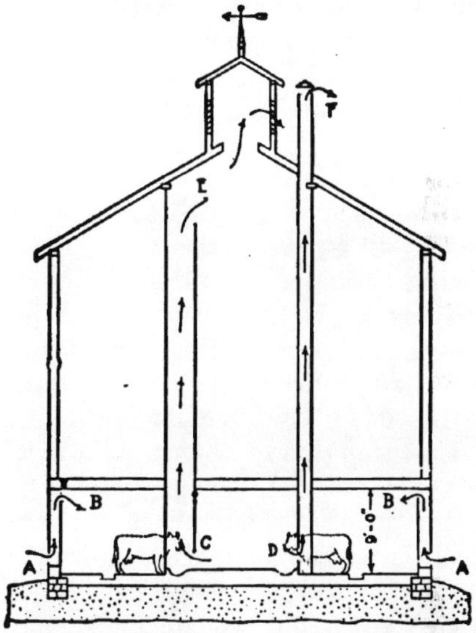

FIG. 153.—Two methods of ventilating a dairy barn. On the right the ventilating flue D F rises straight from the floor, passing out through the roof and rising above the ridge. One, two, or three of these would be used according to number of cattle. The flues should be at one or the other side of the cupola rather than behind it. On the left C E represents how a hay shoot may be used also for ventilating flue. In each of these cases the ventilating flue would take the place of one cow. This method would give the best ventilation but has the objection of occupying valuable space. C, in the feed shoot, is a door which swings out when hay is being thrown down, but is closed when used as a ventilator, the door not reaching quite to the floor. To take air into this stable if it is built of wood with studding, openings would be left at A about 4x12 inches every twelve to sixteen feet, and the air would enter and rise between the sheeting of the inside and the siding on the outside, entering at B as represented by the arrows. If the barn is a basement or stone structure the air intakes could be such as described in figures 150, 151, and 152.

Twenty cows should not be housed in a space much less than 28x33 feet, with ceilings 8 feet in the clear. In warm climates there is no objection, except the matter of cost, to high stables, but where it is cold high ceilings permit the warm air to rise so far above the animals as to leave the stable cold at the floor.

443. Forces Which Produce Ventilation.—The movement of air currents into and from a ventilated stable is caused

1. By the wind pressure against the building tending to force air into the stable.

2. By wind suction on the leeward side of the stable tending to draw air out.

3. By aspiration across the top of the ventilator.

4. By the difference in temperature between the air in the stable and that outside.

When the wind is blowing against a building there is an increase of pressure above that inside which forces air into the stable through any available opening and then out again on the opposite side or up the ventilating flue. At the same time there is a low pressure on the lee side which tends to draw air through any openings on that side.

Where the ventilator rises above the roof as a chimney does the movement of air across its top produces a diminished pressure and the air is aspirated out on the principle of the aspirator used on perfumery bottles.

The difference of temperature causes a difference of pressure because of the expansion making the air in the stable relatively lighter than that outside; and the longer the chimney or ventilating flue the stronger will be the draft, both from difference of temperature and the aspiration across the top of the chimney.

444. Essential Features of a Ventilating Flue.—A good ventilating flue must have all of the characteristics possessed by a good chimney. It should be constructed with air-tight walls so that no air can enter except from the stable. It should rise above the highest portion of the roof so as to get the full force of the wind. It should be as nearly straight as practicable and should have an ample cross section. Stronger currents through the ventilators will be secured by making one or a few large ones than where many small ones are provided, and it is usually best

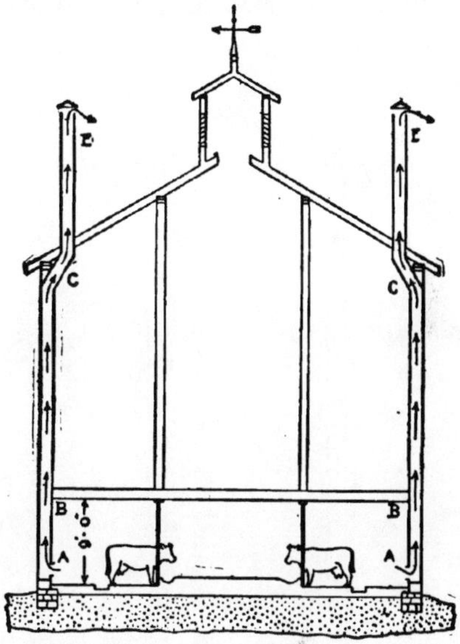

Fig. 154.—Second best method of ventilating an ordinary barn. The air comes in as described in the other figures, and passes out through one or more ventilators rising against the side of the barn and passing out through the roof, as represented at A C E. To make these flues if the barn is a balloon frame, the best method would be to secure the lightest galvanized iron in eight or ten foot lengths, and place the studding where the flues are to be, the right distance apart, so that a width of the metal covers the space between two studs. Sheets of this metal nailed on opposite faces of the stud would make an air-tight flue. On the outside, this metal would be covered with the siding. On the inside in the stable, with the sheeting, but in the barn above nothing would be needed except perhaps an occasional shield to prevent the hay from crushing it in. If it is not desired to carry the flues through the roof, they may end just below the plate, and the air pass out through the cupola. The method represented, however, would give the strongest draft. The width of studding used for the flue would vary with the number of animals to be provided for.

to have as few as practicable and not leave the air impure in distant parts of the stable.

445. Location of Ventilator.—The best location for the ventilating shaft is near the center of the stable where such a position will not interfere with the work. It is not often that this position can be utilized, and when it can-

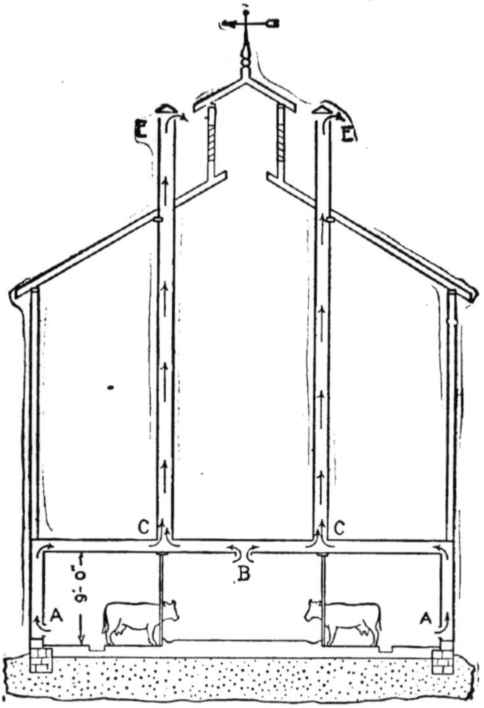

FIG. 155.—Modification of Fig. 157, where the air passes straight out through the roof, instead of being carried in and out through the ridge of the roof. This method would give a stronger current, unless the ventilator passes straight down to the floor between the cows, as represented in Fig. 153.

not it may be located in various places, as indicated in Figs. 153 to 160.

446. Openings to the Ventilator.—The ventilator should reach to the stable floor so that air may enter the shaft from that level. This is very important because: (1) The animals not only stand and lie low down but are so constituted as to breathe the impurities directly to the floor where the carbon dioxide tends to remain, because it is heavier than the rest of the air in the stable, even although its temperature is higher.

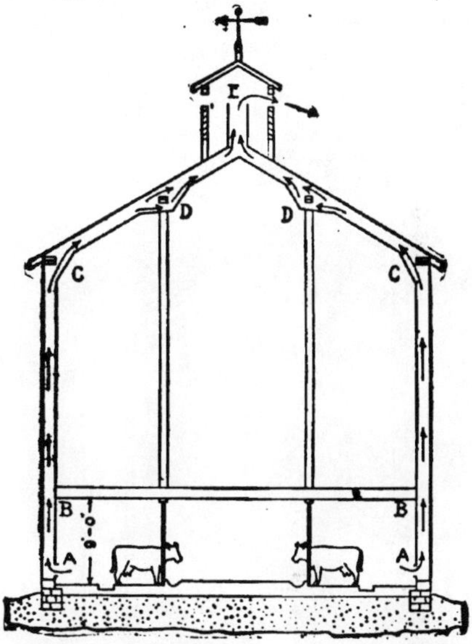

FIG. 156.—Represents a method of carrying the flues up the sides and then along under the roof between the rafters, so as to reach the ridge either under the cupola, or at other places on either side. Such a flue could be made very tight, by nailing the light galvanized iron on the outside and inside of studding, and rafters, having a sufficient width to give the proper capacity for the ventilating flues, and such a system of ventilation would work fairly well but could not be expected to do as effective service as the methods shown in Figs. 153, 154, 158 and 159.

(2) The coldest air is at the floor and the warmest at the ceiling and it is the cold air which should be removed during the winter rather than the warm.

There should be an opening provided at the ceiling for warm air to escape when the stable is too warm and when it is desired to force the ventilation at the expense of the heat developed by the animals.

Both of these openings should be provided with regulating valves so that either or both may be partly or completely closed.

447. Entrance for Fresh Air.—When a stable has been made close and warm, requiring attention to ventilation, provision must be made for air to enter the stable as well as to leave it. This may best be done as represented in Figs. 150–153 and 158–160.

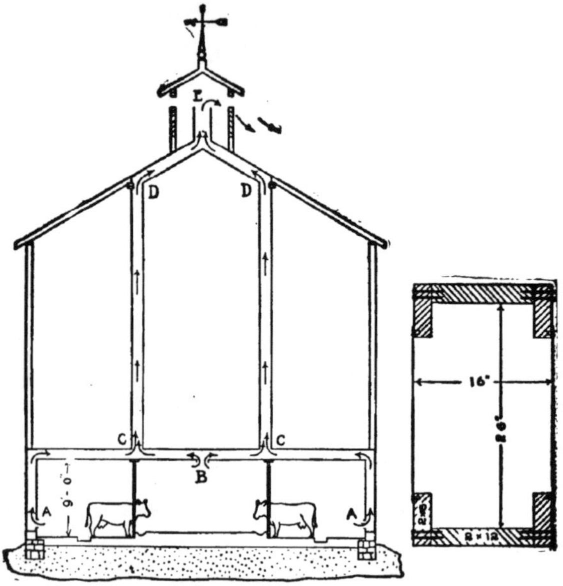

FIG. 157.—Shows method of ventilating an ordinary barn, where the air is taken out of the stable through flues built between the studding and between the joists of the ceiling, the air then rising, through ventilating shafts, made against or as a part of one or more of the purline posts. The air enters at A A and B, following the arrows and passing out along the lines C D E. These ventilators, if desired, can be carried out straight through the roof, or may be terminated inside under the purline plate, or as represented in the figure. The cross section at the right shows how 2x12's and 2x6's may be nailed together and placed so as to constitute a purline post, and at the same time a ventilating flue. The two sides of the purline post or ventilating flue are represented closed with sheets of galvanized iron. They may also be closed with well seasoned matched flooring. The number of bends necessary in this plan is an objection, as they interfere with the draft more or less.

In all of these cases it will be noted that the fresh air enters at the ceiling. This is for the purpose of mingling it with the warmest air of the stable so as to raise its temperature before it falls to the floor. In this way the heat which is wasting at the ceiling is saved and the animals are prevented from lying in cold air.

Provision is further made for the air to enter the intakes outside at a distance of 3 or more feet below the ceiling so as to prevent the warm air being drawn out at these places by suction or to pass out directly as it would if they opened directly through the walls.

These openings should be placed on all sides of the stable if possible so as to take advantage of the wind pressure at all times in increasing the draft. It is better to have many small openings than a few large ones because the cold air is better distributed, lessening drafts.

448. Construction of the Ventilators.—The best form of ventilating flue is that represented in Fig. 160, made of galvanized iron in cylindrical form. Another good form is

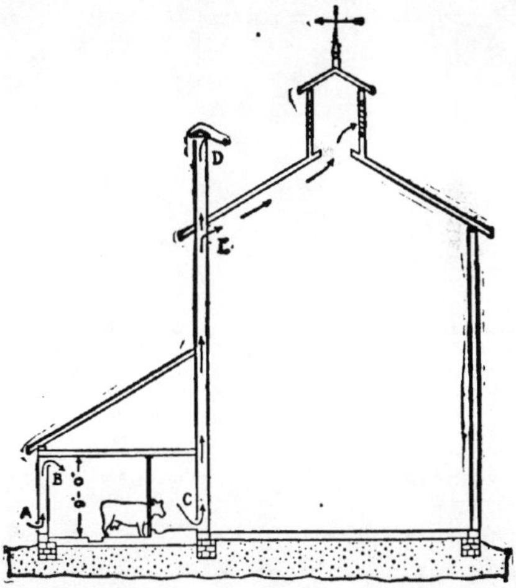

FIG. 158.—Method of ventilating a lean-to stable. The air enters as represented by the arrows at A B and passes out through a flue built on the inside of the upright or main barn. This flue may rise directly through the roof, or it may end at E as shown in the figure, the air passing through a cupola. If the upright barn has a balloon frame, then the space between the studding could be used as ventilating flues in the same manner as described in Fig. 154. These flues could be made tighter by covering inside and out on the studding, with the lightest galvanized iron.

represented in Fig. 157, where the sides are also made of galvanized iron.

As a substitute for galvanized iron in this form of ventilating flue a good roofing paper may be used, such as the ruberoid roofing made by the Standard Paint Company.

449. Ventilation of Basement Stables.—There is a general impression that basement stables are necessarily unhealthful. This idea has grown out of the fact that it has been possible to make these stables much closer and warmer than ordinary over-ground forms, and where ample ventilation has not been provided they have been damp and close.

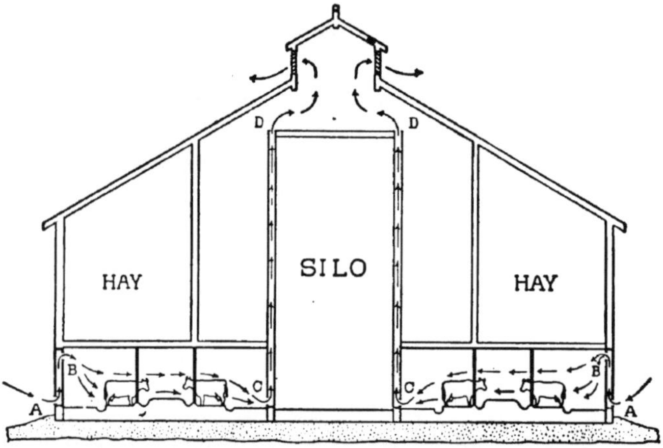

FIG. 159.—Method of ventilating a barn where a silo or granary occupies the central portion. The air enters at A B and the ventilating flues are the spaces between the studding which form the walls of the silo, or other structure. The air entering at C in openings left all around the silo, and passing out at D at the top.

Where basement stables are well lighted and properly ventilated there is no objection to them on sanitary grounds and they have many points in their favor where the conditions admit of their being easily constructed. Methods of introducing the air into these stables are represented in Figs. 150 to 152.

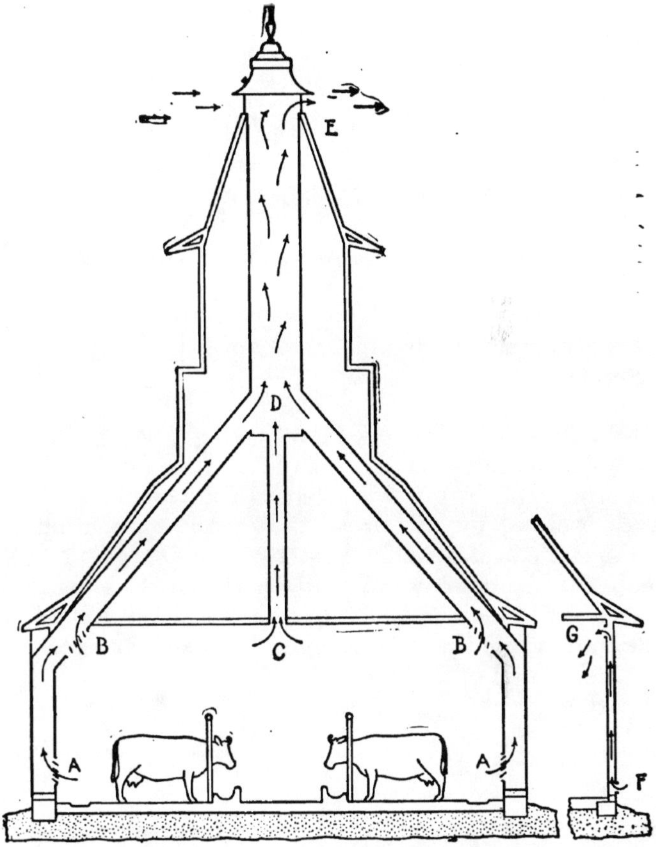

FIG. 160.—Is a section of the cow stable of the dairy barn at the Wisconsin Experiment Station. A single ventilating flue D E rises above the roof of the main barn, and is divided below the roof into two arms A B D, which terminate at or near the level of the stable floor at A A. These openings are provided with ordinary registers, with valves to be opened and closed when desired. Two other ventilators are placed at B B, to be used when the stable is too warm, but are provided with valves to be closed at other times. C is a direct 12-inch ventilator leading into the main shaft, and opening from the ceiling, so as to admit a current of warm air at all times to the main shaft to help force the draft. This ventilating shaft is made of galvanized iron, the upper portion being 3 feet in diameter. The covering on the outside is simply for architectural effect.

CHAPTER XVIII.

PRINCIPLES OF CONSTRUCTION.

RELATION OF COVERING TO SPACE ENCLOSED.

The first cost of a building, when expressed in terms of cubic feet enclosed, is influenced much by its relative dimensions.

450. Relation of Walls to Floor Space.—The form of floor space which can be enclosed by the smallest amount of wall is a circle, and Fig. 161 represents equal amounts of floor space enclosed by the circle, the square and the oblong. If the circle encloses a floor space of 1,600 square feet the length of the outside wall will be about 143.7 feet; the square would then be 40x40 feet and have 160 feet of outside wall; while the oblong would be 20x80 feet and have an outside wall of 200 feet.

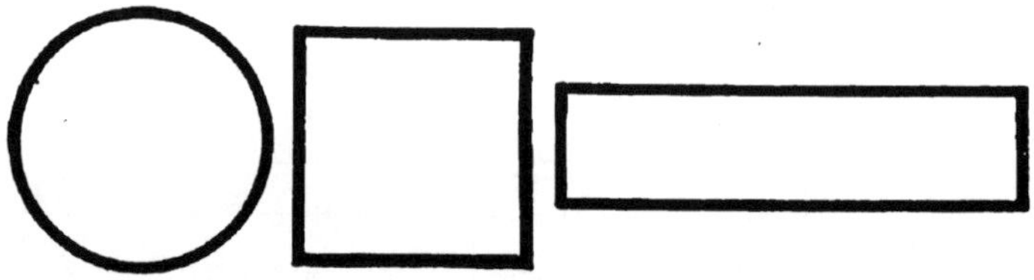

144 ft. 160 ft. 200 ft.

FIG. 161.—Shows equal areas enclosed by three types of walls.

The square which encloses the same floor space as a circle requires 11.44 per cent. more wall, while the oblong whose length is twice the breadth requires nearly 40 per cent. more wall. This means that 40 per cent. more siding, more nails and more paint would be required to cover an oblong building, where the length is twice the width, than would be required for a circular one enclosing the same floor space.

Comparing the square with the oblong building it requires 25 per cent. less wall to enclose it. From these relations it is clear that wherever it is practicable to avoid long narrow buildings there will be not only a saving in materials but the buildings may more easily be kept warm in winter and cool in summer, and in the case of silos there will be less loss of silage.

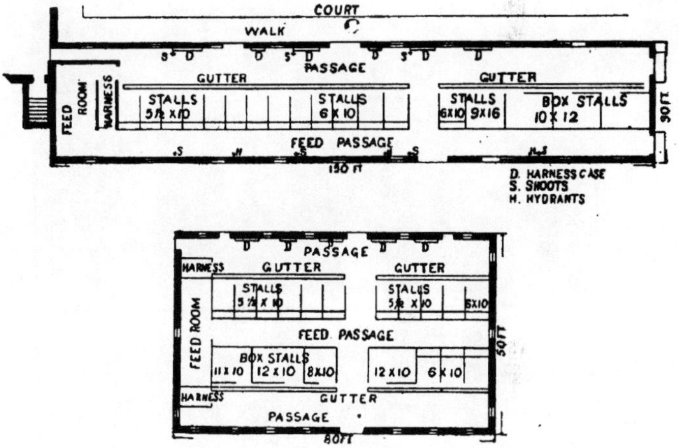

FIG. 162.—Showing the same conveniences in two types of horse barns.

In Fig. 162 are represented two plans for horse barns providing nearly identical accommodations. The longer one is 105 feet 10 inches in length, 30 feet wide with 18 foot posts. The second is 75 feet 10 inches x 44 feet and requires over 8 per cent. less wall and over 6 per cent. less floor space.

451. Relation of Hight to Capacity.—In the building of barns, silos, ice houses, grain bins and root cellars the more depth or hight which can be secured the larger will be the capacity in proportion to roof, ceiling or floor. The material for flooring and roofing a low building is usually no less than is required for a high building and yet the cubic contents are in the ratio of their depth.

In the case of hay barns and silos the capacities increase much faster than the hight because with greater depth of material it is compressed and on this account greater storage capacity is secured.

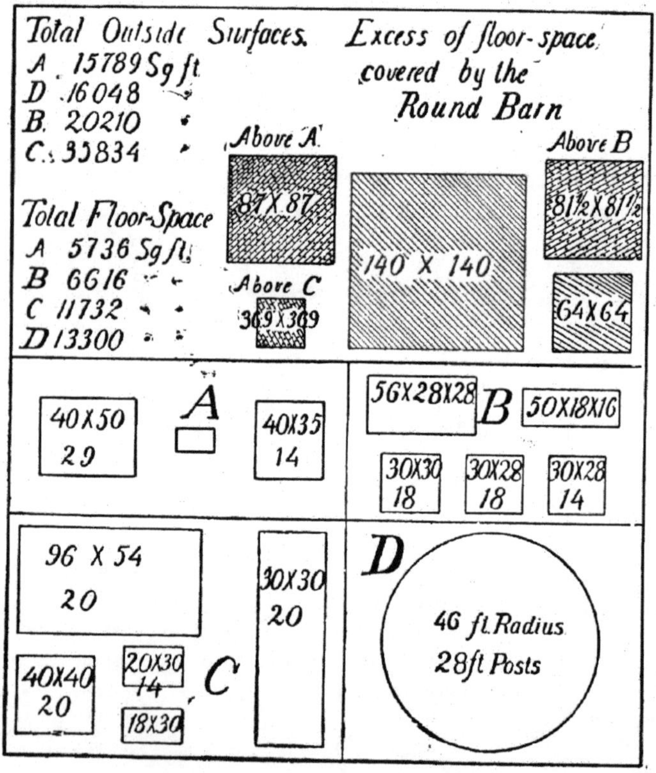

FIG. 163.—Diagram showing the comparative outside surface and amount of floor space in four sets of barns represented in Figs. 164, 165, 166 and 167.

FIG. 164.—Cylindrical barn which accommodates 98 cows and 10 horses, contains a granary and tool house, each equivalent to a floor space 16x40 feet, and a 400-ton silo.

FIG. 165.—Buildings which shelter 37 cows and 15 horses.

452. Combined and Separate Construction.—The amount of capital required to build and maintain in repair a large number of small buildings is greater than that required for a single consolidated structure providing like accommodations. This is clearly illustrated by the comparative chart, Fig. 163, which represents the relations of buildings shown in Figs. 164, 165, 166, 167.

Taking the cylindrical barn as a standard of comparison, it provides shelter for 98 cows and 10 horses, contains a 400 ton silo, a granary 16x40 feet, a tool space 16x40 and storage capacity for all the hay needed; and yet its roof and side area is only 269 feet more than the group of buildings in Fig. 165, which shelters only 37 cows and 15 horses, has no silo, no tool house and not enough space for hay.

FIG. 166.—Group of buildings which shelter 114 cows and 8 horses.

Comparing with the buildings of Fig. 166, their aggregate outside surface exceeds that of the standard by an area 64x64 feet and yet they provide cramped quarters for only 114 cows and 8 horses.

FIG. 167.—Group of buildings which shelter 144 cows and 14 horses with tool house and granary.

In the group of buildings shown in Fig. 167, there is an aggregate outside surface exceeding that of the round barn by 140x140 feet, or more than twice, and they have less floor space by an area of nearly 40x40 feet, and the group of buildings shelters but 36 more cows and 4 more horses. In this last group the buildings are both low and narrow, causing extreme wastefulness of lumber.

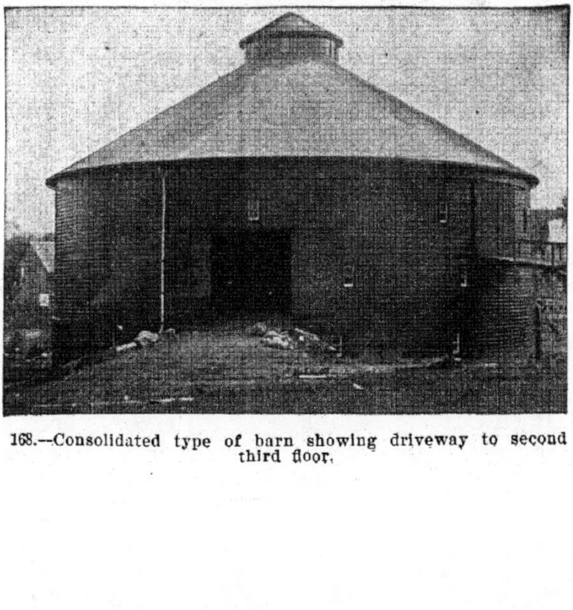

FIG. 168.—Consolidated type of barn showing driveway to second and third floor.

453. Saving of Labor.—It is possible to care for animals with less labor and time where all are brought together under one roof than it is where they are scattered through many buildings and Figs. 164, 168, 169, 170 and 171 represent a consolidated type of barn with composite functions, where all of the stock are brought together under one roof.

Fig. 169.—Consolidated type of barn showing driveway to first and second floor.

Economy in labor is of much greater moment than economy in material because the material simply represents money invested in this case while the extra labor required is a continual expense of a high order.

454. Distribution of Animals in Stables.—The general arrangement of animals in stables must vary in detail in almost endless variety, and individual circumstances must determine just what is best. Three types of arrangement for cows are illustrated in cross-section in Figs. 150 to 159 under the chapter on ventilation, and Fig. 162 represents two convenient groupings for horses. While Fig. 170 shows one plan of division and arrangement of space in a cylindrical barn.

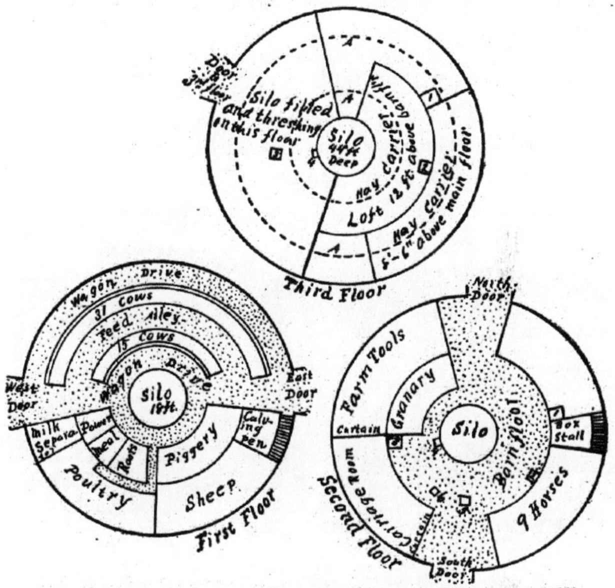

FIG. 170.—Showing plan of the three floors of Figs. 168 and 169.

FIG. 171.—Showing less consolidated type of barn with silo partly outside.

A combined cow and horse barn with silo outside has the arrangement shown in Fig. 172 and permits the work being easily done.

455. Avoiding the Use of Posts.—In cow stables having a second story it will often be possible to carry the floor upon the uprights used to form the stalls or ties for the cows and in this way save lumber by making the same

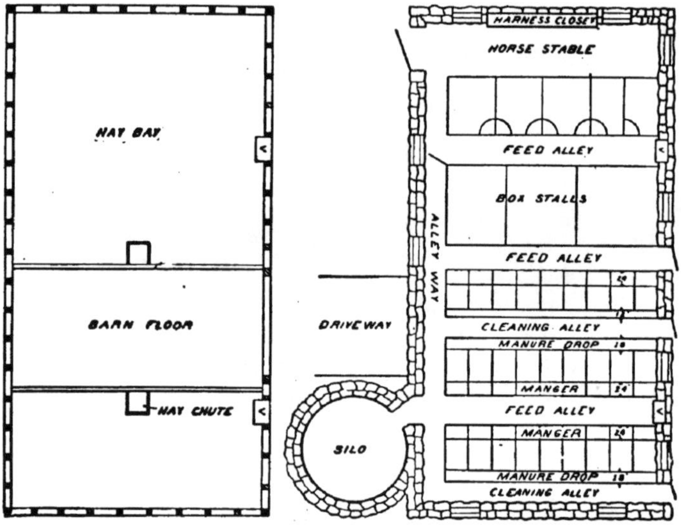

Fig. 172.—Plan of combined cow and horse barn with silo outside.

pieces render double duty, and at the same time avoid the inconvenience of the posts and save the space they would occupy. This plan is illustrated in the various figures showing methods of ventilation.

STABLE FLOORS.

456. Essential Features.—The essential features of a good stable floor are: (1) Imperviousness to water and urine. (2) A surface sufficiently even to be readily and thoroughly cleaned with a small amount of labor. (3) A durability approximating that of the building itself. (4)

FIG. 173.—Rectangular barn showing driveways to second and third floors.

A reasonably low first cost. There are two materials which have been used in the construction of stable floors which fulfill these requirements; they are concretes made either with Portland cement or asphalt. The asphalt is superior to the Portland concrete in being a poorer conductor of heat while the cement has the advantage of less first cost.

457. Cold and Warm Floors.—It is urged against the concrete as compared with wood floors that they are cold. The meaning is that they are better conductors of heat and so serve to carry the heat away from the body of the animal rapidly. It is true that they do convey heat faster than wood and when used in cold climates without bedding are worse than wood from this standpoint. They are not as bad in this respect, however, as many imagine. In the first place the stable ought not to fall below 40° F., and when this is true the floor will only have this temperature and will not lead to inconvenience if other conditions are right. In the second place no animal should be required to lie

FIG. 174.—Rectangular barn with driveway to first and third floors. Same as Fig. 173.

even upon a naked wood floor and when plenty of bedding is provided the cement floor is not too cold for warm stables kept clean.

458. The Use of Bedding.—No farmer who is attempting to maintain the fertility of his land at the standard of best yield can afford to use no bedding or even a scanty supply. He can better afford to overfeed with hay so that the least nutritious portions are rejected and use this for bedding, than go without, because the extra amount of manure made and the greater comfort and cleanliness of his animals will pay a good return for it. The waste roughage of the farm, when used as bedding and mixed with the manure, increases the value of both because it increases the total quantity of manure so much that the fields can be dressed more frequently, thus holding the humus content higher and the soil in better tilth, both essential conditions for large yields. The liability of animals to kick the bedding off from the floor is not a sufficient reason against cement floors. It is only when too little bedding is used or it has not the right texture that the floor is left seriously exposed.

459. All Wood Floors.—These floors are generally laid in one of two ways, either close upon the ground, nailed to stringers bedded in the earth; or else upon joists with an air space between the floor and the earth. When laid in either of these ways they are certain to wear out through the tramping of the animals and the use of the tools in cleaning the stables, but if conditions are favorable so that rotting does not occur they may last as long as 6 to 12 years.

It is oftener true that wood floors give out from decay before they do from wear. Where the floor is kept continually saturated with moisture it will not decay; and when kept continually dry it gives out only through wear, but when it contains the right amount of moisture the growth of moulds, causing the decay, takes place.

When the floor is bedded in a close textured clay soil, where the subsoil is close and all the time saturated with water, decay will go on very slowly; but where the soil is dry and open, and especially if this is the character of the subsoil, decay may destroy the floor in 3 to 5 years. So, too, where the floors are laid upon joists on the ground and a dead air space left beneath, decay is certain to occur in 3 to 5 years, but if the joists are so arranged that there is free circulation of air beneath, destruction from decay is not likely to occur.

460. Making Wood Floors Water Tight.—Wood floors are made so as to prevent water from running through them by using more than one layer with some waterproof composition between them. For heavy floors matched plank are laid and coated with a layer of coal tar roofing composition and then upon this a second layer of plank is laid, painting the joints with the same composition before drawing them together. Lighter floors are made in the same way, using tongued and grooved flooring.

461. Stone Floors.—Thoroughly durable floors for cow and horse stables are made by bedding in clay rounded cobble stone, 4 or 5 inches in diameter, and using upon this an abundance of bedding. The uneven surface holds the bedding so well that the animals are fairly comfortable and neither wear nor decay will destroy them. The most serious objection lies in the difficulty in maintaining cleanliness.

Where a good gutter is made behind the cows and a row of cut stone 10 or 12 inches wide are set for the hind feet to stand upon a durable and quite satisfactory floor is secured.

462. Macadam Stable Floors.—A floor more even in surface than (461) can be made out of carefully constructed macadam work, such as is used in making stone roads, giving it a thickness of 5 or 6 inches. Where this is used there should be provided cement gutters and mangers as represented in Fig. 175.

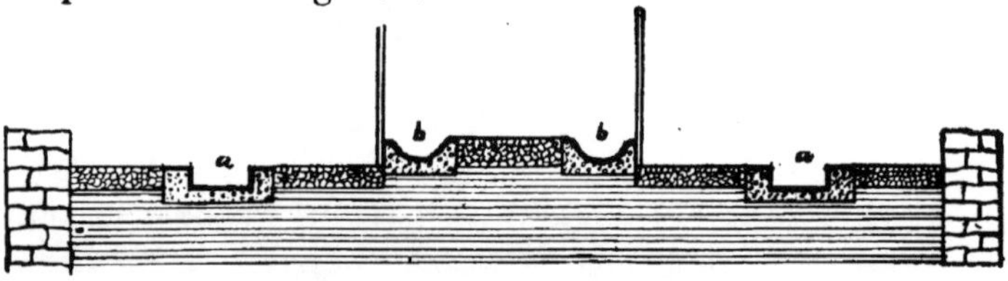

FIG. 175.—Shows method of making a macadam stable floor with cement mangers and gutters.

Before laying such a floor the ground should be shaped and made thoroughly hard by tramping or ramming. The crushed stone should be put on in two layers, thoroughly compacting the first layer and filling the voids with screenings before the surface layer is made. Indeed the method should be the same as that followed in making a good stone road.

463. Macadam Surface for Barnyard.—The paving or flooring the barnyard with macadam surface is perhaps the best solution of the difficult problem of maintaining a hard dry yard. On account of the puddling of the soil by the tramping of feet, surface drainage is all that can be adopted and hence even when the yard has been macadamized it is necessary to scrape the manure into piles so that the water may flow away.

CONSTRUCTION OF CEMENT FLOORS AND WALKS.

464. Kinds of Cement.—There are two classes of cement on the market, Common and Portland. Of the common cements in the United States familiar brands are Akron, Louisville and Milwaukee. They are suitable for laying walls below ground and plastering cisterns but will not answer for stable, cellar or creamery floors, nor for walks, because they do not make a hard enough stone.

For walks and floors some brand of Portland cement should be used. These are American, English or German according to the country in which they are manufactured. American brands are Vulcanite, Alpha, Atlas and Wolverine.

465. Cement Concrete.—The making of cement concrete is in effect the production of artificial stone by binding together pieces of rock and sand with Portland cement. The cement is too expensive to be used by itself for ordinary work and the making of cement concrete aims to produce the largest bulk of strong rock with the use of the least possible amount of the more costly cement. This is secured when only so much space is left between the materials bound together as will leave room for the cement to form a thin layer between the faces of the fragments to be joined together.

466. Materials for Concrete Floors.—The materials used for cement walks and floors should be (1) as large, clean fragments of hard rock as can be readily mixed and worked into the forms and thickness of layer desired; (2) a finer grade of crushed rock or coarse clean gravel which will readily pack into the voids between the larger fragments; (3) a clean, coarse, sharp sand to fill the pores between the fragments of gravel or fine screenings; (4) enough Portland cement to fill the space between the sand and bind the whole together; (5) and finally, water enough to wet all surfaces, fill the pore space of the cement and make the mortar plastic.

467. Presence of Earth, Loam or Dust.—It is of the greatest importance that all of the materials used be perfectly clean and free from dirt or other fine grained material having the texture of the cement itself. If a fine dust is present in the rock, gravel or sand it will tend to form a layer over the surfaces of the fragments which prevents the cement from coming in contact with the pieces which are to be cemented together and a weak concrete results. The fundamental is to have nothing but hard rock fragments large enough to be cemented together and nothing fine present but the cementing material itself.

In the concrete pavements used on the streets of London, and which have a much longer life than the best paving blocks, great care is taken to wash out of the crushed granite and its screenings all dust particles before using them, although the dust may be from the granite itself.

468. Wetting the Crushed Rock Before Use.—There are two important reasons why crushed rock or coarse screened gravel, to be used as the body of concrete, should be wet before mixing with the cement. These are (1) to displace as much adhering air as possible, and (2) so as not to draw out from the cement the water needed to maintain its plasticity and to assist in the setting.

If the coarse materials are mixed with the cement dry a large amount of air will be set free and entangled in the concrete, which will prevent all spaces being filled, but the chief difficulty comes from the air preventing the cement from adhering to the surfaces. So strongly does air adhere to coarse sand that it must be boiled some time under water before it is all removed.

469. Ratio of Ingredients for Concrete.—The amounts of each ingredient required to make a solid concrete with all spaces filled depends upon the pore space in the different materials. Trautwine assumes that for each ingredient the voids are near enough to 50 per cent. so that as a safe working basis this should be taken.

To make a cubic yard of concrete it would be necessary to use, on Trautwine's basis,

Crushed rock.	Gravel or screenings.	Coarse sand.	Cement.
27 cu. ft.	13.5 cu. ft.	6.75 cu. ft.	3.375 cu. ft.

This ratio for pore space is certainly larger than is likely to occur and for farm purposes it will be safe enough to take the ratios of

Crushed rock.	Gravel or screenings.	Sand.	Cement.
27 cu. ft.	12.69 cu. ft.	5.584 cu ft.	2.122 cu. ft.

These figures assume the pore space of the rock to be 47 per cent., of the gravel 44 per cent. and of the sand 38 per cent.

470. Ratio of Ingredients for Finishing.—Where good plastering sand is used for making the finishing surface the pore space to be filled will be about 35 per cent. and this would require a little more than one of cement to two of sand, and unless there is some gravel or screenings to use with the sand it will be safer to make the facing 2 of sand to 1 of cement.

471. Thickness of Floor.—For most stables where the ground has been well firmed and shaped a thickness of 4 inches of concrete and one-half inch of facing will be enough; for house cellars and for the bottoms of silos 3 inches of concrete and one-fourth inch of facing will do. For creameries and milk rooms the concrete better be 4 inches and the facing a full half inch, made richer in cement, in the ratio of one to one.

472. Making the Concrete.—The cement, sand and gravel are put together dry on a mixing board and thoroughly worked over, then enough water added to make a stiff paste. The right amount of crushed rock is thoroughly drenched with water and the whole mixed by shoveling until the rock is thoroughly incorporated with the cement.

473. Laying the Concrete.—The floor of the stable should first be given the proper form and very thoroughly tamped so that no settling shall occur after the floor is laid. The concrete should be laid in blocks four or five feet square, building alternate blocks first, Fig. 176, so as to give time

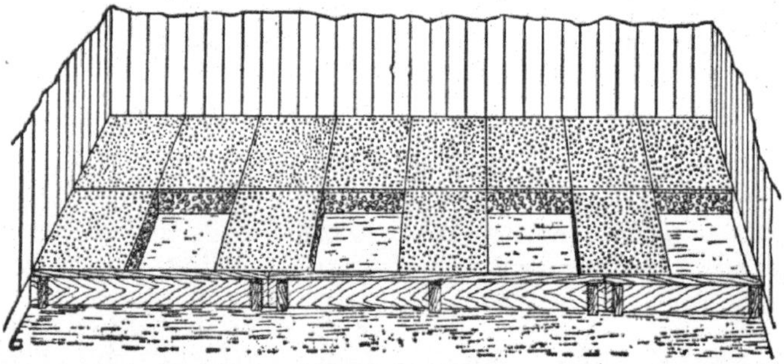

FIG. 176.—Shows method of laying cement floors in blocks to prevent cracking.

for setting and prevent a strong union of the blocks. If the floor is not laid in this manner shrinkage cracks will occur. The concrete should be made only as fast as used and should be thoroughly rammed until the fine cement shows as a layer on the surface. After standing a short time, but before the concrete has set, the finishing surface should be applied and thoroughly troweled until it is even and smooth. Fig. 177 is a cross section of floor and mangers.

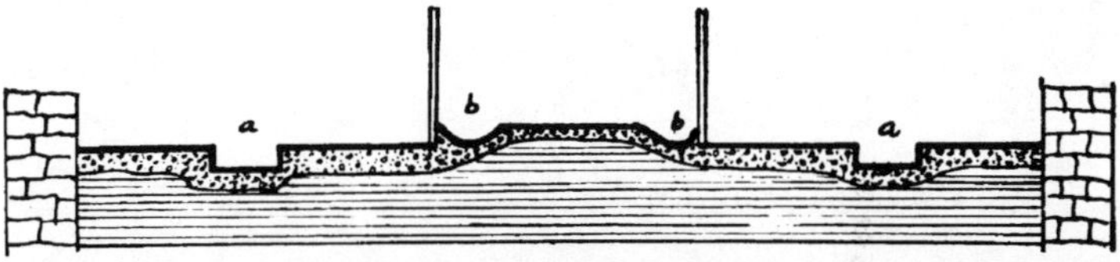

FIG. 177.—Shows cross-section of cement stable floor with mangers and gutters.

For a cellar or creamery floor, where it is desired to have a fine smooth surface, easily cleaned, after troweling, it may be wet with a whitewash brush and some pure dry cement sprinkled over, which is troweled until it is hard, smooth and glossy.

When the second series of blocks in a given tier is made and the surface finished it is necessary to cut through the finishing layer exactly above the joint in the concrete, to prevent cracking, and then neatly round the joint.

474. Cost of Materials for Cement Floor.—Taking materials at the prices given and the concrete 4 inches thick, made in the proportions of **(469)** the cost per 100 square feet of floor, and the amount of materials will be as given in the table below:

The floor made of wood 2 inches thick, laid upon 2x6's, 16 inches from center to center, would cost $4.12 or $4.95 per 100 square feet when the price is $15 or $18 per thousand. This makes the concrete 99 cents per 100 square feet more than the lumber, comparing the lowest prices in each case, and $1.72 more, comparing the higher prices.

Material required for 100 square feet of concrete floor 4 inches thick with one-half inch of facing.

Material.	Amount.	Cost per 100 sq. ft.	
Crushed rock	1.23 cu. yds.	$ .80 per cu. yd.	$ .984
Sand and gravel	.73 cu. yds.	.50 per cu. yd.	.365
Cement	3.76 cwt.	1.00 per cwt.	3.760
		Total	5.109
Crushed rock	1.23 cu. yds.	1.00 per cu. yd.	1.23
Sand and gravel	.73 cu. yds.	.75 per cu. yd.	.55
Cement	3.76 cwt.	1.30 per cwt.	4.89
Total			$6.67

Where crushed rock cannot be had, but coarse gravel and plastering sand are available, a good floor can be made, but more cement must be used, usually 4 of sand and gravel to 1 of cement.

TIES FOR CATTLE.

The methods of tying cattle must vary widely with the taste and objects of the owner. The essential objects to be secured are: (1) comfort for the animals. This is necessary whether the main object is milk, breeding or beef; (2) cleanliness, and (3) economy of time in tying and of space.

FIG. 178.—Wilder swinging stanchion.

FIG. 179.—Scott self-closing swinging stanchion.

475. The Stanchion.—There is no tie for cows, if we except the plain halter or rope, which has been so universally used as one of the forms of stanchions represented in Figs. 178, 179 and 186. It is the simplest, cheapest and most expeditious tie invented and the swinging forms which permit the yoke to turn and to move a little back and forth provide a reasonable amount of comfort; and where the width of the platform is adapted to the size of the animal they secure as high a degree of cleanliness as is practicable.

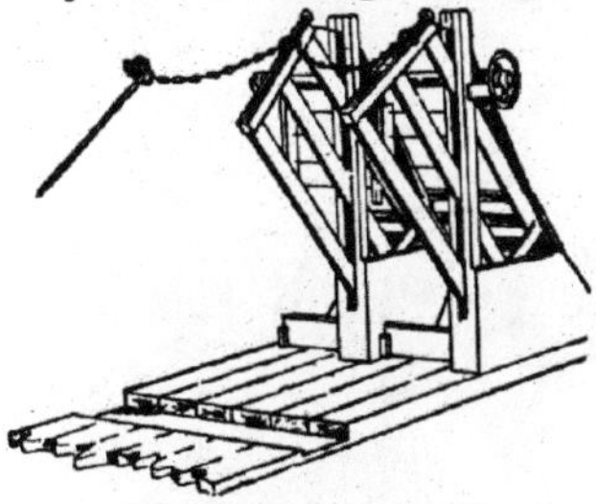

FIG. 180.—Thorp stall.

FIG. 181.--Drown stall.

476. Adjustable Stalls.—The four stalls represented in Figs. 180, 181, 182, and 183 are designed to give the cows the maximum amount of freedom of head movement but to force them to stand close enough to the gutter to prevent the platform being soiled. The manger or the head of the stall is made adjustable so as to crowd the cow back against the chain in the rear which confines her. Practically there is no form of tie which can prevent the cow from soiling the platform upon which she stands on account of the unchangable habit of shortening the body by humping the back when the evacuations occur.

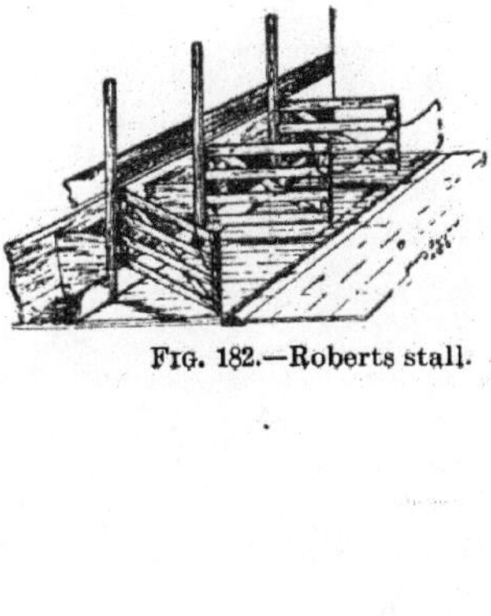

FIG. 182.—Roberts stall.

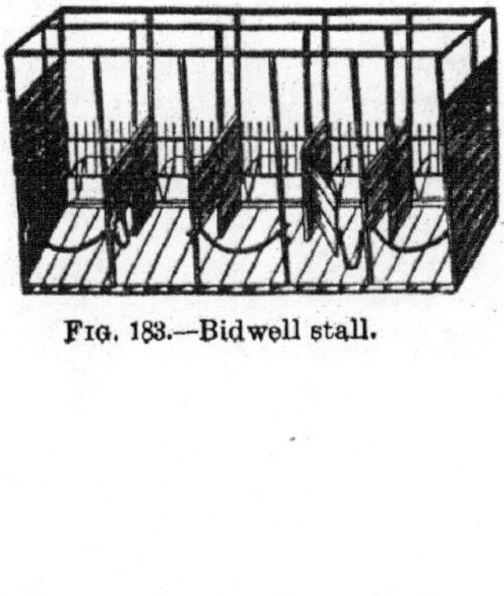

FIG. 183.—Bidwell stall.

The two stalls, Figs. 184, 185, have been designed to secure cleanliness in spite of this habit. In the Newton tie it is expected that while the cow is standing the yoke to which she is tied will force her back sufficiently to prevent the difficulty. In practice, however, there is necessarily so

FIG. 184.--Knapp tie

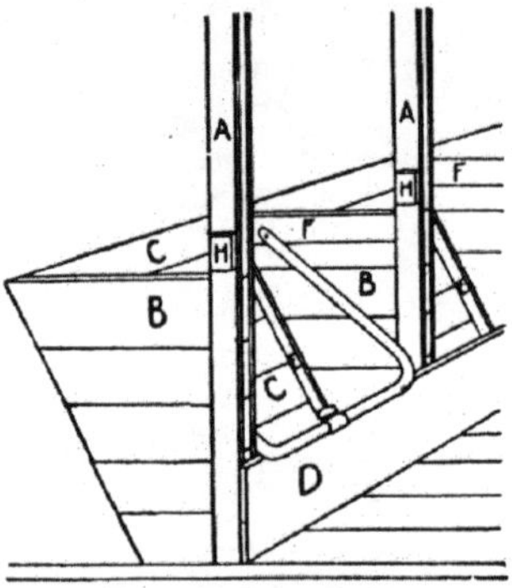

FIG. 185.—Newton tie.

much freedom at the neck that the object is not secured. The "Model tie" provides a bar on the floor, just in front of where the cow's feet are forced to be while standing and feeding, and which is so much of an obstruction that in order to lie in comfort she steps forward enough to lie on the clean bedding.

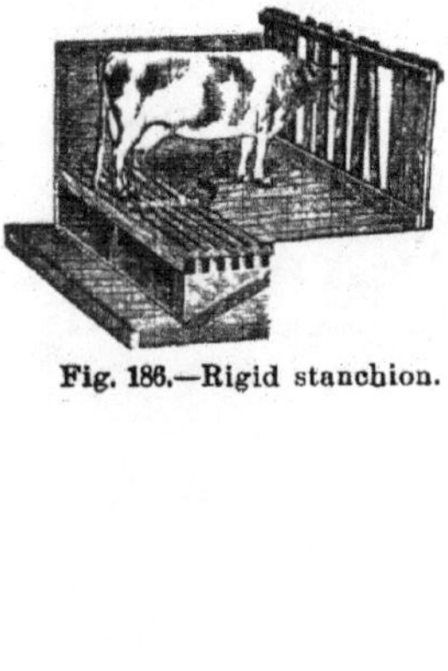

Fig. 186.—Rigid stanchion.

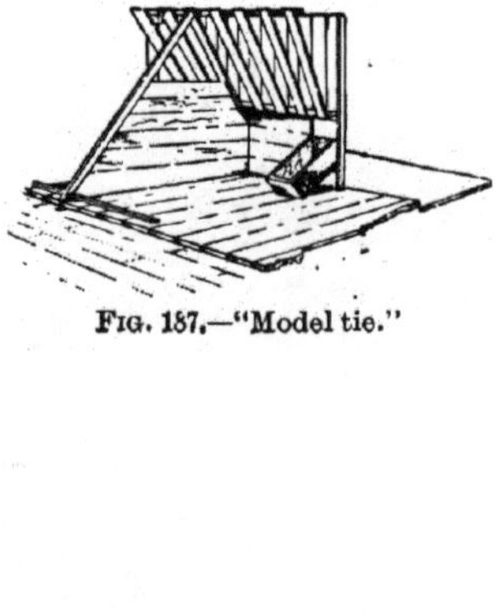

FIG. 187.—"Model tie."

477. Movable Halter Ties.—Another class of ties represented in Figs. 188, 189, attempt to confine the cow in movements forward and backward by using a short chain which slides at the other end in such a manner as to permit freedom of motion up and down.

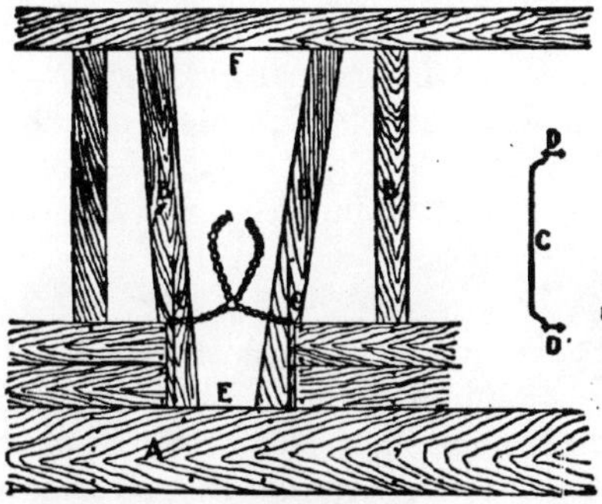

FIG. 188.—Chain tie.

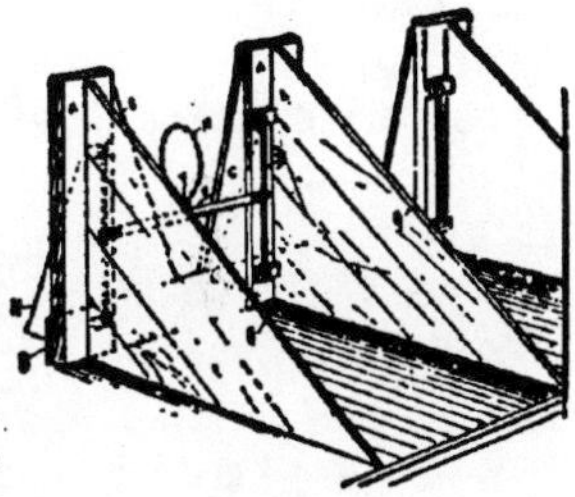

FIG. 189.—Baker tie.

478. Tight Side Partitions.—There is an effort among some feeders to prevent the animal from moving sidewise so as to interfere with the neighbor, either by stepping upon the feet or teats of the cow lying down or of taking the food from the manger. Where such provisions are insisted upon it should be kept in mind that anything which tends to enclose the cow, especially her head, in a tight box tends in a high degree to defeat the purposes of good ventilation by confining the air once breathed about the animal, hence such arrangements should be slatted or else open at the level of the floor.

So, too, wherever box stalls are used these should be slatted or open at the bottom and not "boxes" as they too often are.

479. Tying for Feeding Only.—For calves, young cattle and feeding steers there is perhaps no mode of confining the animals in the stable so good as to give them complete freedom except at the time of feeding, using plenty of bedding on a cement floor which is cleaned as often as needful.

In such cases the stanchion tie is the best as everything is then reduced to the simplest conditions.

480. Mangers.—One of the simplest mangers for feeding cows is represented in Fig. 177, and when made of cement as represented in the cut it is the best for feeding, cleaning and watering, where large numbers of animals are to be handled with the greatest economy. The manger should have an inside width of at least 2 feet, a depth of 8 inches and should have its bottom 3 or 4 inches above the platform upon which the cows stand.

481. The Manure Drop.—This should have a width for adult cows not less than 18 inches and not more than 20 inches. Its depth next to the animals may be 8 inches and on the rear side 6 inches. These dimensions give ample capacity to prevent the walk behind from being soiled and make it easily cleaned.

On some accounts a depth of 6 inches next to the cows and 6 inches in the rear is best; and where a wagon is driven behind the animals to clean the stable a depth behind of only 4 inches gives less hight to lift the manure.

PROVISIONS FOR WATERING.

Where there is a well of ample capacity, and 30 or more cows are kept, the best arrangement, everything considered is to pump the water from the well at the time it is needed. This plan provides water that is both fresh and natural temperature, and does away with expensive storage tanks. In case the power is pumping water faster than is needed it is a simple matter to provide an overflow, returning the water to the well.

482. Watering in the Barn.—In climates having severe winters it is best, if practicable, to water the animals in the barn, and where a good fresh running stream can be maintained the ideal way is to have the water before the cows all the time so that it can be taken when desired.

It is not desirable to keep water standing before the cows continuously as it is certain to become foul; but it may be maintained during the greater part of the day if the drinking basins or troughs are emptied clean each evening.

483. Methods of Watering in the Stable.—We have seen but two reasonably satisfactory methods of watering a large number of cattle in the stable, and these are either to clean the manger and run the water into that or else to have a special long watering trough used for that alone.

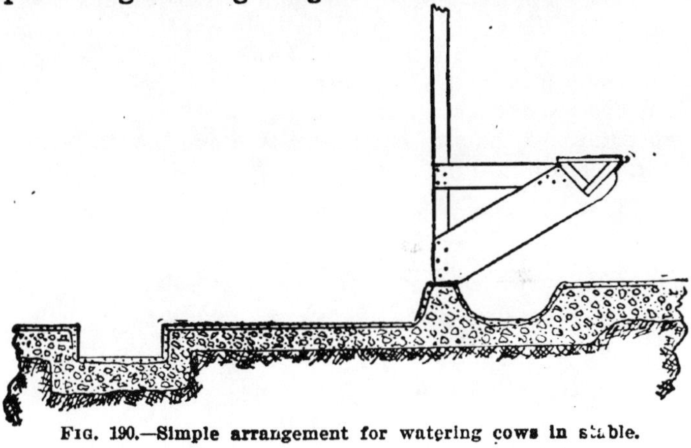

FIG. 190.—Simple arrangement for watering cows in stable.

The simplest arrangement of special trough is represented in Fig. 190, and extends the full length of the stable, the water coming to it from above so that the supply pipe is entirely above ground where it can be gotten at and can be emptied at once after using. The trough is covered its entire length with a hinged lid, but in front of each cow the lid is cut so the cow can raise a section with her nose when drinking, letting it fall when she is through.

484. Storing Water in Tanks.—Where there is a basement barn the best arrangement for a storage water tank is a cement lined cistern beneath the surface in the hill above the barn. Such a cistern is less expensive, is a permanent improvement and will keep the water warm and clean.

We have seen cases where a satisfactory cement lined cistern is built entirely above ground and then covered in by grading a mound of earth about and over it sufficient to make it frost proof. Such a cistern should be provided with a man-hole so that it may be entered if necessary.

485. Watering Trough.—Where stock is watered in the yard a good arrangement for winter, where the ground is porous, is represented in Fig. 191. The tank is a galvanized cylinder 3 or more feet in diameter and 5 feet deep which stands in a dry well 15 or more feet deep and so arranged that the warm air from the bottom of the well all the time surrounds the tank and keeps it from freezing. Water may be pumped into this direct or it may be supplied from the bank cistern. When it is necessary to empty the tank the plug can be removed and the water allowed to drain into the dry well.

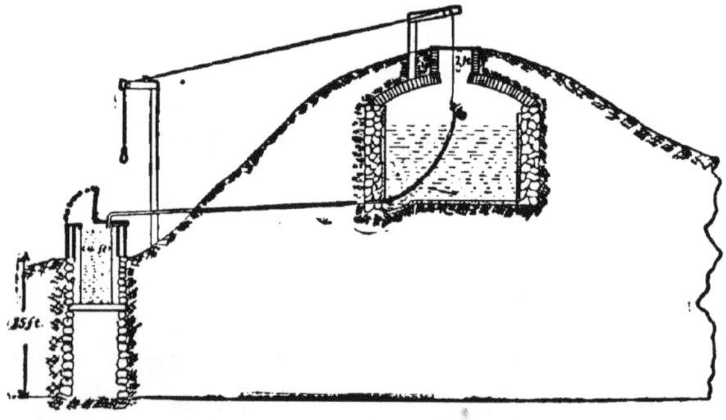

FIG. 191.—Representing a storage reservoir and drinking tank arranged to avoid freezing.

It is of course important to provide a warm jacket about the tank and cover, as represented, so as to assist in keeping the water warm.

ARRANGEMENTS FOR UNLOADING HAY.

486. Unloading Direct from Wagon.—Where the hay is not to be lifted and can be rolled directly from the wagon with the fork into the bay, there is no simpler and more expeditious way; and where the load can be driven to the top of the barn, as represented in Figs. 168, 171 and 173, there is little need of other mechanical arrangements.

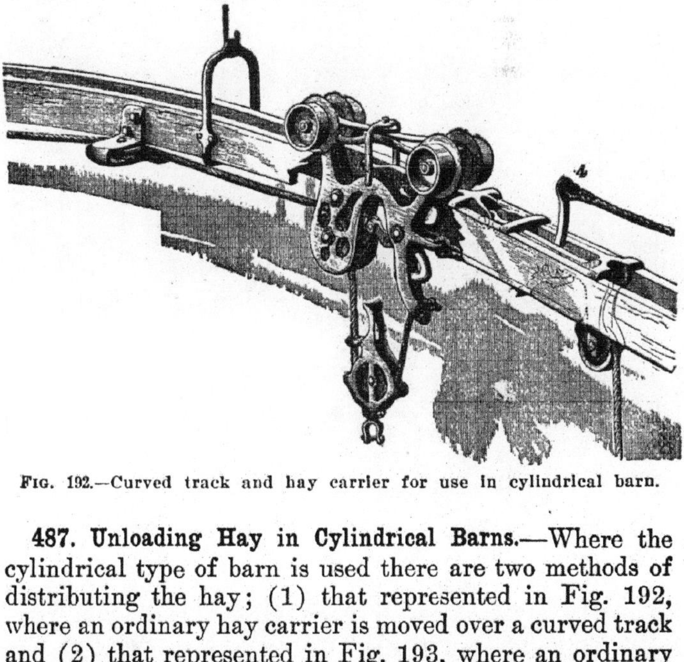

FIG. 192.—Curved track and hay carrier for use in cylindrical barn.

487. Unloading Hay in Cylindrical Barns.—Where the cylindrical type of barn is used there are two methods of distributing the hay; (1) that represented in Fig. 192, where an ordinary hay carrier is moved over a curved track and (2) that represented in Fig. 193, where an ordinary hay carrier delivers the hay upon a central inclined platform, which is turned about by the operator in the bay so as to deliver the hay at any desired point.

488. Tilting Hay Distributor.—It is possible to take advantage of the principle illustrated in Fig. 193 for distrib-

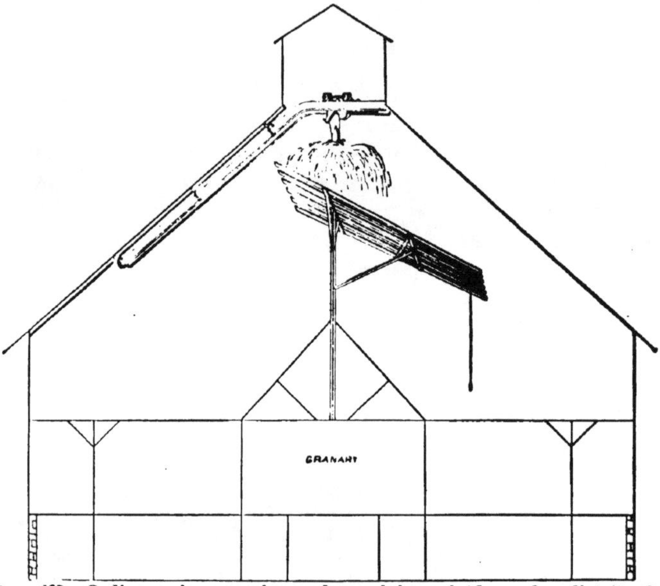

FIG. 193.—Ordinary hay carrier and revolving platform for distributing hay in cylindrical barn.

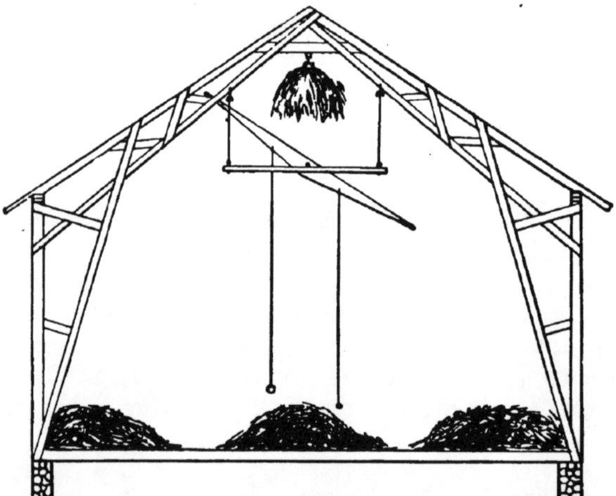

FIG. 194.—Representing a movable, tilting platform for distributing hay in rectangular barn.

uting hay in ordinary rectangular barns, whose timbers are not in the way. Fig. 194 represents a tilting platform, which rocks upon two bars carried by four cables secured to pulleys which roll along tracks or cables secured to rafters, as shown in the cut. With this arrangement hay may be dropped at either side or in the center of the bay, as desired.

CHAPTER XIX.

CONSTRUCTION OF SILOS.

489. Conditions Essential for Preserving Silage.—The only conditions necessary for preserving good corn and clover silage, are close packing in an air tight structure when the materials have reached the right stage of maturity. Whatever means may be adopted to exclude air from these materials will preserve them as silage. If air can find access to it spoiling will be inevitable and the rate and extent will be greater the more readily air can gain access.

490. Depth of Silage.—The depth of silage should be made as great as practicable (1) because in this way the largest amount of feed per cubic foot may be stored. (2) There is less loss relatively at the surface. (3) The strong lateral pressure forces the silage against the walls so closely that less air enters and hence there is less loss.

491. Silo Walls Must be Rigid and Strong.—The outward pressure of cut corn silage when settling, at the time of filling, increases with the depth at the rate of 11 lbs. per square feet for each foot of depth. At a depth of 10 feet the lateral pressure is 110 lbs. per square foot, at 20 feet it is 220 lbs. and at 30 feet 330 lbs.

Because of this great pressure silo walls must be made very strong when they have a depth of 20 or more feet. It is difficult to make deep rectangular silos whose walls will not spread as represented in Fig. 195, and where this takes place the walls are crowded away from the silage so much that air can circulate up and down next to the walls and this results in heavy losses.

In circular silos the pressure is sustained by the tensile strength of the materials in the walls, which gives them the greatest possible advantage.

FIG 195.—Illustrating how the bulging of rectangular wooden silo walls allows air to come down the sides between the walls and the silage, causing it to spoil. The amount of spreading is exaggerated in the figure for clearness of illustration, but it is none the less real.

492. Silo Placed Deeply in the Ground.—In most cases it is best to allow the silo to extend as deeply into the ground as convenience in removing the materials will permit. This can always be as much as 3 feet below the feeding floor and in the case of bank barns where the silo can be placed in the hill a depth of 11 or more feet can easily be secured. Placing the silo deep saves elevating the silage so high when filling and a large portion of it is below frost.

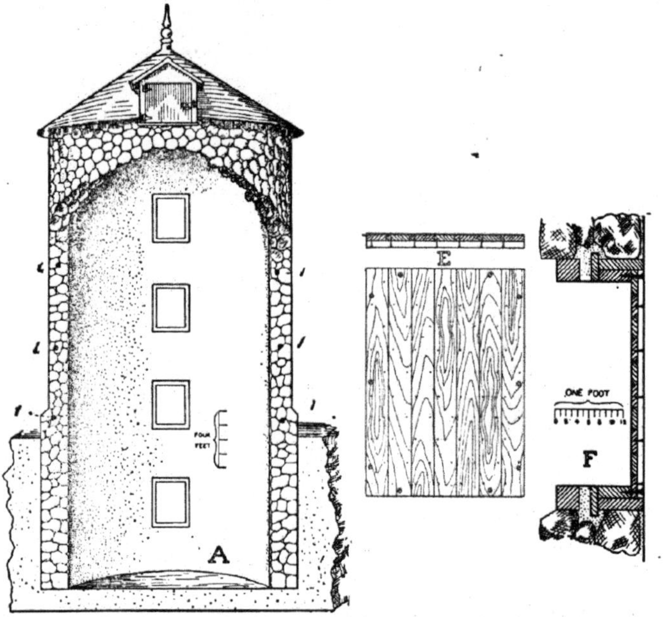

FIG. 196.—Showing an all-stone silo with conical roof and openings for feeding doors; the heavy black dots 1, 1, 1 show where iron rods may be bedded in the wall to prevent cracking from the pressure of the silage. Method of constructing silo door and door jamb for stone silo. E shows cross section of silo door, F shows how the door jamb is made to make it air tight, and how the door is held in place with lag bolts against a gasket of ruberoid roofing.

493. Protection Against Frost.—It is not necessary to build a silo so as to be entirely frost proof in cold climates, but it will pay to build them reasonably warm where they are to be fed from during cold weather. The freezing of silage does not injure it seriously but it is not well to feed it when frozen. If a silo is not to be opened until warm weather no special attention need be given to warmth. If a silo is 10 to 13 feet in the ground and only 20 feet above ground, the settling and the early feeding before severe cold weather will usually have carried the surface of the silage so low that little inconvenience from frost will be experienced even in stone silos. In all the wooden silos, except the questionable stave types, the construction needed for strength and to keep the air from the silage will usually be a sufficient protection against frost.

CONSTRUCTION OF STONE SILOS.

Whenever stone can be had on the farm suitable for building purposes these may be used in silo construction, thus converting idle into active capital. So far as the silo itself is concerned no better or more durable material can be used, and where it can be 10 to 13 feet in the ground the inconveniences from freezing will be small, and the stone silo will be found one of the cheapest of the thoroughly good forms. Great pains should be taken in building the walls to fill all spaces between stones solid with smaller ones and mortar and to have them thoroughly bonded in order to secure strength and prevent cracking.

494. Laying the Wall.—The portion of the silo wall which is below ground better be about 2 feet thick and laid in one of the cheap brands of cement rather than lime, the cement being desirable because lime mortar becomes hard so very slowly in heavy walls, especially below ground. After the wall is two feet above ground good lime mortar may be used, but in this case there ought to be at least two months for the wall to season and set before filling. The upper portion of the silo wall need not be heavier than 18 inches, and if the size of stone permit of it, the outer face of the wall may be drawn in gradually to a thickness of 12 inches at the top.

Too great care cannot be taken in making the part of the wall below and near the ground solid, and especially its outer face, so that it will be strong where the greatest strain will come. It is best also to dig the pit for the silo large enough so as to have plenty of room outside of the finished wall to permit the earth filled in behind to be very thoroughly tamped so as to act as a strong backing for the wall. This is urged because a large per cent. of the stone foundations of wood silos have cracked more or less from one cause or another and these cracks lead to the spoiling of silage.

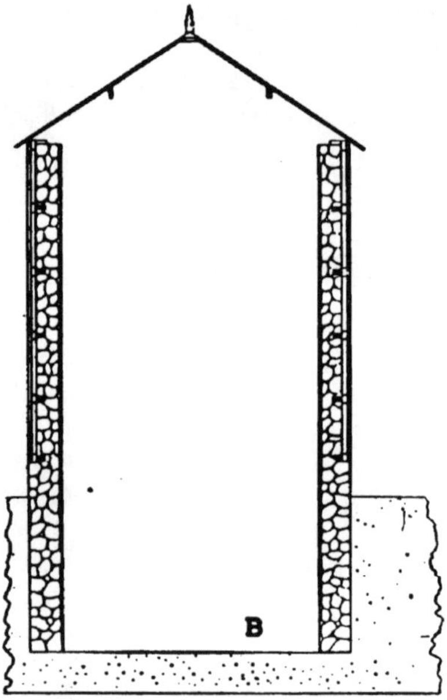

Fig. 197—Shows the method of jacketing a stone silo to protect it against frost; the heavy black squares are blocks bedded into the stone wall to which girts or studs may be nailed to carry the siding.

Flat quarry rock, like limestone, will make the strongest silo wall, because they bond much better than boulders do, and when built of limestone they will not need to be reinforced much with iron rods. It will be best even in this case, however, to use the iron tie rods between the lower two doors.

495. Plastering.—The inner face of the silo wall should be plastered with a thin coat of rich cement not leaner than 1 of cement to 1.5 or 2 of clean sharp sand. If the mortar is not rich and troweled smooth, the acids of the silage will act upon it much more rapidly, dissolving out the lime and leaving it open and porous.

It will usually be prudent also to whitewash these linings every two or three years, especially the lower portion where the silage is longest in contact with the cement, in order to prevent softening, using cement to make the whitewash.

496. Doors.—Doors for filling and feeding should be arranged as represented in A, Fig. 196, and if the lower one is long, cutting out a good deal of the wall, an iron rod should be bedded in the wall above it to prevent cracking between the doors. The rod should be of ⅝ inch round iron bent to the curve of the circle and about 12 feet long. The two ends should be turned short at right angles, so as to anchor better in the mortar.

In deep stone silos, which rise more than 18 feet above the surface of the ground, it will be safest to strengthen the wall between the two lower doors with iron tie rods and, if such a silo is built of boulders, it will be well to use rods enough to make a complete line or hoop around the silo about two feet above the ground, as represented in Fig. 198.

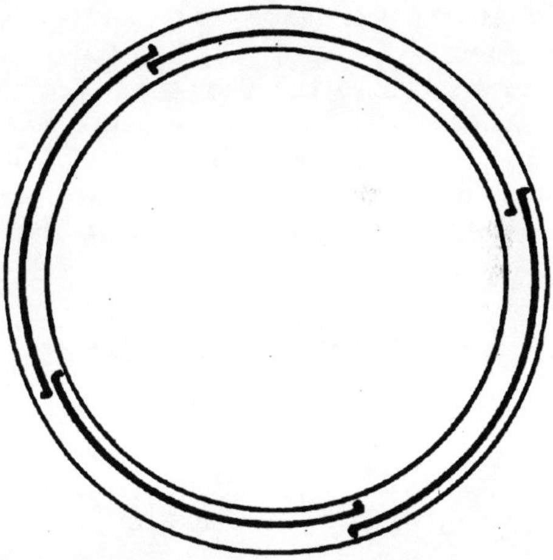

FIG. 198.—Showing method of bedding iron rods in stone, brick or concrete silo walls to increase the strength. The heavy lines with ends bent represent the iron rods.

The door jambs for the stone silo are best made of 4x4's framed together and set far enough apart to give a depth four inches less than the thickness of the wall. This will allow mortar to be filled in between the 4x4's to make an air-tight joint. A 6-inch board may be fitted around the outside of the inner side of the door jambs to form the rabbet for the doors, or the jambs may be made as represented in Fig. 196. There will be slight shoulders left in the round stone silo above and below the doors when these are made flat, and these should be filled out with mortar when plastering, giving a long, gentle slope back to the wall.

The door is best made of two layers of 6-inch flooring, tongued and grooved, crossing at right angles, nailed or screwed together, with a layer of good acid and water proof paper between, as shown at E, Fig. 196. To make the door fit perfectly air-tight there should be tacked to the face of the door jamb, all around, a wide strip of thick roof paper or strips of old worn out rubber belting, and the door drawn up against this with four ½x4 inch lag bolts provided with washers.

If one prefers to do so the door may be made small enough so as to leave a half-inch space between it and the jamb all around, and this space filled with puddled clay after the door is put in place. Either of these methods is better than to tack strips of tar paper over the joints.

CONSTRUCTION OF BRICK SILOS.

Very excellent silos may be made of brick, as represented in Fig. 199, and where brick of a good quality can be obtained at $4.25 to $7.00 per thousand a silo which will last indefinitely may be made at a moderate cost.

497. Foundation.—The foundation of the brick silo is best made of stone, wherever these may be had, carrying the stone work up at least a foot above the ground and beginning below frost line. The brick work will then be set with its inner face flush with the inner surface of the stone work.

If the silo is to be carried 20 or more feet above the stone wall it will be desirable to bed a ⅝-inch round iron hoop into the upper surface of the stone work in order to guard against cracking the wall by the pressure of the first filling before the mortar has had time to thoroughly season, which

FIG. 199.—Shows an all-brick silo with wall 14 inches thick made of three courses of brick, the outer course being set so as to form a 2-inch dead air space as high up as the shoulder.

does not take place until after five or more months. The method of laying the sections of iron rod in the wall is represented in Fig. 198.

498. Walls.—In cold climates it will be best to make the lower portion of the wall, up to within 10 feet of the top, with a 2-inch dead air space, using three courses of brick, thus making the wall 14 inches thick, for all the smaller and medium sized silos. If the silo is to exceed 24 feet inside diameter the lower third of the brick wall should be made of four courses of brick and 18 inches thick, the second third 14 inches thick, and the upper third 8 inches, solid. The dead air space should be next to the outside and this course of brick should be tied to the inner wall as frequently as necessary to make it stable.

499. Strengthening the Walls.—The tendency of the pressure of the silage to crack the walls of round silos increases with the depth and with the diameter of the silo. The tendency of the silage to burst a silo 26 feet inside diameter is twice as great as in one 13 feet in diameter and the same depth, and this makes it necessary to strengthen the walls of the larger brick silos. In all brick silos there should be an iron tie rod bedded in the wall, in the manner illustrated in Fig. 198, between each of the lower doors to compensate for the weakening caused by the doors; and in the larger silos these ties should extend entirely around the silo in the manner shown in Fig. 198.

500. Wetting Brick.—It is very important in laying the brick for a silo wall that they should be wet and especially if the work is done in hot, dry weather. If this is not done the brick will so completely dry out the mortar that it cannot set properly and become strong.

501. Making Walls Air Tight.—There are several ways in which this may be done, and some of these will be given in the reverse order of their effectiveness.

1. After the wall is finished it may be simply given two coats of thick cement whitewash, and this repeated every two or three years as the acid of the silage dissolves it away.
2. The face of the brick wall may be given a good, rich

coat of cement plaster, one-fourth to one-half an inch thick, and then this be kept whitewashed so as to neutralize the acid and prevent it from softening the cement.

3. The wall, or at least the inner portion, may be laid in rich cement mortar, making the horizontal joints about one-fourth of an inch thick and the vertical ones a half inch thick, taking great care to get all joints of the inner tier of brick thoroughly filled with mortar. This method will place the cement where it will not be as readily affected by the acids and frost and does away with the necessity of plastering, care being taken to lay the brick smoothly and to point the joints carefully. Milwaukee cement will answer for this work. Whitewashing the inner face of such a lining will be sufficient for smoothness and tightness.

4. The very best possible lining which could be made would be secured by using the small, thin size of vitrified paving brick. These may be set on edge, to reduce both the cost and the number of cement joints. It will be necessary to tie this course occasionally to the main wall by turning a brick endwise. Rich cement mortar should be used and the joints made thin but thoroughly filled with the mortar. Such a lining would give a surface like a stone jug, thoroughly air-tight and indefinitely permanent.

502. Doors.—The jambs may best be made of 3x6's or 3x8's rabbetted two inches deep to receive the door on the inside. The center of the jambs outside should be grooved and a tongue inserted projecting three-fourths of an inch outward to set back into the mortar and thus secure a thoroughly air-tight joint between the wall and jamb.

The doors are best made as described under the stone silo, of two layers of matched flooring with paper between.

CONSTRUCTION OF BRICK-LINED SILOS.

Next to the all-masonry silos in point of durability and efficiency must be ranked the masonry lined silos, of which there are several types, as follows: (1) Stone silos, jacketed with wood; (2) concrete lined silos; (3) brick lined silos; (4) lathed and plastered silos.

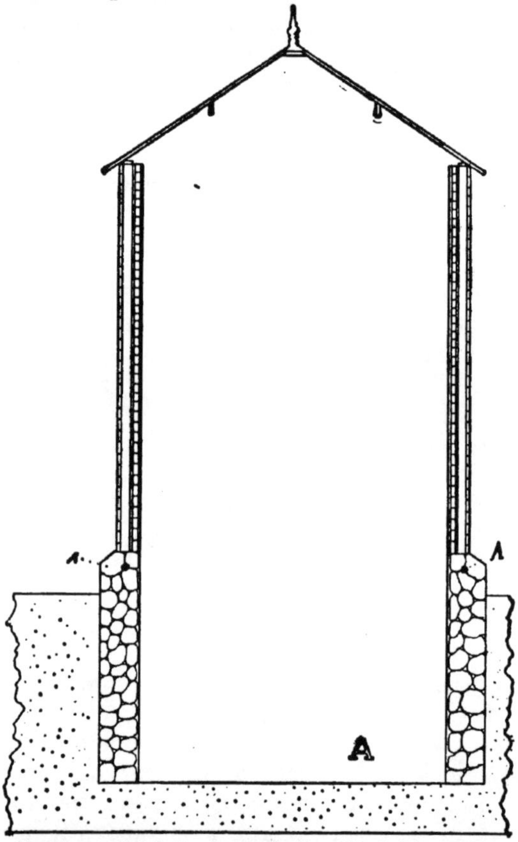

FIG. 200.—Showing a brick lined round silo with bricks set on edge and plastered with cement. Dots A, A show where an iron rod may be bedded in the wall to prevent spreading.

Of these types the brick lined silo is likely to come into the more general use, and its construction will be described first.

503. Foundation and Sill.—Like the brick silo, this form should have a stone foundation, wherever it is practicable to obtain the material for it. Upon this should first be laid the sill made of 2x4's cut in two-foot lengths with the ends beveled so that they may be toe-nailed together and bedded in cement mortar upon the wall in the manner represented in Fig. 201. The sill is set just far enough back from the inside of the wall so that when the brick are laid they come flush with the inside of the silo wall.

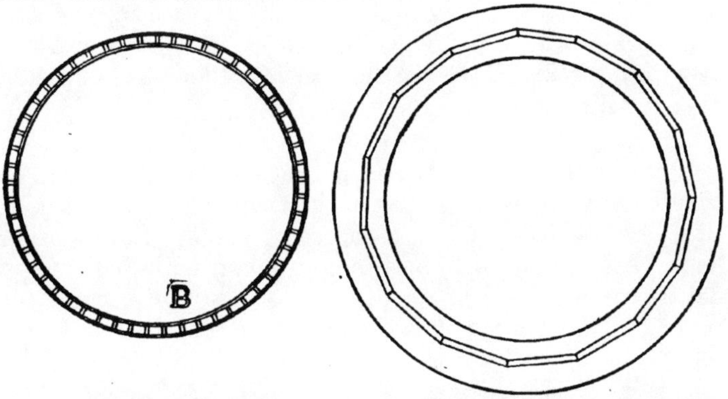

FIG. 201.—Showing method of making the sill of brick lined and of round wood silos. B, plan of studding for all-wood, brick lined or lathed and plastered silo.

504. Setting Studding.—The 2x4 studding are next set up and toe-nailed to the sill. A stud is first set at each angle of the sill, plumbed and stayed from a post set in the center of the silo. After four or five of these are set and plumbed from the center they should be stayed from side to side by tacking to them a strip of half-inch sheeting bent around the outside as high up as a man can reach, taking care to get each stud plumb in this direction before staying. After the alternate studs have been set up in this manner the intervening ones may be put in place, toe-nailed to the sill and stayed to the rib holding the others in place.

505. Sheeting.—The next step should be to put on the outside layer of sheeting which, for all of the silos less than

30 feet in diameter, should be three-eighths inch lumber made by buying a good quality of fencing and taking it to the mill to have it sawed in two. The usual price for sawing fencing in two in this way is $1.00 per thousand. The reason for getting fencing and having it sawed in this manner is to save expense. It is the custom of dealers to charge the same price for half inch as for inch lumber, and hence buying good fencing and having it sawed reduces the cost just one-half, less the cost of sawing. The studding should be covered inside and out with this sheeting, nailing thoroughly with 8-penny nails, two nails in each board at every stud. The object of the boards is to act as hoops and give the silo the needed strength.

506. Siding.—If the silo is out of doors it will need to be covered with house siding with the thick edge rabbetted, or else veneered with a single course of brick. Several silos have been sided with half-inch lumber with both edges beveled at an angle of 45 degrees to take the place of the rabbet. This method gives greater strength, but is not likely to keep out rain as thoroughly.

507. Lining.—The brick lining of the silo should be laid in rich Milwaukee, Akron or Louisville cement mortar, the bricks being previously wet. The most rigid lining will be secured by laying the brick flatwise, making the layer 4 inches thick, but with one-half the amount of brick they may be set on edge, thus considerably lessening the cost. If set on edge, as represented in Fig. 200, a row of spikes should be driven into the studding through the joints of every fourth course to hold the brick more securely in place until the cement has had time to season.

The mortar should not be made more than one-fourth of an inch thick and great care should be taken to leave no open space anywhere. The necessity of plastering the wall may be avoided by filling behind each brick with one-half an inch of mortar, which will keep out the air as well as if on the front side and there will be the additional advantage of the cement not coming in direct contact with the silage juices. If care is taken in setting the brick so as to secure a smooth face, pointing the joints carefully, it will not be necessary to even whitewash the wall and a permanent lining requiring no attention will thus be secured.

In this form of silo the brick may have one face filled with coal tar, or the vitrified paving brick may be used, giving a lining wholly air tight and permanent.

ROUND PLASTERED SILO.

Where brick are high, lumber low, and clean, sharp sand may be readily obtained, a cement plastered lining may be made to take the place of the brick lining, using the Milwaukee, Akron, Rosendale or Louisville cement in making the mortar. The first coat is usually made with hair and a little lime to make it hang to the wall better.

There are a good many of these lathed and plastered cylindrical silos in Racine and Kenosha counties in Wisconsin, and across the line in Illinois. Some of these have been in use since 1889 and have given good satisfaction.

508. Construction.—The frame work of the silo should be made exactly like that of the silo with brick lining except that there should be two layers of half-inch sheeting on the inside with a layer of 3-ply Giant P. and B. paper between, or other of as good quality.

After the woodwork of the silo has been completed it should be lathed and plastered with a cement mortar made of 1 of cement to 2 of sand.

If wood lath are used there should be furring strips of lath nailed to each stud up and down and the lath nailed through these. If metal lath is used this may be nailed directly to furring strips of lath nailed to the studding over the lining and the plastering then done.

It should be understood that it would not do to lath and plaster a rectangular wood silo because the springing of the walls would crack the cement. It should be understood further that on account of the fact that the layer of cement is so thin it is a matter of greater importance to keep the

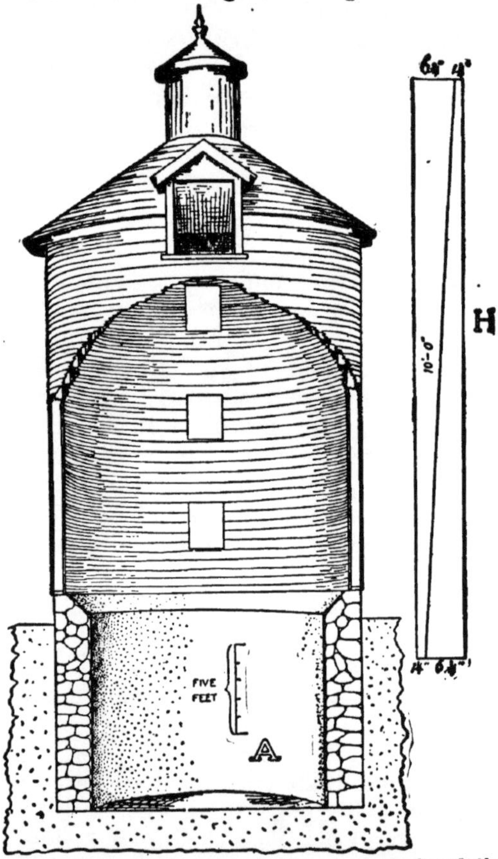

FIG. 202.—Showing an all-wood round silo on stone foundation. H represents a method of sawing boards for the conical roof.

surface whitewashed to prevent the acid from softening the cement and rendering it porous. It is because of this also that two layers of lining with paper between are recommended.

CONSTRUCTION OF ALL WOOD SILOS.

Up to the present time more silos have been built of wood than of any other material, and since 1891, the majority of wood silos built have been after the model represented in Fig. 202. Very few silos of the rectangular type are now built unless they be of stone.

509. Foundation.—There should be a good, substantial masonry foundation for all forms of wood silos and the woodwork should everywhere be at least 12 inches above the earth to prevent decay from dampness. There are few conditions where it will not be desirable to have the bottom of the silo 3 feet or more below the feeding floor of the stable and this will require not less than 4 to 6 feet of stone, brick, or concrete wall. For a silo 30 feet deep the foundation wall of stone should be 1.5 to 2 feet thick.

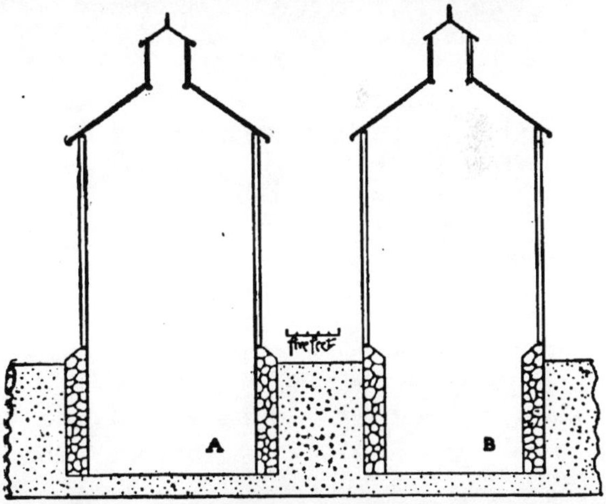

FIG. 203.—Showing two methods of placing the wood, brick lined or lathed and plastered silo on a stone foundation. A shows the silo set with upper portion flush with the inside of the stone wall, and B shows the upper portion flush with the outside of the stone wall.

The inside of the foundation wall may be made flush with the woodwork above, as represented in Fig. 203 A, or the building may stand in the ordinary way, flush with the outside of the stone wall, as represented in Fig. 203 B. In both cases the wall should be finished sloping as shown in the drawings.

510. Cementing the Bottom.—After the silo has been completed the ground forming the bottom should be thoroughly tamped so as to be solid and then covered with two or three inches of good concrete made of 1 of cement to 3 or 4 of sand and gravel. The amount of silage which will spoil on a hard clay floor will not be large, but enough to pay a good interest on the money invested in the cement floor. If the bottom of the silo is in dry sand or gravel the cement bottom is imperative to shut out the soil air.

511. Tying Top of Wall.—In case the wood portion of the silo rises 24 or more feet above the stone work and the diameter is more than 18 feet it will be prudent to stay the top of the wall in some way.

If the woodwork rises from the outer edge of the wall, then building the wall up with cement so as to cover the sill and lining as represented in Fig. 207 will give the needed strength, because the wood-work will act as a hoop; but if the silo stands at the inner face of the wall, it will be best to lay pieces of iron rod in the wall near the top to act as a hoop.

Where the stone portion of the silo is high enough to need a door it is best to leave enough wall between the top and the sill to allow a tie rod of iron to be bedded in this portion. So, too, the lower door in the woodwork of the silo should leave a full foot in width below it of lining and siding uncut to act as a hoop, where the pressure is strongest.

512. Sills and Studding.—The sill in the all-wood silo may be made of a single 2x4, cut in 2-foot lengths, in the manner represented in Fig. 201 and described under the brick lined silo.

The studding of the all-wood round silo need not be larger than 2x4 unless the diameter is to exceed 30 feet, but they should be set as close together as one foot from center to center, as represented in Fig. 201, B. This number of studs is not required for strength but they are needed in order to bring the two layers of lining very close together so as to press the paper closely and prevent air from entering where the paper laps.

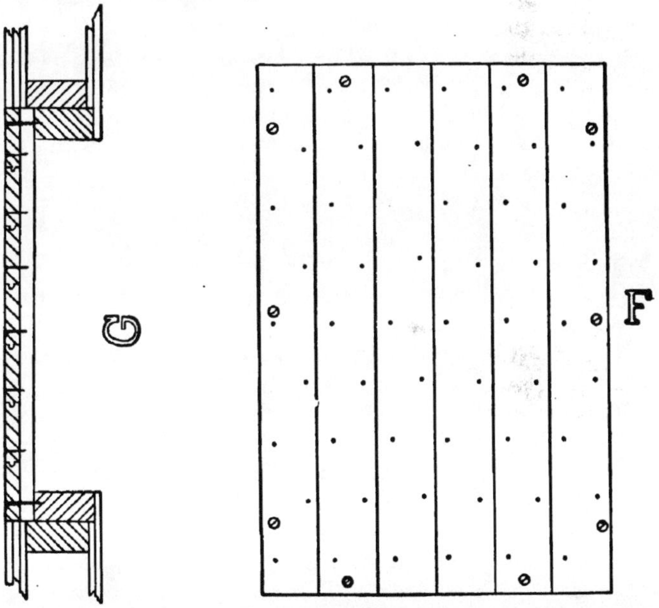

FIG. 204.—Showing the construction of the door for the all-wood silo G is a cross-section of the door resting against the door jamb, which is provided with a gasket of three-ply ruberoid roofing and held in place with four lag bolts and washers, the door opening on the inside. F is a front view of the door made of two layers of four inch or six inch tongued and grooved flooring with a layer of three-ply acid and water proof P. & B. paper between.

To stay the studding a post should be set in the ground in the center of the silo long enough to reach about 5 feet above the sill and to this stays may be nailed to hold in place the alternate studs until the lower 5 feet of outside sheeting has been put on. The studs should be set first at the angles formed in the sill and carefully stayed and plumbed on the side toward the center. When a number of these have been set they should be tied together by bending a strip of half-inch sheeting around the outside as high up as a man can reach, taking care to plumb each stud on the side before nailing. When the alternate studs have been set in this way the balance may be placed and toe-nailed to the sill and stayed to the rib, first plumbing them sideways and toward the center.

On the side of the silo where the doors are to be placed the studding should be set double and the distance apart to give the desired width. A stud should be set between the two door studs as though no door were to be there and the doors cut out at the places desired afterwards. The construction of the door is represented in Fig. 204.

513. Sheeting and Siding.—The character of the siding and sheeting will vary considerably according to conditions and size of the silo.

Where the diameter of the silo is less than 18 feet inside and not much attention need be paid to frost, a single layer of beveled siding, rabbetted on the inside of the thick edge deep enough to receive the thin edge of the board below, will be all that is absolutely necessary on the outside for strength and protection against weather. This statement is made on the supposition that the lining is made of two layers of fencing split in two, the three layers constituting the hoops.

If the silo is larger than 18 feet inside diameter, there should be a layer of half-inch sheeting outside, under the siding.

If basswood is used for siding care should be taken to paint it at once, otherwise it will warp badly if it gets wet before painting.

In applying the sheeting begin at the bottom, carrying the work upward until staging is needed, following this at once with the siding. Two 8-penny nails should be used in each board in every stud, and to prevent the walls from getting "out of round" the succeeding courses of boards should begin on the next stud, thus making the ends of the boards break joints.

When the stagings are put up new stays should be tacked to the studs above, taking care to plumb each one from side to side; the siding itself will bring them into place and keep them plumb the other way if care is taken to start new courses as described above.

514. Forming the Plate.—When the last staging is up the plate should be formed by spiking 2x4's, cut in two-foot lengths, in the manner of the sill, and as represented in Fig. 205, down upon the tops of the studs, using two courses, making the second break joints with the first.

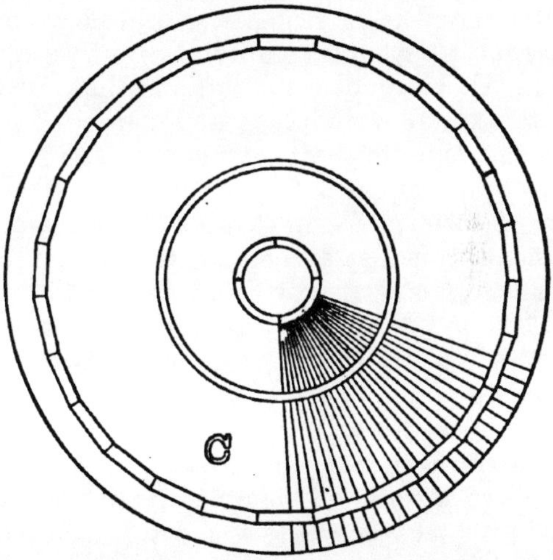

FIG. 205.—Showing construction of conical roof of round silo where rafters are not used. The outer circle is the lower edge of the roof, the second circle is the plate, the third and fourth circles are hoops to which the roof boards are nailed. The view is a plan looking up from the under side.

515. Lining for Wood Silos.—There are several ways of making a good lining for the all wood round silo, but whichever method is adopted it must be kept in mind that there are two very important ends to be secured with certainty. These are (1) a lining which shall be and remain strictly air tight, (2) a lining which will be reasonably permanent.

Galvanized Iron in Silo Lining.—The tightest lining for a wood silo may be made with a light weight of galvanized iron, No. 28 to No. 32. Where the silos are 18 feet in diameter or less this may be put directly upon the studding, buying the strips 8 feet long and 36 inches wide, so as to be nailed on up and down and exactly cover the space between three or four studs. Headers should be put in every 8 feet to nail the ends of the sheets to between the studs, and these are best when sawed to the curve of the silo. The metal should be put on with roofing nails, nailing close so as to make the joints tight.

After the metal is in place it should be given a heavy coat of asphalt paint, taking special care to make it heavy where the nails and laps come so as to shut out the air.

When the metal is in place and painted it should be covered with a layer of sheeting made the same as that used outside, by splitting good fencing in two. The object of this layer of sheeting is, first to take the pressure of the silage; second, to act as a hoop for strength, and third, to keep the silage from softening and wiping the paint from the metal lining. Were it not for the fact that the heat of the silage tends to soften the paint, and its settling to wipe it off, it would be better to let the metal come next to the silage.

Where the silo is more than 18 feet in diameter it will be best to use two layers of fencing split in two, placing the galvanized iron between the two layers. In these cases the sheets of metal may be put on horizontally, using those 36 inches wide.

All Wood Lining of 4-inch Flooring—If one is willing to permit a loss of 10 to 12 per cent. of the silage by heating, then a lining of tongued and grooved ordinary 4-inch white pine flooring may be made in the manner represented in Fig. 206, where the flooring runs up and down.

When this lumber is put on in the seasoned condition a single layer would make tighter walls than can be secured with the stave silo where the staves are neither beveled nor tongued and grooved.

In the silos smaller than 18 feet inside diameter the two layers of boards outside will give the needed strength, but when the silo is larger than this and deep there would be needed a layer of the split fencing on the inside for strength; and if in addition to this there is added a layer of 3-ply Giant P. and B. paper, a lining of very superior quality would be thus secured.

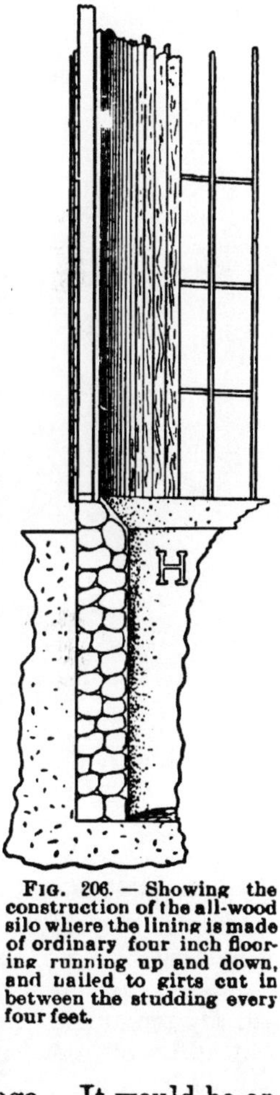

Fig. 206. — Showing the construction of the all-wood silo where the lining is made of ordinary four inch flooring running up and down, and nailed to girts cut in between the studding every four feet.

Lining of Half-inch Boards and Paper.—Where paper is used to make the joints between boards air tight, as represented in Fig. 207, it is extremely important that a quality which will not decay and which is both acid and water-proof be used. A paper which is not acid and water-proof will disintegrate at the joints in a very short time and thus leave the lining very defective.

Great care should be taken to have the two layers of boards break joints at their centers, and the paper should lap not less than 8 to 12 inches.

The great danger with this type of lining will be that the boards may not press the two layers of paper together close enough but that some air may rise between the two sheets where they overlap and thus gain access to the silage. It would be an excellent precaution to tack down the edges of the paper closely with small carpet tacks where they overlap, and if this is done a lap of 2 inches will be sufficient.

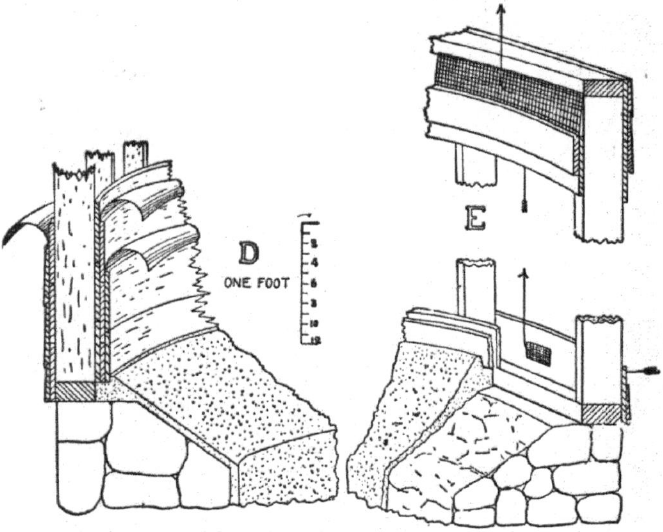

FIG. 207.—D, Showing method of constructing the all-wood round silo and connecting it with the wall flush with the outside. This figure shows the most substantial form of construction with three layers of half-inch lumber and two layers of three-ply acid and water proof P. & B. paper between them. A very excellent silo is made after this plan omitting the inner layer of lining and paper and the layer of paper on the outside. With small silos 15 feet in diameter only the siding on the outside is necessary for strength and protection against weather. E, Showing method of construction for ventilating the spaces between the studding in all-wood and lathed and plastered silos. The lower portion shows the intakes of fresh air from the outside at the bottom, and the upper portion shows where the air enters the silo at the plate to pass out at the ventilator in the roof.

Such a lining as this will be very durable because the paper will keep all the lumber dry except the inner layer of half-inch boards, and this will be kept wet by the paper and silage until empty and then the small thickness of wood will dry too quickly to permit rotting to set in.

A still more substantial lining of the same type may be secured by using two layers of paper between three layers of boards, as represented in Fig. 207, and if the climate is not extremely severe, or if the silo is only to be fed from in the summer, it would be better to do away with the layer of sheeting and paper outside, putting it on the inside, thus securing two layers of paper and three layers of boards for the lining with the equivalent of only 2 inches of lumber.

516. Construction of Roof.—The roof of cylindrical silos may be made in several ways, but the simplest type of construction and the one requiring the least amount of material is the cone, represented in Figs. 202 and 205.

If the silo is not larger than 15 feet inside diameter no rafters need be used, and only a single circle, like that in the center of Fig. 205. This is made of 2-inch stuff cut in section in the form of a circle and two layers spiked together, breaking joints.

517. Ventilation of Silos.—Every silo which has a roof should be provided with ample ventilation to keep the underside of the roof dry and in the case of wood silos, to prevent the walls and lining from rotting. One of the most serious mistakes in the early construction of wood silos was the making of the walls with dead-air spaces which, on account of the dampness from the silage, lead to rapid "dry rot" of the lining.

In the wood silo and in the brick lined silo it is important to provide ample ventilation for the spaces between the studs, as well as for the roof and the inside of the silo, and a good method of doing this is represented in Fig. 207, E, where the lower portion represents the sill and the upper the plate of the silo. Between each pair of studs, where needed, a one and one-fourth inch auger hole to admit air is bored through the siding and sheeting and covered with a piece of wire netting to keep out mice and rats. At the top of the silo on the inside the lining is only covered to within two inches of the plate and this space is covered with wire netting to prevent silage from being thrown over when filling. This arrangement permits dry air from outside to enter at the bottom between each pair of studs and to pass up and into the silo, thus keeping the lining and studding dry and at the same time drying the under side of the roof and the inside of the lining as fast as exposed. In those cases where the sill is made of 2x4's cut in 2-foot lengths there will be space enough left between the curved edge of the siding and sheeting and the sill for air to enter, so that no holes need be bored as described above and represented in Fig. 207 E. The openings at the plate should always be provided and the silo should have some sort of ventilator in the roof. This ventilator may take the form of a cupola to serve for an ornament as well, or it may be a simple galvanized iron pipe 12 to 24 inches in diameter, rising a foot or two through the peak of the roof.

518. Painting Silo Lining.—It is impossible to so paint a wood lining that it will not become wholly or partly saturated with the silage juices. This being true, when the lining is again exposed when feeding the silage out, the paint greatly retards the drying of the wood work and the result is decay sets in, favored by the prolonged dampness. For this reason it is best to leave a wood lining naked or to use some antiseptic which does not form a water proof coat.

THE STAVE OR TANK SILO.

We have examined personally 19 stave silos and have made a careful study of the unavoidable losses in one of these. We have also studied the unavoidable loss in two kinds of small stave silos. As a result of these observations it has been demonstrated that there are several very serious objections to stave silos intended as permanent buildings out of doors. Some of these are stated below:

1. When the silo is empty the staves shrink and loosen the hoops and in this condition the wind racks the building, getting it out of round, out of plumb, and out of place upon the foundation. It is much more easily blown down than

other forms of silos. Two of the fourteen out-of-door silos visited had been blown down; one of these was abandoned and the hoops sold to another farmer; the other was set up again at the expense of a day's drive for new staves and getting the carpenters to set it up, the accident happening just as they were ready to fill the silo.

A third silo of the fourteen out-of-doors we visited had moved on the foundation so much that I could put my arm up through between the stone wall and the outside of the staves. This silo had been stayed to the end of the barn, using fence wire for guy rods.

Three others of the fourteen out-of-door stave silos had been found so unsatisfactory that they were subsequently lined on the inside to prevent the silage from spoiling, and in two of these three the inner lining has rotted out on account of the dampness which the outside staves confines.

2. There is great danger of the hoops being broken by the intense pressure of the silage increased by the swelling of the staves. In one of the silos visited eight out of ten hoops on one side of the silo and six out of ten on the opposite side had sheared in two the 2x4's used for lugs; but, by a fortunate coincidence, two of the ten hoops remained intact to hold the silo up, assisted by some half-inch boards which had been bent around the inside of the silo at the top to prevent the staves from falling in.

In another silo where 4x4 oak pieces had been used as lugs, the 2-inch iron washers had been crushed their full depth of one-half inch into the hard wood and two of the pieces of wood had been badly injured by the severe strain upon them.

In a fourth silo where the hoops were provided with iron lugs the staves on one side had been thrown into the silo by the swelling of the wood.

It is urged by the advocates of these silos that with a little care and judgment the nuts of the hoops may be tightened or loosened as needed and such accidents averted. There is enough truth in this statement to induce many farmers with limited means to take the risk, but life is too short and there are too many other things to engross the attention of good farmers for them to lie awake nights wondering whether the silo hoops are too tight or too loose.

3. Staves do not contain the same amount of sapwood in all parts and for this reason shrink unequally, with the result that after 3 or 4 years' use there are places which do not close up tightly on swelling and which open again on the sunny side of the silo, and thus admit air, even where the silage is in contact with them.

Three of the silos visited showed these peculiarities, and in one of them visited last winter we could see through between several staves on the south side of the silo close to the silage surface, on the inside.

4. The expansion and contraction of the staves during wetting by the silage and drying when the silo is empty makes it difficult to securely anchor a permanent roof and impossible to connect the staves permanently with the foundation, so as to be air-tight. Something must be done each season to cement the joints between the staves and foundation or air will enter.

5. There is no reason to hope that good silage with small losses in dry matter can be made in the stave silos which are not carefully constructed of good lumber with the staves both beveled, and tongued and grooved. It is really more difficult to make a stave silo air tight than it is to make a tank water-tight, and we have found by careful tests that the unavoidable losses in a new stave silo next to the walls were as high as 24 to 28 per cent.

510. Construction of Stave Silos.—There are three methods adopted in the construction of these silos. The best and only one which should be used in the permanent silo is that represented in Fig. 208, where the staves are both beveled and tongued-and-grooved; the second is where the staves are beveled so that the flat surfaces fit together accurately as water tanks are made; the third plan uses the lumber without either beveling or tonguing-and-grooving, and this both observation and principles of construction indicate should be adopted with very great hesitation and as a temporary makeshift only until more experience and exact knowledge has been obtained regarding their permanent efficiency.

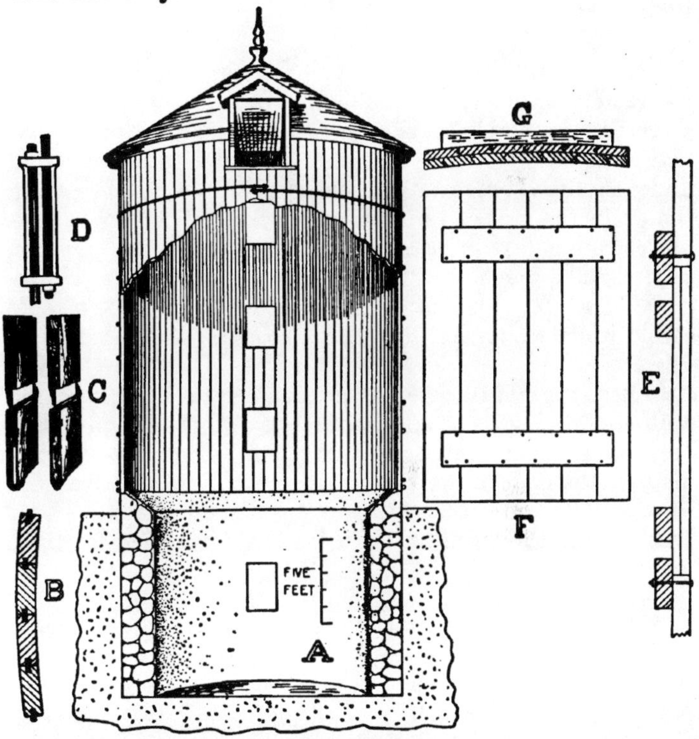

FIG. 208.—Showing the construction of the stave silo. A shows the silo complete on stone foundation with four feeding doors. B is cross-section of four staves showing how they are tongued and grooved to make them air tight. C shows a method of splicing staves. D shows iron lugs for tightening hoops. F is front view of door viewed from outside. G cross-section of same. E is a vertical section showing the shoulder against which the door rests, and upon which should be a gasket of three-ply ruberoid roofing. The door should also be drawn tight against it with four lag bolts and washers, opening from the inside.

This third plan has been recommended because the first cost is relatively low and because it is assumed that the pressure due to the swelling of the wood and the rigidity of the hoops will result in crushing the edges of the staves together so as to make a sufficiently tight joint to preserve the silage.

520. Lumber for Staves.—The lumber selected for the staves of this type of silo should be of the grade known commercially as "tank stuff," and lumber freest from knots and straightest grained is best. Wood is quite air-tight under low pressures in directions across the grain but along the grain the air passes more or less freely. The Washington cedar appears to be an excellent wood for this purpose, as it shrinks much less than the pine after the silage is removed and, for this reason, the building will be much more stable when empty and less liable to burst the hoops when filled.

Where the silo is to be deeper than can readily be secured with single lengths of lumber the staves may be spliced in the manner represented at C, Fig. 208, where a saw-cut is made in the ends of the two staves and a piece of galvanized iron, a little wider than the stave is slipped into it. This crushes into the wood on the sides and forms a water tight joint.

521. Foundation of Stave Silo.—On account of the tendency of the stave silo to work off from the wall when empty a flat cement floor has been recommended, made of sand and gravel or crushed rock, forming a bed of concrete about 12 inches thick. This is perhaps as good as can be done under the circumstances but it precludes the extension of the silo into the ground.

If the silo stands upon a stone wall, as represented in Fig. 208, it will be prudent to have a shoulder jutting into the silo as much as 2 inches and a similar amount on the outside, to permit of some movement on the foundation.

522. Hoops for Stave Silo.—Five-eighths inch round iron rods, in about 16-foot lengths, form the best hoops and they should be provided with long threads and joined with iron lugs and nuts, as represented in D, Fig. 208. The iron lugs should always be used in preference to the 2x4's or 4x4's because they are better in every way. So, too, should they be used in preference to posts set up against the silo outside or shaped to act as a part of the staves as has been recommended. In visiting over 100 silos in 1891 it was found that wherever a silo lining had a heavy timber back of it, the holding of dampness caused rotting there in three or four years, and it is quite certain that the use of iron lugs is the safest way to avoid this danger in stave silos.

523. Doors for Stave Silos.—A good method of constructing doors for the stave silo is represented in Fig. 208. Two inch lumber is bolted to the staves on the outside, projecting two inches into the doorway all around, thus forming a rabbet against which the door may rest. A strip of thick ruberoid roofing should be used on the rabbet under the door and the door drawn down tight with four lag bolts and washers.

A common way of making these doors is to cut the staves out on a bevel and make the door fit into this beveled cut directly. If the work is carefully done and then, at the time of filling, if the face of the bevel is plastered with a thick coat of puddled clay and the door forced tightly into this a fairly close joint may be secured.

524. Pit Silos.—In localities where both lumber and masonry are expensive or cannot be had, and where the soil is of such a character that a pit 15 to 20 feet deep may be sunk in the ground, a good silo may be made in this way. The most serious objection to such a silo is the inconvenience of removing the silage to feed.

If the soil is of such a character that it will not cave in the pit may be made circular in form, of the desired size and depth and then plastered with cement in the manner of a cistern. If there is a little difficulty in the walls standing the pit may be made with sloping sides, smallest at the bottom.

In using such a silo, especially when filling it, care should be observed in going into it when there is a possibility that carbonic acid has accumulated to a dangerous extent. There need be no danger in using such a silo if caution is observed as stated on page 427.

525. Weight of Silage per Cubic Foot.—The weight of corn silage increases with the depth below the surface, with the amount of water in the silage, and with the diameter of the silo. In silos of small diameters the amount of surface in the wall is so much greater in proportion to the silage contained that the friction on the sides has more influence in preventing the settling of the silage. In the following table will be found the weights of silage per cubic foot in round silos given for different depths and the mean weight of silage above the given depth:

Table showing the computed weight of well matured corn silage at different distances below the surface, and the computed mean weight for silos of different depths, two days after filling.

Depth of silage.	Weight of silage at different depths.	Mean weight of silage per cu. ft.	Depth of silage	Weight of silage at different depths.	Mean weight of silage per cubic foot.	Depth of silage.	Weight of silage at different depths.	Mean weight of silage per cu. foot.
Feet.	Lbs.	Lbs.	Feet.	Lbs.	Lbs.	Feet.	Lbs.	Lbs.
1	18.7	18.7	13	37 3	28.3	25	51.7	36.5
2	20 4	19.6	14	38 7	29.1	26	52.7	37.2
3	22.1	20.6	15	40.0	29.8	27	53.6	37.8
4	23.7	21.2	16	41 3	30.5	28	54.6	38.4
5	25.4	22.1	17	42 6	31.2	29	55 5	39 0
6	27.0	22 9	18	43.8	31 9	30	56.4	39.6
7	28.5	23.8	19	45.0	32 6	31	57.2	40.1
8	30.1	24.5	20	46.2	33 3	32	58 0	40.7
9	31.6	25.3	21	47.4	33.9	33	58.8	41.2
10	33.1	26.1	22	48.5	34.6	34	59 6	41 8
11	34.5	26.8	23	49.6	35.3	35	60.3	42.3
12	35.9	27.6	24	50.6	35.9	36	61.0	42.8

526. Capacity of Silos.—The amount of silage which may be stored in a silo increases in a higher ratio than the depth increases. A silo 36 feet deep will store nearly 5 times the amount of feed that one 12 feet deep will.

Doubling the diameter of a silo increases its capacity more than fourfold and a silo 30 feet in diameter will hold more than 9 times as much as one 10 feet in diameter and of the same depth. It is clear from this that small silos must be relatively more costly that those of larger diameter.

Table giving the approximate capacity of cylindrical silos for well matured corn silage, in tons.

Depth, Feet.	Inside Diameter in Feet.											
	15	16	17	18	19	20	21	22	23	24	25	26
20......	58.84	66.95	75.58	84.74	94.41	104.6	115.3	126.6	138.3	150.6	163.4	176.8
21......	62.90	71.56	80.79	90 57	100.9	111.8	123.3	135.3	147.9	161.0	174.7	189.0
22......	67.35	76.52	86.38	96.84	107.9	119.6	131.8	144.7	158.1	172.2	186 8	202.1
23......	71.73	81.61	92.14	103.3	115.1	127.5	140.6	154.3	168.7	183.6	199.3	215.5
24......	76.12	86.61	97.78	109.6	122.1	135.3	149.2	163.7	179.0	194.9	211.5	228.7
25......	80.62	89.64	103.6	116.1	129.8	143.3	158.0	173.4	189.5	206.4	223.9	242.2
26......	85.45	97.23	109.8	123.0	137.1	151.9	167.5	183.8	200.9	218.8	237.4	256.7
27......	90.17	102.6	115.8	129 8	144.7	160.3	176.7	194.0	212.0	230.8	250 5	270.9
28......	94.99	108.1	122.0	136.8	152.4	168.9	186.2	201.8	223.3	243.2	263.9	285.4
29......	99.92	113.7	128.3	143.9	160.3	177.6	195 8	214.9	234.9	255.8	277.6	300 2
30......	105.0	119.4	134.8	151.1	168.4	186.6	205.7	225.8	246.8	268.7	291.6	315.3
31......	109.8	124.9	141.1	158.2	176.2	193.2	215.3	236.3	258.2	281.8	305.1	330.0
32......	115.1	135.9	147.8	165.7	184.6	204.6	225.5	247.5	270.5	294.6	319.6	345.7

In this table the horizontal lines give the number of tons of silage held by a silo having the depth given at the left of the column.

527. Horizontal Feeding Area.—In the construction of silos it is very important to have the horizontal dimensions such that the rate of feeding shall be rapid enough not to permit moulding to occur on the exposed or feeding surface. It is also important to have the horizontal dimensions as large as possible because the larger the silo is the less it costs in proportion to the feed it stores. Then, too, narrow, small silos do not allow the silage to settle as well, and hence in them the necessary losses are proportionally greater than in the larger ones.

Observations indicate that if silage is fed down at a rate slower than 1.2 inches daily, moulding is liable to set in. This is more likely to be true in the upper half of the silo than in the lower half but it will be prudent to have the silo of such a diameter as to lower the surface more rapidly in feeding than is necessary rather than less rapidly.

A silo 30 feet deep will allow 1.5 inches in depth of silage per day for 240 days, and one 24 feet deep will allow 1.2 inches for the same time. From the table on page 424 it will be seen that the mean weight of silage per cubic foot for a silo 30 feet deep is 39.6 lbs., and allowing 40 lbs. of silage per cow per day it is seen that a cubic foot of silage on the average will feed a cow one day. But from the same table it will be seen that if the silo is 24 feet deep there will be required 1.114 cubic feet of silage to give the desired weight.

Table giving the inside diameter of silos 24 feet and 30 feet deep which will permit the surface to be lowered in feeding at the mean rate of 1.2 to 2 inches per day, assuming 40 lbs. of silage to be fed to each cow daily.

No. of Cows.	Feed for 240 Days.						Feed for 180 Days.					
	Silo 24 feet deep. Rate 1.2 in. daily.			*Silo 30 feet deep.* Rate 1.5 in. daily.			*Silo 24 feet deep.* Rate 1.6 in. daily.			*Silo 30 feet deep.* Rate 2 in. daily.		
	Tons.	Inside diameter.		Tons.	Inside diameter.		Tons.	Inside diameter.		Tons.	Inside diameter.	
		ft.	in.		ft.	in.		ft.	in.		ft.	in.
10.....	48	11	11	48	10	2	36	10	4	36	8	9
15.....	72	14	7	72	12	5	34	12	8	54	10	9
20.....	96	16	10	96	14	4	72	14	7	72	12	5
25.....	120	13	10	120	16	0	90	16	4	90	13	10
30.....	144	20	8	144	17	6	108	17	10	108	15	2
35.....	168	22	4	168	18	11	126	19	4	126	16	4
40.....	192	23	10	192	20	3	144	20	8	144	17	6
45.....	216	25	7	216	21	5	162	21	11	162	18	7
50.....	240	26	8	240	22	7	180	23	1	180	19	7
60.....	288	29	2	288	24	9	216	25	8	216	21	5
70.....	336	31	6	336	26	9	252	27	4	252	23	2
80.....	384	33	8	384	28	7	288	29	2	288	24	9
90	432	35	9	432	30	4	324	30	11	324	26	8
100....	480	37	8	480	31	11	360	32	8	360	27	8

Using these data the inside diameter of cylindrical silos 24 feet and 30 feet deep which will hold feed enough for different numbers of cows may be computed and such results are given in the preceding table.

528. Danger in Filling Silos.—It never should be forgotten in connection with the filling of silos, that carbon dioxide is generated very rapidly the first few days after silage is put into the silo, and it sometimes happens if the air is very still over night, and if the surface of the silage is a considerable distance below any door, that carbonic acid accumulates in sufficient quantity over the silage to make it impossible for a man to live in it. Cases are on record where people have been suffocated by going into a silo under these conditions. If the doors in a silo are so close together that a man standing on the silage will have his head above an open door the carbonic acid gas will flow out of the door and not accumulate to such an extent as to be injurious.

In cases where the silage is below any opening far enough to leave a man's head below the opening care should be taken not to go into the silo in the morning after filling has begun until after the machinery has been started. After the silage has been dropping into the silo for a few minutes it will stir the air up sufficiently to render it pure enough for a man to work in it without danger. Ordinarily the air currents outside are sufficiently strong to prevent the carbonic acid from accumulating, but it should be kept in mind that it is possible on still nights for this accumulation to take place.

FARM MECHANICS.

CHAPTER XX.

PRINCIPLES OF DRAFT.

If it were possible to construct a perfect road its length would be the shortest distance between the places connected, and it would offer no resistance to movement over it. A pair of parallel, level, smooth and rigid steel rails, well bedded, constitutes the nearest approach to the perfect road yet devised, and how vastly superior the steel track of the railroad is to the best paved street is shown by the enormous loads moved and high speed attained over them.

529. How the Draft Increases With the Grade.—A pull of 2,000 lbs. is required to lift a ton vertically, but to simply move it horizontally only the friction of the carriage and the resistance of the air need be overcome. The more nearly level that roads are built, therefore, the heavier and the faster may loads be moved over them. If the road-bed rises one foot in 100 feet it is said to have a one per cent. grade, and this amount of slope will increase the draft one per cent. of the weight of the load over what it would be on the same road-bed level. A two per cent. grade rises two feet in every 100 feet and the draft is increased by it two per cent. of the load; a ten per cent. grade rises ten feet in every 100 feet and will increase the draft of a ton 200 lbs. over what it is on a level road of the same character. The heavier the loads to be moved, therefore, the more objectionable becomes any grade in the road. This is why with all railroads the heavier their freight the more they overhaul their tracks and lower the grade.

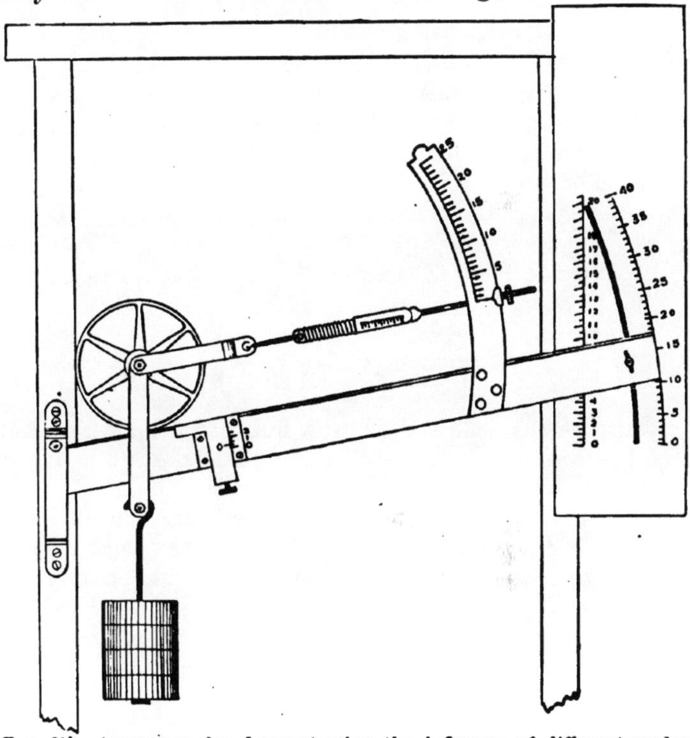

Fig. 209.—Apparatus for demonstrating the influence of different grades and of obstructions on the draft of wagons on roads.

530. Experimental Demonstration of Influence of Grade on Draft.—In Fig. 209 the steel bar may be set so that it represents any grade from one to twenty per cent., and by setting the road-bed at these different grades the spring balance shows the force necessary to sustain the load in the several cases. If the load with the carriage is made equal to 60 lbs. then the scales will read .6, 1.2, 1.8, 2.4, etc., up to 12. lbs. for the 20 per cent. grade. If now the road-bed is set for a 10 per cent. grade and then the load, including the carriage, varied it will be found that the draft on the scale will be always 10 per cent. of the load.

531. The Mechanical Principle Involved in the Relation of Draft to Grade.—It is a general truth or principle in overcoming any resistance or in doing work of any kind that the force or power doing the work, when multiplied by the distance through which it moves, is always equal to the resistance or work multiplied by the distance through which it is moved. Stated mathematically the equation stands

$$\text{Power} \times \text{Power Distance} = \text{Weight} \times \text{Weight Distance}$$

or

$$P. \times P.\,D. = W. \times W.\,D.$$

Suppose the road-bed in Fig. 209 has a length of 100 and the grade is 10 per cent., then if a load of 60 is drawn along the length of the road the power will have passed over a distance of 100, acting parallel with the road-bed, but, leaving friction out of consideration, the work done is to lift the load vertically through a distance of only 10, and since the distance which the weight is raised is only $\frac{1}{10}$ of that over which the power has acted it is only necessary that the power shall be $\frac{1}{10}$ of the weight or

$$P. \times P.\,D. = W. \times W.\,D.$$
$$P. \times 100 = 60 \times 10$$
$$\text{whence } 100\,P. = 600$$
$$\text{and Power} = 6 \text{ lbs.}$$

532. The Steepest Grade Admissible.—When it is asked what is the steepest grade which should be permitted on a given road there are many factors which must be considered, but the most general rule is to make the grade as small as practicable on roads where horses are expected to carry all they can well handle on good, nearly level roads, and the better the level part of the road, the longer the haul and the more teams to pass over it, the less steep should the grade be. On all well designed roads a great effort is usually made to keep below a rise of seven feet in 100 feet.

Just why low grades are so necessary will be readily understood from the following considerations:

About the maximum walking draft of a horse on a good level road is measured by one-half his weight. Trials have shown that a 1,634-lb. horse can exert a steady pull of 800 lbs. while walking 100 feet, and that an 836-lb. horse may maintain through the same distance a steady draft of 400 lbs. It would not be safe, however, to repeat such strains often nor maintain them long. Even a draft equal to one-fourth the weight of the animal is a heavy and exhaustive pull. Indeed a steady pull equal to one-tenth of the weight of the horse for a ten-hour daily service at the walking pace of 2.5 miles per hour is an average of effective service and the work of a 1,000-pound horse would equal

$$\frac{5,280 \times 2.5 \times 100}{60 \times 33,000} = \tfrac{2}{3} \text{ H. P.}$$

Taking this as the safe rate of work for a team on the road an 800-pound horse may pull steadily 80 lbs.; he may pull over hills at the rate of 200 lbs. and in emergencies 400 lbs. A 1,600-pound horse at the same rating may pull steadily 160 lbs., up hills 400 lbs. and in an emergency 800 lbs.

It has been found that to move a gross ton over a good level dirt road requires a traction of about 140 lbs. A team of 800-pound horses may therefore come to a hill with a load of

$$\frac{160}{140} \text{ tons} = 2,285\tfrac{5}{7} \text{ pounds.}$$

Up how steep a grade may such a team carry this load with a steady exertion of 200 lbs. per horse? To overcome the resistance the road-bed offers to the load requires a steady pull of

$$\frac{2,285\frac{5}{7}}{2,000} \times 140 = 160 \text{ lbs.}$$

and this leaves the reserve draft to go up the grade

$$(200 \times 2) - 160 = 240$$

The load to be carried up the grade is the weight of the team plus that of the load or

$$(800 \times 2) + 2,285\tfrac{5}{7} = 3,885\tfrac{5}{7} \text{ lbs.}$$

Up how steep a grade will 240 lbs. carry $3,885\frac{5}{7}$ lbs.? Solving this problem by applying the principle of **(531)** we shall have

$$\text{P.} \times \text{P. D.} = \text{W.} \times \text{W. D.}$$

$$\text{or} \quad 240 \times 100 = 3,885\tfrac{5}{7} \times \text{W. D.}$$

$$\text{whence} \quad \text{W. D.} = \frac{24,000}{3,885\frac{5}{7}} = 6.176 \quad \text{or}$$

a rise of about 6.2 feet per 100 feet, which is a 6.2 per cent. grade.

By taxing the team to its utmost capacity its effective power to ascend the grade would be

$$(400 \times 2) - 160 = 640 \text{ lbs.}$$

Proceeding as in the other case we shall have

$$\text{P.} \times \text{P. D.} = \text{W.} \times \text{W. D.}$$

$$\text{and} \quad 640 \times 100 = 3,885\tfrac{5}{7} \times \text{W. D.}$$

$$\text{whence} \quad \text{W. D.} = \frac{64,000}{3,885\frac{5}{7}} = 16.47$$

or about a 16.5 per cent. grade. That is, a grade of 16.5 feet in 100 feet is the steepest dirt road a team can be expected to carry the load over which it was able to bring over a level dirt road to it.

These results have been computed from the standpoint of an 800-pound horse, but since the ability of a team to work is in a general way proportional to its weight the same results would have obtained had we taken the 1,600-pound horse with a proportional load.

533. Good Roads Make High Grades More Objectionable.— When the good macadam road-bed is substituted for the common dirt road then the same draft, 140 pounds, which draws a ton on the dirt road will draw

$$\frac{140}{60} = 2\tfrac{1}{3} \text{ times as much or } 4,666\tfrac{2}{3} \text{ lbs.} = 2\tfrac{1}{3} \text{ tons.}$$

on the level macadam road. Since it requires but 60 lbs. to move a ton on a macadam road the team may come to the hill with a load of

$$\frac{160}{60} = 2\tfrac{2}{3} \text{ tons} = 6,933\tfrac{1}{3} \text{ lbs.}$$

The effective power of the team will be

$$400 - 160 = 240 \text{ lbs.}$$

Up how steep a grade will 240 lbs. carry the team and $2\frac{2}{3}$ tons? Solving this as we did the other we get

$$240 \times 100 = 6,933\tfrac{1}{3} \times \text{W. D.}$$

$$\text{whence W. D.} = \frac{24,000}{6,933\frac{1}{3}} = 3.46$$

or a little less than a 3.5 per cent. grade. That is to say, when a dirt road is improved so as to reduce the draft from 140 lbs. per ton to 60 lbs. per ton then, in order to utilize this improved road with equal effectiveness under the conditions assumed, the 6.2 per cent. grade should be reduced to 4 per cent.; and the highest grade could not exceed 9.23 per cent.

DRAFT OF WAGONS ON THE LEVEL.

There are many factors which modify the draft of a wagon over a level road and some of the most important of these are:

1. Smoothness of the road-bed.
2. Rigidity of the road-bed.
3. Width of the tire.
4. Diameter of the wheel.
5. Distribution of the load on the carriage.
6. Direction of the line of draft.
7. Rigidity of the carriage.

534. The Smoothness of the Road-bed.—When the road-bed is not smooth and has numerous ruts, stones or other obstructions upon its surface, the draft of the load is increased and the wear on the vehicle and on the road-bed is also greater so that much effort and care should be exercised to have the road smooth. The increase in the mean draft of the load is not so great, however, as the other difficulties which result for the reason that when the wheel enters a rut or passes down off from an obstruction there is a push forward which tends always to give back a portion of the energy expended in raising the load upon the obstruction or out of the rut.

535. Rigidity of the Road-bed.—A yielding road-bed is perhaps the most serious defect of roads, and the one which increases the draft more than any other. If a wheel is steadily cutting into its road-bed it is continually tending to rise over an obstruction or out of a rut, or it is doing what is in effect all the time passing up a grade, as represented in Fig. 210, the hill being steeper in proportion as the wheels are smaller.

In Fig. 209 is represented a method of measuring the increase in draft due to the wheel rising over an obstruction whose hight is a stated per cent. of the radius of the wheel.

The arrangement at C is provided with a screw and graduated so that the block may be raised or lowered at will, setting it so as to represent the wheel passing over an obstruction, 3, 4, 5, etc., per cent. of the radius of the wheel. By setting the road-bed inclined as shown in the figure, the draft is first noted and then the thumb screw at D is turned until the wheel rises upon the block and the difference between the two readings of the scale expresses the increased draft due to the obstruction.

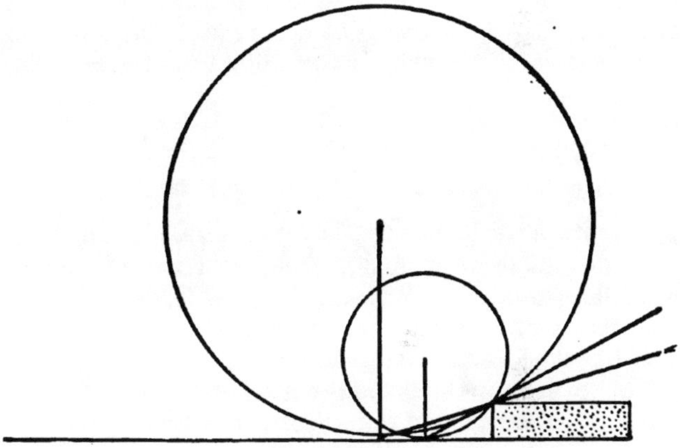

FIG. 210.

When the obstruction is only four per cent. of the radius of the wheel the draft is increased more than two-fold. That is to say, if a wheel is 48 inches in diameter, an obstruction of four per cent. would be only .96 of an inch, and yet the draft is made by it more than twice as heavy.

When the wheel cuts in one inch the draft would not increase quite so much because the wheel never rises quite out of the rut, but the difference between the draft on the macadam and dirt road is due mostly to the difference in the yielding, or cutting in of the wheels.

An experiment conducted by the United States Department of Agriculture, testing the draft of ordinary wagons on a steel wagon road, showed that a single small horse easily drew 11 tons, or 22 times the weight of the animal, and it is stated in the report that the horse could readily have hauled 50 times his own weight. This would be, for a 1,000-pound horse, 25 tons, but of course with such a load the road must be practically level, for a grade of one per cent. would increase its draft 500 pounds.

536. Draft of Wagon Shown by English Trials.—The power required to draw a four-wheeled wagon over roads of different characters has been tested and the following expresses the results in pounds per 2,000 lbs. of gross load:

On cubical block pavement	28 to 44 lbs. per ton
On macadam road.....................	55 to 67 lbs. per ton
On gravel road	75 to 140 lbs. per ton
On plank road.........................	25 to 44 lbs. per ton
On common dirt road	75 to 224 lbs. per ton

537. Draft With Different Widths of Tire.—Prof. J. H. Waters[1] has made an extended series of trials to test the effect of the width of tires on the draft of loads under different conditions of road. He used always a net load of one ton, but the 6-inch tired wagon was 245 pounds heavier than the 1.5 inch, making the gross loads 3,225 and 2,980 pounds respectively, when the wagons were free from mud. The following are his results:

On macadam streets, wide tire 26 per cent. less than narrow tire.

On gravel road, wide tire 24.1 per cent. less than narrow tire.

On dirt roads, dry, smooth, free from dust, wide tire 26.8 per cent. less than narrow tire.

On clay road, with mud deep, and drying on top and spongy beneath, wide tire 52 to 61 per cent. less than narrow tire.

On meadow, pasture, stubble, corn ground and plowed ground from dry to wet, wide tire 17 to 120 per cent. less than narrow tire.

On the other hand he found that when the roads were covered with a deep dust, or with a thin mud but hard below, the narrow tired wagon gave the lightest draft. Also when the mud was thick and so sticky as to roll up on the wheel, loading it down, and again when narrow tired wagons had made deep ruts in the road which the wide

---

[1] Bull. No. 39, Missouri Agr. Exp. Station.

tired wagon tended to fill up, the narrow wheeled wagon gave the lightest draft.

538. Size of the Carriage Wheel.—It is plain from what has been said, that on yielding road-beds the draft must necessarily be heavier, other things being the same, the smaller the wheels of the vehicle. This must be so both because small wheels present less surface to the road-bed to sustain the load, and because when the wheel has depressed the surface it must move its load up a steeper grade than the large wheel. It follows also from these statements that wagons with small wheels must be more destructive to the road itself, whether this be of dirt, gravel, stone or iron.

Some unpublished data bearing upon this point are given here by permission of Prof. T. J. Mairs of the Agr. Exp. Station, Columbia, Mo.

Wagons with three sizes of wheels were used in these experiments:

1. High, 44 inch front wheels and 56 inch hind wheels.
2. Medium, 36 inch front wheels and 40 inch hind wheels.
3. Low, 24 inch front wheels and 28 inch hind wheels, all having tires 6 inches wide.

The total load including the wagon was: For 1, 3,762; for 2, 3,580, and for 3, 3,362 pounds.

The drafts in his trials are stated in the table below:

Description of Conditions.	High wheels.	Medium wheels.	Low wheels.
	Lbs per ton.	Lbs. per ton.	Lbs. per ton.
Dry gravel road; sand 1 inch deep; some small, loose stones	84.48	90.45	110.2
Gravel road up grade 1 in 44; covered with one-half inch wet sand; frozen beneath	123.0	132.1	173.1
Dirt road frozen; thawing one-half inch; rather rough; mud sticky	100.6	119.2	139.1
Timothy and blue grass sod, dry, grass cut	131.9	145.2	178.8
Timothy and blue grass sod, wet and spongy	172.9	202.6	281.1
Cornfield, flat culture, with spring-tooth cultivator; across rows; dry on top	178.5	201.2	265.1
Plowed ground not harrowed, dry and cloddy	252.5	302.8	373.6

For use on the farm the advantage of truck or low wheels comes in the saving of labor in high lifts in placing manure and other materials upon the wagon, and here a sacrifice of strength of the horse may advantageously be made to save that of the man. A lighter draft and lower lift in handling loads are secured by using the low down carriage bed in the upper part of Fig. 211, than are possible with the very low wheeled wagons shown in the same cut.

539. Distribution of Load on the Carriage.—When there is nothing to prevent doing so, the load carried by the wagon should be so distributed upon the wheels as to be divided proportionately to the surface the wheels present to the ground, and when the front wheels are smaller they should carry a smaller load. When care is not exercised

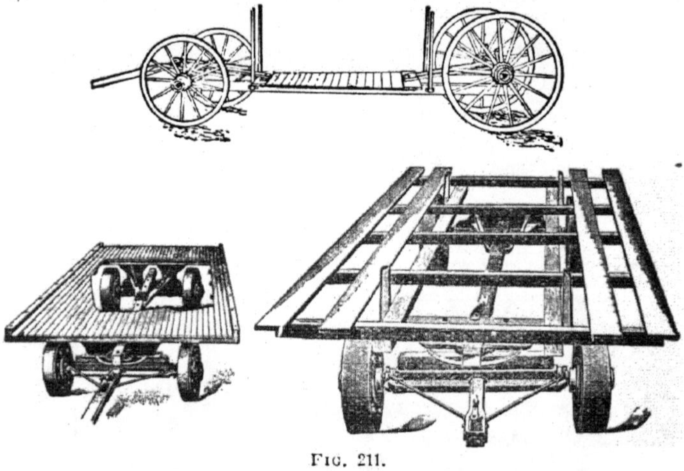

FIG. 211.

in this matter there is danger, especially on soft roads and in the field generally, of very materially increasing the labor of hauling. When the load is heaviest on one side the wheels of that side are unduly depressed, thus increasing the draft. The tilting of the wagon in this way throws the center of the load to one side still further and to a very serious degree if the load is high, as is the case in hauling hay or cord-wood.

540. Heaviest Load on the Hind Wheels.—In loading the ordinary wagon the heaviest load should be placed on the hind wheels for three important reasons: First, because they are larger and will not depress the road-bed so much and will draw easier if they do; second, when the wheels track, the front wheels make a road, by firming the ground, over which the balance of the load may be more easily drawn; third, when the axle of the front wheel is free to be turned, as in the common wagon, the slight inequalities of the road-bed tend all the time to keep the tongue vibrating, so that there is a strong tendency, by this to and fro swinging, to cause the front wheels to cut more deeply into the ground and thus increase the draft. On a very rigid road-bed this matter is not as important as in doing field work, but the differences are large enough on earth roads so that they should never be overlooked.

In the following table some observed differences are recorded:

	Dry sheep pasture.	Dry meadow.
	Lbs. per ton.	Lbs. per ton.
Load equally on four wheels	110.4	174.0
Load heaviest on one side	120.0	187.5
Load heaviest on front wheels	129.3	229.9
Load heaviest on hind wheels	101.8	190.9

These statements may appear to contradict the common practice of hauling logs butt end forward and the general tendency of placing the heaviest portion of the load forward. The conditions, however, are quite different from those where there is a real advantage in placing the heaviest load forward. The reason for this will be better understood from the considerations of the next paragraph.

541. Direction of the Line of Draft.—In drawing a load over a plane surface which remains unchanged during the movement the least draft is required when the line of draft is maintained parallel with the road as shown at A. B., Fig. 212, where the apparatus may be used to clearly demonstrate this principle. It will be seen that as the spring balance is moved up upon its arc the line of draft is such that it tends partly to lift the load off the road and so much that if it were pushed around until the direction

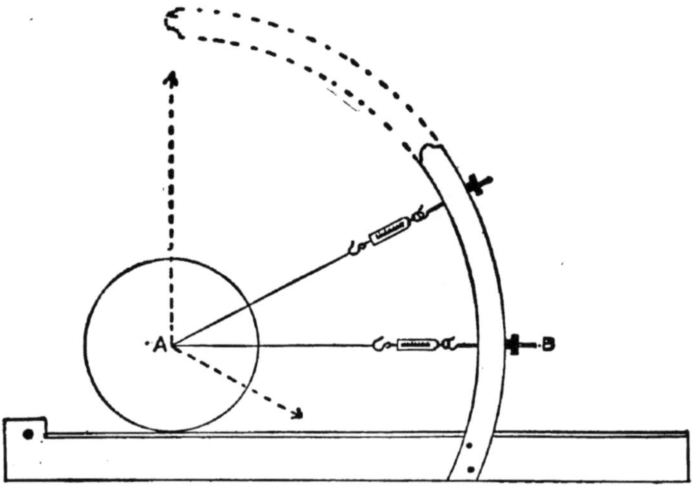

FIG. 212.—Apparatus for demonstrating the influence of the direction of the line of draft on the draft of wagons.

became vertical the whole weight of the load would come upon the spring balance. Then, too, if the line of draft is carried below a parallel to the road-bed the draft must increase because then it is partly downward upon the bed, tending to practically increase the weight of the load by the lost portion of the force of traction, for it is clear that were the scales carried downward until the draft became vertical to the road the whole effect would be lost in producing pressure.

In the movement of cars by the locomotive over the

smooth unyielding bed of the steel rail the line of draft is always parallel with the rail.

542. Line of Draft on Road Wagon.—The statements of the last paragraph may appear to be contradicted by the general practice of having the traces nearly always slope decidedly backward and downward. The former statements, however, are not incorrect, neither is the common practice fundamentally wrong. The apparent contradiction grows out of the fact that the road is seldom either smooth or rigid so that the wheels on the average are in effect continually rolling up an inclined plane.

The principle is clearly shown in Fig. 213 where the wheel is rising over the obstruction which in effect makes

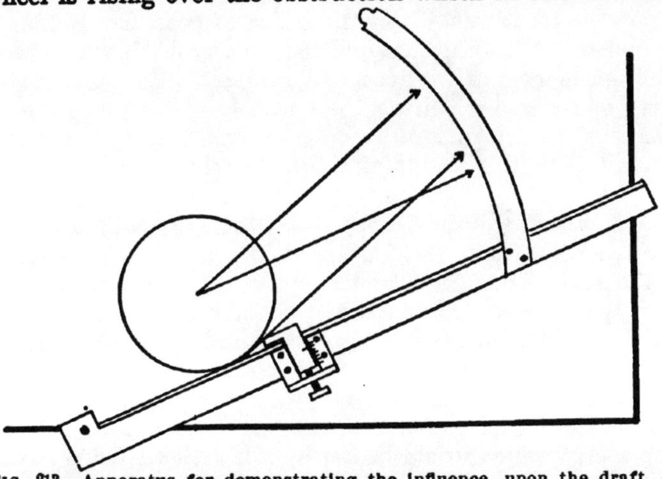

FIG. 213.—Apparatus for demonstrating the influence, upon the draft, of the direction of the line of the draft of a wagon when the wheels are passing over an obstruction or cutting into the road or ground.

an inclined road upon the general road-bed. If now the draft required to bring this load upon the obstruction is measured when the line of draft is parallel with the general road-bed and then the line of draft is made more and more slanting until the direction finally becomes parallel with the secondary road made by the obstruction, it will be found that the draft decreases until this direction is reached, but that passing beyond it again increases. In other words, the draft is least when the direction of the traces is parallel with the effective road-bed.

It is clear, therefore, that in teaming with wagons on the field and on any but rigid, smooth roads the least draft is secured when the traces incline more or less downward, the amount increasing the more yielding and the more uneven the road.

In regard to the division of the load between the front and hind wheels it is clear that the hind wheels are drawn by the reach from the king-bolt, the line of draft being nearly horizontal, and, this being true, it may fairly be concluded that on ordinary roads and upon the field the load must draw harder if the heaviest portion is not placed upon the front wheels where the line of draft can be more inclined. It is quite possible and even probable that when the unevenness of the road is considerable the least draft may be secured when the front wheels are carrying more than half the load. More observations, however, are required along this line to establish the whole truth.

543. Rigidity of the Carriage.—Where the road is not perfectly smooth and where the speed is faster than a medium walk, springs under the load diminish the draft and the advantage of elasticity increases with the roughness of the road and with the speed. For small and rigid inequalities in the road the maximum advantage is secured in the use of the elastic tire, and especially with the pneumatic form, where the load is not too heavy, because in these cases the energy which would be lost by concussion is prevented, the tire quickly and effectually conforming to the road. Where the loads must be heavy, and where the inequalities are larger, then springs under the load carried by the axles respond in rapid transit and relieve the concussions and thus lessen the draft, diminish the strain upon the carriage, and permit less injury to the road.

544. Results of General Morin's Experiments in France.—General Morin, after a series of experiments carried on under the French government, reached the following conclusions regarding the draft of carriages on roads:

1. The traction is directly proportional to the load, and inversely proportional to the diameter of the wheel.

2. Upon paved or hard macadam roads the traction is independent of the width of the tire when this exceeds three or four inches.

3. At a walking pace the traction is the same for carriages with springs as for those without springs.

4. Upon a macadam or paved road the traction increases with the speed above a velocity of 2.25 miles per hour.

5. Upon soft roads of earth or sand the traction is independent of the velocity.

6. The destruction of the road is in all cases greater as the diameter of the wheels is less, and it is greater by the use of carriages without springs than of carriages with them.

CHAPTER XXI.

CONSTRUCTION AND MAINTENANCE OF COUNTRY ROADS.

Having outlined the principles underlying the draft of wagons on roads the next consideration should be how to make and maintain the road for the given locality which, everything considered, is the most economical.

545. Establishing the Grade.—For ordinary country roads the road-bed will generally conform with the natural slope of the surface over which it passes; steep hills, however, should, if possible, always be avoided either by turning to one side or by grading and filling.

Where the hills are short and steep they may usually be graded down to better advantage than to pass around them, but when the hill is both long and high then it may be best to reduce the grade by passing obliquely up the hill, or in mountainous countries where ranges are crossed through passes it often becomes necessary to pass down the long steep slopes by a series of zigzags, having short and steep rounded turns.

546. Factors to Be Considered in Establishing the Grade.—There are many factors which must be considered in deciding the particular grade a road over a given hill may be permitted to have. If the road for the main travel is generally excellent and level, with a good deal of traffic over it, then it is important to keep the grade as low as practicable. Where the country is generally rolling, so that there are many hills which must in any event have a high grade, it will not be as important to cut other hills down as much as a more level country would warrant.

The better the more level portions of the road are, where heavy teaming is done, the more important it is to reduce the grade to a low per cent. because it is important to be able to go over any hill readily which can be approached with the largest load the team is able to handle without injury to itself. The great importance of this point will be readily understood when it is stated that the steepest grade admissible on an average macadam road is 10.5 per cent., and on a dirt road in good condition 16 per cent. But as these grades will tax the team to its utmost the hills should not be permitted to rise if practicable faster than 4 feet in 100 feet for the ordinary macadam and 6.2 feet in 100 feet for the earth road in good condition.

In thinly settled sections people must be content to improve the roads gradually, but if the end finally to be reached is kept in mind all the time it will usually be possible to make each year's work count as permanent improvement and avoid tearing down one year the work of the years preceding.

ROAD DRAINAGE.

The keeping of the road dry, both above and below. is the most fundamental necessity of a good permanent highway. Fill any soil, however hard and firm, completely with water, and a child walking over it will mire; and to completely drain and dry any soft and marshy place will leave it so that heavy loads may be moved across it readily and safely. Drainage is one of the first requisites of a good road.

In some places only surface drainage requires attention. Where the surface is more or less rolling and underlaid with coarse porous materials, so that standing water in the ground does not occur within 10 to 20 feet of the surface, under drainage will not be necessary; but wherever the adjacent fields would be improved by drainage, wherever the ground is springy, and wherever the ground water at any season of the year rises to within three or four feet of the surface there the road-bed should be drained.

In humid climates provisions should be made to surface drain every road.

547. The Relation of Water to Roads.—When a soil is completely filled with water the individual soil grains are invested by water and tend to float in it so that there is the greatest freedom of motion of the particles. On the other hand let all water be removed from the soil and the ground, while hard, easily frets into fine, loose, separate dust particles, which not only increase the draft but are easily drifted away by the wind, thus injuring the road much as it would be were the top washed away by running water.

There is a medium condition or amount of water in the soil which gives it power to withstand the eroding tendency of the tramp of the horses' feet and the rolling of the wheels. When sand is just wet enough its surface is hard and will carry a heavy load, the grains being bound together by the surface tension of the water films. So, too, with the clay roads and those of the best of loam, the right amount of water always present, so as to keep the surface damp and dark without making them soft, greatly improves the quality and lengthens their life. So valuable is the right amount of water on earth roads that sprinkling them in arid and semi-arid climates and in dry times in humid climates, is one of the most effective means of maintenance.

548. Depth of Under Drainage.—Where under drainage is needed the drain should not be less than three to four feet deep, and this is especially true if heavy traffic is to be maintained over it.

No one thinks of walking on the yielding surface of the water of a lake or stream, but let it be covered with a sufficiently thick layer of ice and it then makes the best kind of a road-bed. The drained ground beneath the road surface must be sufficiently thick to float, on the soft soil beneath, any load which may be driven along it, just as the ice floats its burden.

549. Place For the Drain.—In the narrow roads of eight to sixteen feet, where the water to be removed is that which may be raised by hydrostatic pressure vertically upward beneath the road-bed, the best place for the drain is directly beneath the center of the drive-way.

Where the main source of the water causing the trouble is an underflow through sands and gravels from adjacent higher lands then the drain should be placed upon the side of the road from which the water comes.

Where the ground is marshy on all sides, and particularly if the road is wide, it may then be necessary to lay two lines of tile, one on each side.

If springy places occur under or near the road-bed drains must be connected with the spring itself, so as to effectually remove the excess of water.

550. Fall of the Drain.—The fall of the drain will usually conform somewhat nearly to the grade of the road-bed, but should not be less than two inches in 100 feet, if this can be secured. It will, however, be necessary sometimes to lay the drain on a slope less than this, even as low as ½ an inch in 100 feet. In all cases care should be exercised to lay the tile on a true grade, not allowing them to drop anywhere below or rise above a rigidly maintained grade line. If they are not laid in this manner water will stand in the sags and behind the bends, and in these places the tile may become filled with silt.

It may sometimes occur that the road is so nearly level that there is no fall for the drain. In such cases it may be necessary to lay the beginning end of the drain nearer the surface of the ground by as much as six or even twelve inches. In this way there could be given a fall of one inch in 100 feet over a distance of 1,200 feet, but of course the upper portion of the road could not be as well drained and the plan should be followed only where there is no other alternative.

551. Outlet of the Drain.—The drain should be turned out to the side of the road whenever there is an opportunity for doing so, that is, whenever there is a natural line of drainage leading across the road which will answer for the purpose. The free end of the drain is best made of one length of cast iron sewer pipe eight feet long, because this will not be injured by freezing nor be easily broken. There should be a free fall at the end of the drain, and it is better that the opening should be protected by some sort of metal grating or screen to prevent animals from running in in dry times.

552. Size of Tile.—Tile three inches in diameter is the best to use for the reason that, in case the grade is very small, slight errors in laying the line cannot carry the entire opening of the tile above or below the grade line and hence permit the drain to be entirely closed by silt.

553. Kind of Tile.—Where the tile can be laid two feet or more below the surface of the road ordinary drain tile which are well burned, straight, smooth inside and having the ends cut squarely off so that they may fit closely together are best. Great care should be taken in placing the tile to turn them until the ends fit very closely all the way around, and then to fix them rigidly there. This care is needed in order to prevent silt from being washed in at the joints.

Where the tile must come less than two feet below the surface it will be safer either to use the vitrified drain tile or else second quality sewer tile not likely to be disintegrated by frost.

554. Surface Drainage.—The quick removal of water from the surface of a road and the prevention of seepage down through the road-bed are the most important points to be secured in the matter of maintenance. The surface of every road, therefore, should be so shaped as to act like a roof in throwing all rains quickly and completely off, permitting only a little moisture to be drawn downward by capillary attraction to moisten the material and lessen the formation of dust. If the compacted material of the road and the road-bed beneath it can be kept with only a small per cent. of capillary water in them the danger of injury from frost is greatly lessened and the liability to soften during wet periods is also largely removed.

Water should under no conditions be permitted to stand either upon the surface nor along the side of the road, the shape being sufficiently rounded to throw the rains quickly to either side, and the surface ditches deep enough, clean enough and possessing sufficient capacity to carry all water rapidly away.

555. Slope of the Road Surface.—In order to have quick, complete surface drainage it is necessary to so arch the face as to make a road twelve feet wide three inches higher in the center than at either margin, a slope of about four per cent. or four inches in 100 inches. But if the road has itself a considerable grade, then the slope must be made enough greater than four per cent. to force the water to the side ditches rather than to permit it to flow down the center of the road. But evenness or smoothness of surface is the most important condition to be secured and maintained in order to afford perfect drainage. If the road surface is left uneven, or is permitted to become so, no amount of slope which can be tolerated will secure the drainage.

The road must not be made too rounding or sloping for the reason that then teams all drive in one place on the surface and wear it into ruts and this prevents drainage.

556. Water-Breaks.—On steep grades where the hill is long it is a common practice to throw a ridge obliquely across the road at intervals to turn the water to the side. This is a bad practice and should be avoided wherever possible, and in all but the steepest grades this may be done by making the slope of the road higher than the grade.

If the water cannot be turned off in this way it is better to make two paved gutters meeting V-shaped in the center of the road with the point up the grade. The paving will prevent washing and making the gutters meet in the center does not tip the wagon in passing across them.

Whenever it becomes necessary to carry water across a road on a hill from one gutter to the other it is much better to carry it under the road than above it, as is so often done with the aid of water-breaks. A culvert is of course necessary but it should be used.

TEXTURE OF ROAD MATERIALS.

Closeness of texture is necessary to the building of a solid road. The more completely all pores can be obliterated and the road given the close texture of iron the better and more durable will it be.

Field soil in its natural condition may have from 30 to 50 per cent. of space unoccupied by anything but water and air, and in this condition it cannot form a good road. It is too yielding to pressure and water percolates through it too rapidly. When it is properly rolled and tamped the pore space is very greatly reduced, giving it so close a texture that water does not enter it readily, and so large a portion of the grains are in actual contact that it approaches the character of a rock. Of whatever material a road is built it should permit the parts to pack so closely as to resemble a solid rock.

557. Roads Should Be Built in Layers.—Whether a road is to be built of crushed rock or earth it is indispensable that the materials used shall be put on in layers. The thickness of the layers will depend primarily upon the size of the pieces of material used, the layers being thicker the coarser the material. With crushed rock having pieces 2 to 2 1-2 inches in diameter the layers will need to be 3 to 4 inches thick; with smaller pieces the layers should be thinner. If thicker layers than these are made the effect will be the formation of a closely packed crust, a little thicker than the diameter of the material used, over a loose and open structure below.

The hardest and best earth road can be built only by spreading the material on very uniformly in thin layers and thoroughly compacting each layer before the next is put in place; the thickness of these layers should be 2 inches and less, rather than more.

558. Uniformity of Size of Material Used.—It is impossible to crush rock into sizes varying all the way from fine dust to pieces 1.5 inches in diameter and then use this material unsorted to make a solid, unyielding road. The materials when laid down at once with all sizes mixed will not pack so as not to work up loose with the travel upon it; and this is the main reason why more solid roads cannot be built from earth.

Crushed rock must be carefully separated into nearly uniform sizes by means of screens and the different grades applied to the road in layers.

When a layer is made of only a single size of pieces these may be brought together by packing so that all touch and press firmly against one another. If now a grade is used of smaller pieces such as will work readily into the pores left between the angles of the larger ones, pressing hard upon all sides, a still more stable layer will be formed. If it were practicable to follow this method step by step there would be reproduced a nearly solid rock from the fragments made and the most substantial of roads built.

559. Shape of Fragments.—The shape of the materials used in road building has important bearings on the quality of the road. The best form is that which approaches most closely to the cube with broad, flat faces, sharp angles and having the same diameter in three directions. Fragments of this form pack most readily and, as the broad, flat faces set against each other, the fragments do not so readily turn under the wheel or horses' feet and withstand a heavier load without crushing.

Where sands and gravels are used in road building those of glacial origin which are much sharper and more angular than water worn types are much to be preferred, for the simple reason that when packed together they give a more rigid body and stronger binding. Beach gravels and sands cannot be held rigidly by any ordinary cementing material because, with the round, smooth surfaces, there is little opportunity for any locking.

560. Cleanness of Material.—Where crushed rock is used in the building of roads it is important that these materials be clean and free from dirt, clay and rubbish of any sort. So with gravel or sand, when these are called for they should be clean. In general, anything which works against uniformity of material should be avoided.

EARTH ROADS.

In the country in most parts of the United States the greatest number of miles of travel for a long time to come must be made over earth roads. It is therefore of great importance that they should be built in the best possible manner. The proper construction of earth roads is made the more important through the fact that when well built and well maintained there is no road easier on the team, the carriage or the parties riding, where speed is an important consideration, than an earth road.

561. Forming the Road-bed.—After the grade has been established and under-drainage provided where necessary, all organic material and stone should be cleared out of the way and the road given the form and width desired by a road machine such as represented in Fig. 215, or by other means.

The road itself should have a width of 16 or 18 feet bordered on either side by a strip of grass three feet wide, outside of which should be the surface drains, where needed, five feet wide at the top, two feet at the bottom and 24 inches deep, making a total width of 32 or 34 feet as represented in Fig. 214.

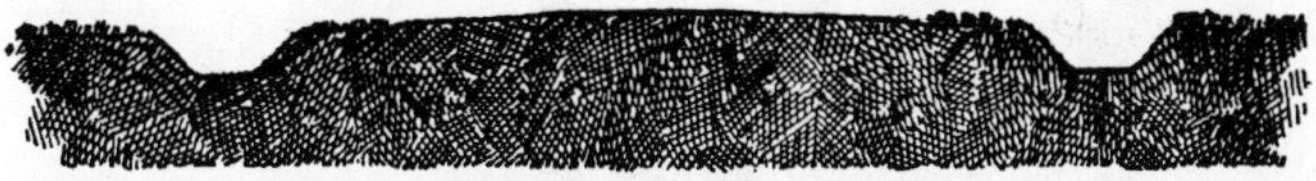

FIG. 214.—Showing cross-section of an earth road 18 feet wide; bordered on each side with 3 feet of grass, outside of which are placed the surface drains when needed. The center of the road is three inches higher than the sides at the grass.

The center of the road-bed should be thoroughly rolled with as heavy a roller as practicable in order to compact it and to discover in it any soft places. If soft places are found these should be filled and brought to the proper level. If the soft place is due to a different kind of material this should be removed and replaced by other and better.

The center of the finished road should be two to six inches higher than the margins at the grass border, varying with the width of the track, in order to give quick, complete surface drainage, and this should be built up in thin successive layers of as uniform material as possible. If earth is brought in from the sides and ditches great care should be exercised in distributing it evenly, and thoroughly harrowing it ahead of the roller, so as to secure the necessary uniformity of texture. This is of the utmost importance in order to prevent the formation of ruts. Thorough rolling should follow the addition of each layer of material and should be kept up until a hard, even surface has been secured.

In making earth roads it is particularly important not to make them wider than necessary because the narrow road is always more quickly and better drained and lack of drainage more than anything else will destroy the earth road.

FIG. 215.—View of one type of road machine, Champion road grader.

If the soil contains cobble stones everything larger than one inch in diameter should be thrown out, otherwise they will form ruts.

If, in establishing the necessary grades on the earth roads, fills must be made, this filling should be done systematically, distributing the earth in uniform layers which are thoroughly firmed with the roller as the work progresses.

562. Utilizing the Old Road as a Road-bed.—In cases where the grade does not require changing and where natural under-drainage is adequate the old road-bed may be utilized in its already tramped and packed condition upon which to build the new road. This may be fitted with the road machine by throwing the loose and uneven portion of the surface outward to form the shoulders. Then if there are still low places these should be filled in and thoroughly packed with the roller, the use of which is necessary even where no leveling is needed, in order to discover any soft spots, quite certain to exist, and in order to give the foundation a more thorough packing than the wagons have secured.

563. Preparing the Road-bed a Year or More in Advance.—It will generally be found advantageous to get the road-bed into proper shape to receive the surfacing material, whether this be gravel or crushed rock, a year or more in advance, utilizing the weathering of rains, the frost of winter and the traffic to settle the road-bed, but directing and assisting these agencies by a timely and judicious use of the harrow, road machine and roller. It is particularly important to allow time to intervene where there has been much filling necessary.

564. Roads on Gravelly Loam.—Where the soils are a gravelly loam the best earth roads are possible. The reason for this is found in the fact that a gravelly loam is made up of large and small grains in such proportions that when they are thoroughly worked and compacted the coarser sand particles work in between the gravel, and the fine clay particles between those of sand, in such a way that there is left almost no open space; under these conditions the water is shed the most rapidly and completely so that the road is less liable to soften under the travel over it and it is less liable to be injured by frost.

565. Roads in Fine Clay Soil.—Where the soil is a fine adhesive clay it is hardly possible to make a good road without the aid of foreign material. Of course by grading it into proper form so as to secure the needed drainage the road will be good when it is not wet, and under these conditions it will remain fair much longer than if not so prepared because, when this soil has been once thoroughly compacted and dry, water enters it very slowly, so that it is only during long wet spells and when the frost is going out that the most serious injury to the road comes.

566. Clay Roads Surfaced With Gravel.—Where gravel of suitable quality is available a covering of three or four inches, thoroughly rolled and packed, will very greatly improve the surface of a clay road, preventing it from softening so readily with every rain and with the action of frost. Even sand and good loam, where nothing better is available, will improve the quality.

In some cases burning the clay has been practiced so as to render it less plastic and sticky, but this practice will be one of the last to be resorted to at this time of cheap transportation and high price of fuel.

567. Sandy Roads.—The making of good roads in a country of very sandy soil is extremely difficult on account of the nearly complete absence of binding properties in the sand when dry. If there were any cheap method of keeping the surface wet, sand would make an excellent road. Even the rounded grains of beach sand for a short time after the waves have withdrawn are so tightly bonded that a horse may canter along the beach, making but little impression upon it. The water, however, drains away so rapidly from the coarse clean rounded grains that there is no longer anything to bind them together, and the foot or wheel easily sets them aside. When, however, there are a sufficient number of much finer particles commingled with the coarse sand grains a loam is the result whose water holding power is increased so that for a longer time the grains are bonded together by it, enabling the loam to form the better road. On the other hand, the amount of water may be too great to permit it to act as a binding material and as the water-holding power of the clays is greater than the loams, they more quickly come into the condition of over saturation during long rains and so the loam which is intermediate between the two extremes makes the best earth road, sand tending most of the time to retain too little water and the clay retaining too much for tight binding.

With this principle to direct practice it is clear that if the right amount of finer soil particles can be obtained to incorporate with the sand of sandy roads their firmness will be increased. It is unfortunately too often true that in districts where sandy roads prevail there is no clayey or loamy material available, either to incorporate with the sand or to place above it.

568. The Use of Straw, Sawdust and Tan Bark on Sandy Roads.—It is well known that these materials when applied to sandy roads have temporarily a beneficial effect. The fundamental principle underlying this improvement is that stated in the last paragraph; that is, in the power they have of maintaining a higher per cent. of water in the sand, which is necessary in order to bind the grains together. The sawdust, tan bark and straw act in two ways to maintain the needed amount of water in the sand. At first they act as a mulch, lessening the rate of evaporation from the surface. Later, when they begin to disintegrate, they form a humus-like material, in its physical effects, which increases the capillary power and diminishes the rate of percolation downward after rains.

The reason why these materials are only temporary in their effect is because they rapidly decay, being converted into soluble salts and gaseous products which finally leave the sand as if nothing had been added.

569. Road Gravel.—It occasionally happens that natural gravel beds are found which possess the right characteristics for making roads, and when the gravel is just right excellent roads may be made from it.

There are several important features which a good road gravel must possess:

1. There must be one prevailing size of pebble in sufficient quantity so that when thoroughly rolled they press against one another.
2. There must be enough of the finer sizes of coarse sand and fine gravel to fill the voids between the coarser gravel.
3. There must be enough of fine loam to fill the voids between the coarse sand and fine gravel and retain a sufficient amount of water to bind the sand grains together and prevent their rolling.
4. The coarse and fine gravel and the sand must be made up of more or less angular fragments in order that flat faces of rock may set together and thus lessen the danger of rolling and of crushing under the weight of the load.

It is not possible to give specific, concise directions for identifying a good road gravel, but a man who has seen and worked with it readily recognizes it.

570. Clean White Gravel Not Suitable.—It will be apparent at once that the several characteristics which have been pointed out are not likely often to occur together in just the right ratios; and so there will be all possible gradations from the ideal gravels to those which will not answer at all. Indeed it must be said that most gravel beds have had the finer materials so completely washed out that only clean sand and gravel remains; and when this is true it is useless to try to make a road with it. Such materials can only be used to temper a road which is too clayey in its texture, by reducing its water capacity.

571. Texture of Gravels Altered by Crushing and Screening.—It happens in the majority of cases that much of the gravel is too large and too rounded to permit close packing and fast binding. When this is true much better qualities may be secured by using either the crusher or the screen or both together, one form of which is represented in Fig. 216. It will be at once apparent that where much of the gravel is too coarse, to run it through the crusher so as to reduce the material to a more uniform size and at the same time to increase the angularity of the fragments will make a much better road material to use either by itself, or as a tempering material.

FIG. 216.—Champion rock crusher and screen.

572. Some Gravels Contain Too Much Clay.—There are many deposits of gravelly clay which it might appear would make a good road material, but the principle must be kept always in mind that too much of a too fine material will take in and retain so much water that the binding quality of the water is lost. These gravelly clays occur in many of the hills of the glaciated portions of the United States and through which roads are often cut.

573. Gravel Roads.—In the construction of a gravel road, as in that of a stone road, it is of prime importance to secure first of all a properly shaped and thoroughly rolled and firmed road-bed before any gravel is laid on. When this has been done, and a suitable gravel has been found, the next step is to spread evenly over the surface and thoroughly roll a layer which, when finished, will measure three inches thick.

In the rolling it will be important to firm the outer edges of the gravel first in order that the rolling may not force it outward and destroy the slope. Should the gravel be too dry to pack it must be moistened or the work be suspended to take advantage of the rains.

To make a good road there should be not less than three 3-inch layers, and usually four will be better. Of course a road 6 inches thick will be a great improvement, and often where the travel is light and the road-bed thoroughly made, three inches of good gravel, well placed, will make a great improvement in the road, serving as a wearing surface.

Where the gravel must be crushed and screened to secure the proper sizes the revolving screen represented in Fig. 216 should be used and should have two sizes of holes 1.5 to 2 inch and 3 to 4 inch in diameter. The coarser size of gravel will form the body of the road while the finer will have to be discarded unless it happens to be of the right quality to use as a binding material or in making a bicycle path along one side of the road.

574. Roads in Swampy Places.—It occasionally happens that roads must be built in places which cannot be drained and which are too soft to permit of the construction of a solid earth foundation. A common way to meet this type of conditions is to lay a foundation of logs, poles or even brush, having the desired width of the road and of sufficient body to enable an earth or gravel road to be built upon it. When such roads are built in situations where the wood is kept constantly beneath the water it does not decay and a road of considerable permanence and solidity is secured.

Where logs are used care is taken to arrange them at right angles to the direction of the road, parallel with one another and like sizes side by side. The depressions between the logs are filled with smaller logs or poles, whole or split, while these in turn may be covered with twigs and limbs forming a mat upon which the earth or gravel road is built. Upon this mat of wood is usually first thrown the material taken from ditches on either side made for drainage, building the earth or gravel road upon this after it has first been well spread and firmed.

STONE ROADS.

Stone roads of one form or another date back to and possibly beyond Roman times; and Fig. 217 represents two types of the extremely massive and substantial roads which were built ten or fifteen centuries ago, some of which still survive. These roads had a width of 30 feet and pavements of heavy stone at the bottom and often one or more layers of stone bedded in cement to make the road water proof. One type of construction which they followed made the road consist of four layers:

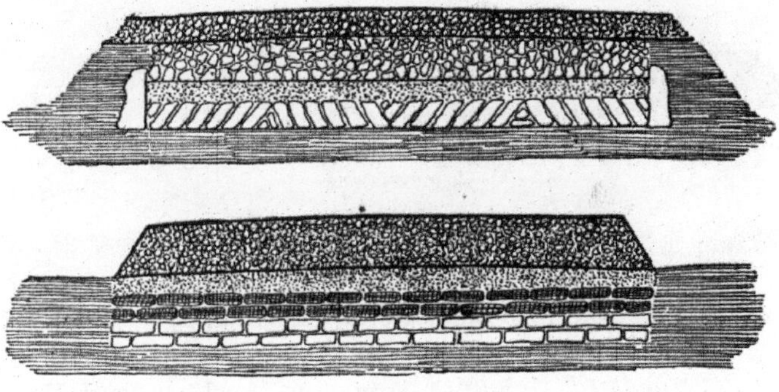

FIG. 217.—Two types of Ancient Roman stone roads. (After Shaler.)

1. Two or three courses of flat stone or, if these were not obtainable, of other stone, generally laid in mortar.
2. A layer of rubble masonry or coarse concrete.
3. A finer concrete upon which was laid
4. A layer of paving blocks jointed with the greatest nicety.

It is stated that with many of the great roads the paved portion had a width of 16 feet bordered by raised stone causeways outside of which, on each side, were unpaved side-ways each eight feet wide, and the paved way sometimes had an aggregate thickness of three feet.

575. Macadam Roads.—The use of crushed rock in road building is at least as old as Roman history; but as, during the dark ages, little road building of a permanent character was practiced, the art had to be revived in modern times and about 1764 the French engineer Tresaguet appears to have introduced into France the type of road represented in Fig. 218, consisting of a stone pavement covered with two or three inches of crushed rock as a facing material. After being introduced into England and Scotland, where the details were modified and perfected by Telford about 1820, this type of stone construction came to be known as the Telford road.

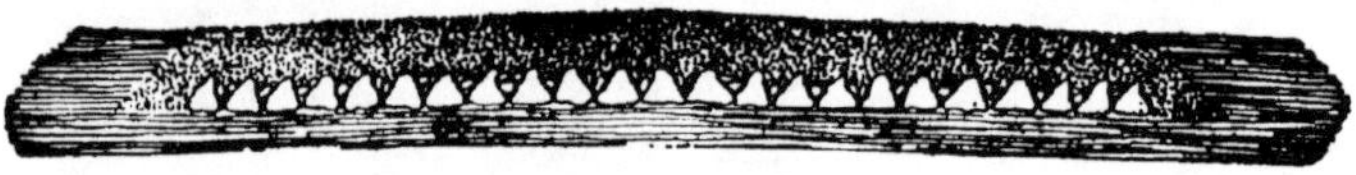

FIG. 218.—Type of road introduced into France by Tresaguet about 1764. (After Shaler.)

Macadam's work began somewhat earlier than Telford's in 1816, and to him apparently is due the idea that when any road-bed is thoroughly under-drained, so as to remain permanently hard, then crushed stone alone may be used, the pavement of Roman practice becoming unnecessary.

576. Construction of Macadam Roads.—After the foundation for the stone road has been completed the border is left with a shoulder of earth on each side as represented in Fig. 219, between which the road-bed is covered with a layer of crushed rock as nearly one size as possible and three or four inches thick. This layer is next thoroughly rolled and then covered with enough of finely crushed rock to fill the voids between the larger fragments. This material is worked in with the roller and water until a solid bed has been formed.

After the first layer has been placed the second is applied in the same manner, rolled, and the binding material applied and again rolled, until thorough consolidation has been secured.

FIG. 219.—View showing the road-bed, in the foreground, shaped with road grader and receiving the foundation layer of crushed rock 4 inches thick.

577. Fitting the Road-bed.—It is of the utmost importance to have a thoroughly firmed and seasoned road-bed put into proper form and well drained before the stone surface is to be applied, and to do this most economically it is well to do all of this preliminary work a year or more ahead so that traffic, rains and frosts shall have an opportunity to do the work of consolidation, and to discover the soft places which may exist. In short, the formation of a good earth road to be used for a number of years as such will generally be found the best and most economical preparation for the stone road.

578. Forming the Shoulders.—The formation of the shoulders represented in the foreground of Fig. 219 is best done with a road grader or road machine. With this tool the surface of the road-bed is prepared at the same time and the shoulders left in such shape that very little hand labor will be required for the finishing touches. After the shoulders have been roughly formed and before the finishing touches are given the roller should go over the road-bed to make sure that it is properly firmed and that there are no soft places.

579. Kinds of Rock for the Road.—Practical experience has demonstrated that the best rocks for road making are the dark green, black and dark gray trap or igneous rock such as are known in common language as "nigger heads" in glaciated countries where large boulders are common in the fields and cuts of roads. They are tough, fine grained rock, much less brittle than most others, which yield when grinding upon themselves and under the wheel a fine rock flour whose texture is such that it holds the needed amount of moisture to make it bind together well, and consequently a road built from these fragments sets sooner than almost any other crystalline rock and hence is subject to less internal wear.

Next to the trap rock in value for road building purposes stand the closer grained hornblend-bearing syenites and gneisses which are species of granite where hornblend takes the place of mica of the true granites. It is the class of dark minerals allied to hornblend composing much of the trap rock referred to above which makes that the best road stone.

Next in order stand the true granites made up of quartz, feldspar and mica, and their gneissoid varieties. The best of this class of rocks are the close fine-grained varieties having the least tendency to break into thin layers, giving flat instead of cubical blocks.

To the granites and syenites with their banded or gneissoid varieties belong the lighter colored and flesh colored boulders which are usually associated with the "nigger heads" of glacial drift.

The chief difficulty with syenites and granites for road metal is their brittle, unyielding quality and coarse crystalline structure which makes them grind and pound up into a coarse sand without a sufficient amount of the finest dust to give it the needed water-holding power to permit it to properly bind the pieces together. The road-bed fails to

FIG. 220.—View showing where four inches of crushed rock for wearing surface is being built upon four inches of road-gravel as foundation layer.

set quickly and the internal wear is larger while there is a greater tendency for ruts to form in wet weather and for the surface to ravel or throw out loose pieces in a dry time.

Next to the syenites and granites in general availability for road metal stand the close grained hard limestones which break into hard, clean blocks and fragments with sharp edges and little material which will rub off under the fingers. Any rock which crushes readily into an earth-like or sandy material will not answer for road work.

When a good road limestone wears down under the wheels, the horses' feet or the roller, a loam-like powder is formed which holds the right amount of water for good binding, and besides this it appears more quickly to pass into that cementing stage which in nature cements beds of loose fragments into rock.

The chief objection to limestone as a road metal is its softness, which permits it to wear away rapidly, leaving the surface dusty in dry and muddy in wet weather.

The extremely hard and brittle quartzite which throws off angular bits under the blows of horses' feet and the rolling of wheels makes one of the poorest road materials because it too nearly possesses glass-like brittleness and the dust is too coarse and sand-like to hold the needed water for binding.

580. Foundation and Surfacing Stone May be Different.—Where there is in the locality a rock which does not make a good wearing surface but which binds well, like limestone, this may be used to advantage for the foundation of country roads, thus making it necessary to import only the wearing surface layer.

581. Sorting Boulders Before Crushing.—In localities where there are many boulders available for road work it will often be practicable to sort these when hauling them to the crusher in such manner as to use the lighter colored varieties for the foundation, reserving all of the "nigger heads" for the surface layer, and in this way increase the efficiency of the material.

582. Using Limestone for Binding.—Where only granitic rock and quartzite are available for road work and these do not bind well, it will often happen that the limestone of the locality may be crushed fine to form screenings and used to great advantage as a binding material to hold the harder rocks more securely in place. This practice would be especially desirable for the foundation layer where it could not be converted into dust. But in localities where both limestone and the harder rock are available, but where the limestone can be obtained at much the less cost, this may be used alone for the foundation and as a binding material for the surface layer.

583. Roads Made Without Binding Material.—It was Macadam's practice in road building to strictly forbid the use of all binding material whatsoever. He preferred to wait for the general traffic over the road to develop from the wear of the crushed stone, both superficial and internal, the necessary amount of rock flour to do the work of filling and cementing. While this work was in progress the road was given constant supervision to keep it in proper form. At the same time the filling and binding material was being slowly produced there was brought upon the road with the wheels and horses' feet a considerable amount of earth which slowly worked downward and united with the rock flour to complete the consolidation. Macadam certainly secured in the end a better road by this method than was usually secured with the use of the then available binding material.

It must be remembered, however, that in his time rock were crushed by hand and little fine material was made to use for binding, whereas with the modern rock crushers a large amount of this material is produced which must be a dead loss if it cannot be used for binding and surfacing, and it is quite certain that had Macadam used our modern rock crushers he would have availed himself of the screenings.

584. Use of Sand for Binding.—The great readiness with which clean dry sand works into and fills the voids between the stone of a road, the ease with which it may be handled and the readiness with which it may often be obtained, leads to its occasional use as a binding material in macadam road. The coarse silicious sands, however, have very little cementing quality, they do not retain water well enough either to make the road shed the rains nor give the surface tension of water much opportunity to bind the grains to-

FIG. 221.—View showing the binding material or screenings being applied to the foundation layer of crushed rock.

gether firmly; consequently the best results cannot be secured when it is used.

If loam is used there is danger that it will pack in the upper surface of the layer of stone and prevent even the combined use of water and the roller from working it to the bottom so as to completely fill the voids. There is the still further danger that it will work in between the flat surfaces of the crushed rock, holding them apart to such an extent that heavy loads will produce too much rocking of the pieces and quickly lead to the formation of ruts. If the loam could be had in a dry condition, such as is usually the case with the screenings and the sand, it would be possible with dry rolling to nearly completely fill the voids so that the subsequent use of water would, with the roller, lead to good results.

585. Limestone for Stone Roads.—There is no doubt that crushed limestone although a soft rock will make an excel-

FIG. 222.—View of distributing cart being raised to spread crushed rock.

lent country road where the traffic is not heavy and the use of it should be encouraged wherever suitable quality of rock is available. There is no rock which breaks in better form or which binds as well and sets as quickly. It is readily quarried and put in shape for the crusher; and the power required for crushing being small makes it less burdensome for towns to invest in the necessary machinery.

It is true that the road wears rapidly under heavy traffic and the surface becomes dusty in a dry time, but not more so than clay roads do. It is true that careful road engineers advise against its use, but it is usually from the standpoint of city and suburban traffic rather than from that of the purely country road.

FIG. 223.—View of distributing cart spreading crushed rock on the road.

586. Spreading the Rock on the Road-bed.—It is important that the crushed rock should be laid down on the road-bed in a sheet both of uniform thickness and uniform density and where this is not done the road is quite certain to roll to an uneven surface which will make it necessary to add more material in some places and remove it in others. But this will unnecessarily add to the cost of the road.

Not only this, but when a wagon-load of stone is all dumped in one place, leaving it for a man to spread, it is certain to occur that all of the dust and fine materials not removed by the screen will drop into the voids at the place where

FIG. 224.— View of surfacing crushed rock as left by the distributing cart on the road. The watch, 2 inches in diameter, serves as a scale to show the size of the rock fragments.

the load was left and this will give rise to a spot more compacted than the balance of the road and hence when it comes into service two ruts or depressions are liable to form one on either side of the harder spot.

To avoid these difficulties and to save time in spreading the material the distributing cart represented in Figs. 222 and 223 has been devised. In it can be placed two cubic yards of rock, and after tilting the box as shown in Fig. 222 the end board may be opened to such a width as to deposit

FIG. 225.—View of the surfacing rock after it has been packed by the roller.

a uniform layer of any desired thickness while the team travels along at a slow and uniform pace. Fig. 224 is a view showing how the surface was left by the distributing cart and the watch is a scale by which the size of the pieces may be judged, its diameter being a trifle less than two inches.

587. Thickness of Layer.—The thickness of a layer placed at one time should vary somewhat with the size of the pieces, the depth being greater with the larger fragments. With pieces of the size shown in Fig. 224 the layer when packed should not be greater than four inches and three inches will pack more quickly and closely than four inches. A too thick layer tends to form a crust on the surface, making it difficult to fill all the voids below completely.

588. Rolling.—The function of rolling is to arrange the fragments in the positions of the greatest stability with reference to the rolling of wheels and the tramping of horses. The first effect of the roller is to bring the pieces nearer together and to reduce the size of the voids. This is clearly brought out by the two photo-engravings, Figs. 224 and 225.

There is one other important thing which rolling should secure and that is to put the several pieces of stone together in the positions of the most stable equilibrium; that is, in positions such as to make certain that they shall not tip or turn when the stress of the wagon or team is brought upon them.

589. Size and Weight of Roller.—The diameter of the roller should be large to prevent it from shoving the stone forward as it moves and in order that the thrust may be as nearly directly downward as possible. It will be observed that even the front wheel of a loaded wagon often slides rather than rolls when coming upon the unpacked layer of rock on the road, and such movement cannot do proper packing.

There appears to be a lack of agreement between practical men regarding the proper weight of the roller, some advocating a roller of 3.5 to 5.5 tons, while others hold that only one of 15 to 20 tons weight will serve the purpose. Others advocate a light weight to begin with and a heavier one at the close.

FIG. 226.—View showing horse roller at work compacting the road metal.

590. Amount of Rolling.—The only general rule which can be given in regard to the amount of rolling a given layer should receive is that the work should be continued until the stones cease to move in front of the roller or until the roller no longer sensibly depresses the bed and it has become hard and smooth. It should be kept in mind, however, that the road may be rolled too much, or until the stone again begin to move. This is most likely to occur when the stone is too dry.

591. Manner of Rolling.—The rolling should begin at the outer sides of the road, packing the stone first against the shoulder. If this is not done the fact that the road-bed is highest in the center will lead to flattening the slope and thinning out the rock in the center through a side creeping of the material from under the roller.

592. Kind of Roller.—There are three methods of consolidating the layers of stone put into a road. The first, now largely abandoned as being too expensive and too uncertain, is to allow it to be done by the natural traffic. The second, also being abandoned as too expensive, is the use of a 3.5 to 5-ton horse roller; and the third, which is regarded the cheapest and best, is with the aid of an 8 to 20-ton steam roller.

The safest indications seem to point to the use on country roads of an 8 to 10-ton steam roller as most satisfactory; although good work can be done with the horse roller of half this weight which may be made heavier or lighter by taking on and laying off weights Such a roller as this is represented in Fig. 226 which, naked, weighs 3.5 tons, but by the addition of castings to the inside of the roller may be increased to 5.5 tons. This roller has the frame and tongue so constructed that the team may be turned without reversing the roller, a very important feature.

It will be readily seen that the use of two men and two teams must make the service of this roller very expensive, and when the disturbing effects of the horses' feet are recalled it becomes clear that the steam roller easily managed by one man is much better.

593. Rock Crushers.—Until recently all rock crushing for road work has been done by hand and hammer, and in the days of slave labor when the man was a machine which managed, fed, cared for and reproduced itself, it is clear how such Herculean tasks as the ancient Roman roads could be accomplished. But happily, the use of steel and inanimate forces is freeing man from such drudgery; and in Figs. 227 and 228 are two views of a rock crusher at work, breaking stone, sorting it and delivering it into bins where it may easily be dropped into wagons for delivery upon the road.

FIG. 227.—View of No. 3 Austin Crusher, with revolving screen breaking boulders for road; and wagon loading coarsest grade of broken stone.

At the time these views were taken the crusher was being driven by a 22 H. P. traction engine and was crushing rock at the rate of 100 wagon loads per day. The material is separated into three sizes, the coarsest used for the foundation, the intermediate for the wearing surface and the finest as binding and surfacing material, and Fig. 227 shows a wagon loading with the foundation size, and Fig. 228 with the screenings or binding material.

There are various forms of crushers on the market and Fig. 216 represents another type.

594. Revolving Screen.—The revolving screen is an indispensable attachment to a rock crusher, because a good road cannot be made with the unsorted material, for with this method of putting the crushed rock upon the road the fine materials are certain to work downward and the coarser fragments to come to the surface. It should be thoroughly understood too that the chute screen will not do the work.

FIG. 228.—Side view of No. 3 Austin Crusher and wagon loading

595. Earth and Stone Road Combined.—Where it is desired to cheapen the construction of stone roads it is practicable to make the central portion 8 feet wide of this material and then have on one or both sides an earth road of eight feet, giving a total width of 16 or 24 feet to the margin of grass and 30 feet to the side ditches. The most serious objection to this combined plan is the securing at all times of sufficient and quick surface drainage.

The chief difficulty which will arise in the carrying out

of this plan will come from the tendency of summer traffic on the narrow earth road to go so persistently in one track as to develop wheel and foot ways deep enough to prevent surface drainage. The fact that the stone road may come into service when the ground is wet will only lessen the tendency to develop the evil pointed out but not prevent it. For winter service in cold climates it seems clear that the earth road will be likely to ensure better sleighing.

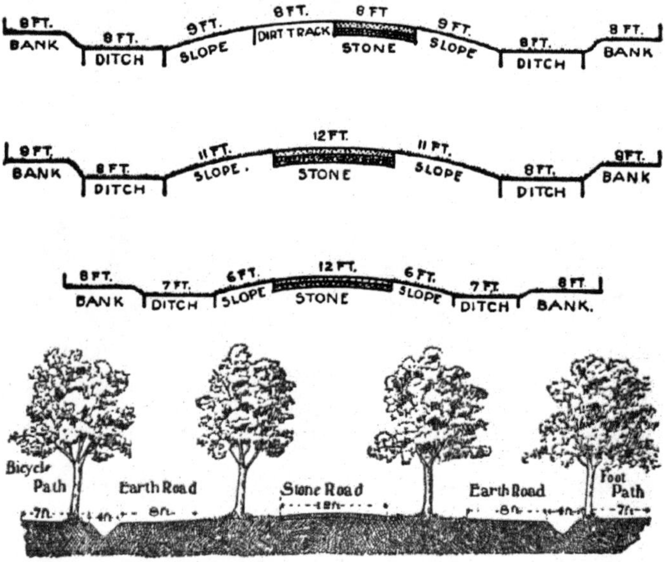

FIG. 229.—Diagrams showing profiles of earth and stone road combined.

596. Telford Foundation.—When it is necessary to build the road where the ground is soft then it may be best to lay a foundation of larger stone as was the general practice with the Romans and with the English engineer, Telford, whose name is now attached to this type of road foundation. The paving blocks should be uniform in size, laid in rows across the road after it has been given the proper slope, the pieces breaking joints. The stones should not exceed 10 inches in length, 6 inches wide on the bottom and 4 inches at the top, the thickness being 4 or 5 inches for a road 8 inches thick. The surface of the pavement foundation should be as even as practicable and the voids filled with broken stone. It is necessary to have each piece thoroughly bedded before the macadam material is added so as not to be tilted on the surface.

FIG. 230.—View showing road with the stone portion in the foreground nearly completed.

597. Culverts.—Culverts are necessary for carrying under a road the water from adjacent fields which collects as surface drainage in times of heavy rains and melting snows. The permanent forms are made of sewer tile, cement tile, cast iron sewer pipe or of stone. Wood should only be used as a temporary expedient.

Where the amount of water to be conveyed is small so as to demand only one, two or three 12-inch sewer, or cement tile, it will usually be cheapest to use these, but where a water-way demanding a cross-section of more than 8 square feet is necessary and where stone are available, it will be cheapest to make it of arched masonry.

Where the culverts are of sewer pipe there should be not less than 18 inches of earth in the road above them to prevent crushing.

The cast iron pipe is the safest to use and cheaper than either sewer or cement tile when diameters above 16 inches are required.

MAINTENANCE OF COUNTRY ROADS.

Important as the matter of construction of good roads is, it is, or should be, secondary to that of maintenance; when a good thing has been made which is designed for permanent service it is clearly a matter of sound business policy to provide whatever economic means is practicable for keeping it in order.

598. Section Men Necessary.—In the maintenance of railroads it was early learned that two or more men provided with proper tools must be employed by the year, permanently or as long as they rendered efficient service, to care for and keep in order a certain number of miles of road. It is the business of these men to daily go over their section and keep it in first class repair and their tenure of office is only conditional upon their doing this satisfactorily.

It is self-evident that good country roads can only be maintained by adopting and keeping in force a system which is equivalent to that found indispensable in railroad maintenance. That is, men competent to do the work, provided with the necessary authority, tools and materials, must have constant employment at a price which will permit them to devote their time to it, and they must be made responsible for the maintenance of a certain number of miles of road 365 days in a year.

FIG. 231.—View of country stone road with foot path on one side, near Maybole, Ayrshire, Scotland. From photo in 1895.

599. Road Master.—In the country road service it will be necessary to have one man who corresponds in duties and responsibilities to the "Section Boss" of the railroad. He must be competent, temperate and in every way reliable and trustworthy. He must have a practical knowledge of the principles and details underlying the maintenance of good roads and at his command the necessary authority, assistance and appliances for doing the work required.

600. Width of Tires Controlled.—When we come to have a system of good roads and the means for maintaining them it will be necessary to have ordinances regulating the width of tire and diameter of wheel which may be used on the roads when carrying specified loads. In Europe, where better roads are found and a better system for maintenance exists, there are ordinances which fix the width of tire to be used with given loads. In Bavaria the regulations are as follows:

Two wheel carts with two horses, 4.133 inch tires.

Two wheel carts with four horses, 6.180 inch tires.

Four wheel carts with two horses, 2.596 inch tires.

Four wheel carts with four horses, 4.133 inch tires.

Four wheel carts with five to eight horses, 6.180 inch tires.

Carts with more than four and wagons with more than eight horses are not allowed to use the roads without a special permit from the authorities.

Other countries of the Old World have found similar ordinances necessary and it is clearly rational and just that such matters should be regulated, for otherwise one man may easily put in jeopardy the interests of a whole community.

601. Maintenance and Repairs.—A sharp distinction should always be made between the maintenance of a road and its repairs. It is only when some accident has occurred to seriously injure a road or when, from long neglect, it has become well nigh worn out that repairs are needed, but the daily touching up of slight defects and places of evident wear constitutes maintenance.

602. Good Maintenance.—Good maintenance will consist in daily attention to all the details which are necessary to keep a section of road up to the standard of perfection practicable to its type, influenced by its local surroundings and conditions. It must consist in (1) keeping the road in proper form; (2) in adding materials to the wearing surface where needed; (3) in keeping the road surface and drainage channels clean; (4) in keeping the road sides free from weeds and otherwise neat; (5) in caring for and maintaining road trees if they are grown; (6) in maintaining the proper conditions in winter in regard to snow.

603. Maintenance of Earth and Gravel Roads.—The first requisite for the maintenance of any road is the knowledge which can be gained by going over the road while or immediately after it rains. Observations at this time will show the road master where the most serious defects exist and he should make careful note of them to use in directing his efforts as soon as the weather permits. It should therefore be the business of the road master to study his roads in wet weather and he should be equipped with clothing, etc., in a way which will permit him to do this without risk of injury to health.

FIG. 232.—View of French country road 20 feet wide, showing piles of crushed limestone used in maintenance. Photo. in 1895, near Grignon.

Whenever ruts or saucers begin to show in the road they should be corrected immediately, provided the moisture conditions permit of doing so, but on the earth roads the soil may be either too wet or too dry to allow this to be done well, and the highest success will be attained when the road master comes to know and understand his conditions and then is alert to move at just the right time. The ruts will be formed chiefly in both the very wet and the very dry weather, and in the country where sprinkling the roads cannot be afforded, everything must be planned to take advantage of every shower heavy enough to bring the road into condition for working with grader, shovel, rake and roller.

FIG. 233.—View on the same road showing the tool house where appliances for caring for the road are kept. Photo. in 1895, near Grignon.

The intelligent use of the grader and roller at the right time after the rains of a wet period and after a dry period will make marvelous changes in the character of earth roads of all classes and particularly in those which are proverbially bad.

We cannot too strongly emphasize that to drive up one side of the road with a road machine and back on the other, scraping a lot of loose, heterogeneous rubbish and earth into the middle of the road, to be tramped out again by the traffic, is neither repairing nor maintaining the road. The material brought upon the road should be well distributed and harrowed until an even, uniform layer has been secured and then the roller should be thoroughly applied when the earth is in just the right condition to pack well. Work of this sort will count and will be appreciated.

31

CHAPTER XXII.

FARM MOTORS.

The tendency of modern civilization is toward the adoption of methods and appliances which free man from the necessity of expending his strength in developing mere mechanical power such as a horse, a windmill or an engine may create, and thus to leave him greater freedom to devote a larger share of his time and energies to lines of mental activity, the necessity for which becomes greater and greater as competition becomes wider and more intense. As a result of this tendency farm machines are steadily increasing, becoming more complicated and demanding more and more the employment of one or another form of motor or engine to drive them. This in turn makes it necessary for the farmer to know more of mechanical principles, and how to handle and care for machinery than was formerly necessary.

604. Farm Motors.—The sources of energy which are used on the farm to drive machinery are (1) animal motors, (2) wind motors, (3) water motors, (4) steam motors, (5) oil motors and (6) electric motors.

All of these motors are machines designed to utilize the energy of (1) chemical action, (2) moving air and (3) running water. The horse, the steam engine and the oil engine each derives its power from the chemical action of the fuel consumed or food eaten and may therefore be called chemical engines; the windmill and the water wheel get their power by arresting the motion of wind or water, actuated by the force of gravity, and these may be called gravitation engines. The chemical engines use the energy derived from the collision of molecules and atoms, while the gravitation engines use the energy derived from the movement of streams of air or water traveling as a body.

ANIMALS AS MOTORS.

When animals are viewed from the standpoint of machines they are wonderful mechanisms. Not only are they self-feeding, self-controlling, self-maintaining and self-reproducing, but they are far more economical in the energy they are able to develop from a given weight of fuel material, than any other existing form of motor.

While they are like the steam engine in requiring carbonaceous fuel, oxygen and water for use in developing energy these are made to combine in the animal body at a much lower temperature than is possible in the steam engine, and a much smaller proportion of the fuel value is lost in the form of heat, when work is being done.

605. The Horse as a Motor.—The essential elements which constitute the horse a machine for developing power are (1) a system of rigid levers united by ligaments and capsules at the joints which are automatically lubricated by a synovial fluid; (2) a system of muscles, each one of which is a motor, corresponding in function to the piston and cylinder of a steam engine; (3) a fuel supplying and waste removing system, consisting of the digestive, excretory and respiratory organs; (4) a co-ordinating and regulating mechanism, consisting of the nervous system, which throws the different motors or muscles into and out of action at the times needful to secure the results; (5) a protecting and insulating system, consisting of the skin and hair, which keeps all of the working parts free from dust and reduces the waste of heat.

606. Muscles Are Motors.—Muscles are made up of bundles of fibers which can be stimulated by the nervous system and made to shorten, thus exerting a pull of greater or less intensity as desired. All muscles do their work by shortening and a pull and they are arranged in systems of pairs designed to produce movements in opposite directions, as illustrated by the biceps and triceps muscles which move the fore arm as represented in Fig. 234. Just how the shortening of the muscle fibers is accomplished under the nervous stimulus sent to it is not clearly understood, but it is known that, while in action, the muscle fibers are in a state of vibration which gives rise to sounds known as the muscular murmur.

When muscles are in action and are producing mechanical movements their temperature changes but little, but if the muscles are held in a state of rigid contraction without producing motion as the result, then the temperature rises, showing that the energy which normally would be changed into mechanical motion is changed into heat; this is exactly what occurs in a steam engine. When it is working hard a large portion of the heat energy of the steam is transformed into mechanical work and the heat generated in the fire box thus disappears but the moment the engine is stopped and the steam is held so that it is unable to produce motion of the piston the temperature rapidly rises.

607. Strength of Muscles.—The strength of individual muscles is often very great and more than at first seems possible. Taking the case of the biceps and triceps muscles in the arm, represented in Fig. 234, it is possible to measure their power with a spring balance. If a loop of strong cord is fastened to each end of a spring balance and the foot put through one, while the hand is put through the other, the strength of the muscle can be measured by lifting against the balance, bending the fore arm so as to make a right angle with the arm and holding it horizontal with the elbow against the edge of the desk.

A man of average strength will exert a pull of 50 pounds in this way; and as the lever arm upon which the muscle acts is only one-sixth of the length of the weight arm the pull of the muscle must have been 300 pounds. When the strength of the triceps muscle is measured in a similar way it is found to be able to exert a pull of 25 to 30 pounds; and as the lengths of the lever arms in this case are in the ratio of 1 to 20 or 1 to 24 the power of the muscle must equal 500 to 600 pounds.

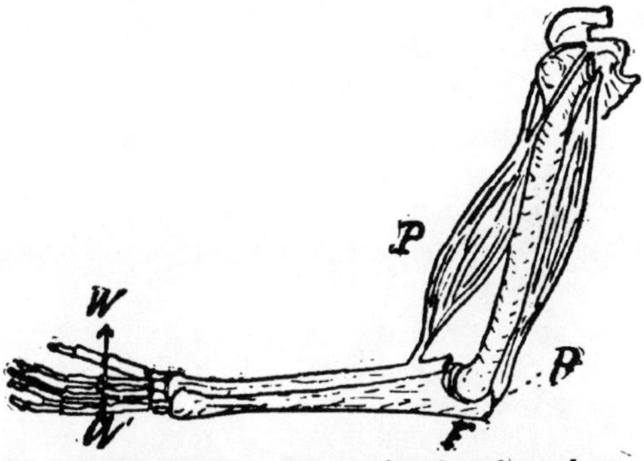

Fig. 231.—Showing the mechanical action of muscles.

It is clear from these measurements that the power of the larger muscles in a horse must be very great indeed.

608. Need of Great Muscular Strength.—It is because the rate at which muscles are able to change their length is relatively quite slow and because they are only able to contract through short distances, that it is necessary to have them act upon the short ends of levers in order to secure the rapid movements through long distances which animals are obliged to make. The horse as an engine consists of a large number of very powerful motors acting through a system of levers.

609. Rate at Which a Horse Can Generate Energy.—It is recorded in (532) that about the maximum walking draft of a horse is one-half his own weight; pulling with this intensity and traveling at the rate of 2.5 miles per hour the ability of a 1,600-pound horse would be

$$\frac{2.5 \times 5280 \times 800}{60 \times 60 \times 550} = 5\tfrac{1}{3} \text{ horse power.}$$

It is not safe, however, to have a horse repeat such strains as this often nor maintain them long at a time. Even when a horse is pulling with an intensity of one-fourth its weight this is too heavy for steady work and represents

For a 1,600-lb. horse, 2⅔ H. P.
For a 1,200-lb. horse, 2 H. P.
For a 1,000-lb. horse, 1⅔ H. P.
For an 800-lb. horse, 1⅓ H. P.

Indeed, it is commonly allowed that for steady and continuous work 10 hours per day at the rate of 2.5 miles per hour a horse should not be asked to pull more than ⅛ to 1/10 of its own weight. At this rate the work of horses of different weights would be

For a 1,600-pound horse, 1.06 to 1.33 H. P.
For a 1,400 pound horse, .93 to 1.17 H. P.
For a 1,200-pound horse, .80 to 1.00 H. P.
For a 1,000-pound horse, .67 to .83 H. P.
For an 800-pound horse, .53 to .67 H. P.

610. Horse Power Required to Haul Loads on a Wagon.—Taking 1 H. P. equal to 550 foot-pounds per second and the data in the table of (538), the number of horse power required to haul two tons, including the weight of the wagon, under the conditions there stated, are given in the table below:

Table giving the number of H. P. required to haul 2 tons on wagons under different conditions, when the rate of travel is 2.5 miles per hour.

Conditions.	High wheels.	Medium wheels.	Low wheels.
Dry gravel road; sand 1 inch deep; some small, loose stones	1.13	1.21	1.47
Gravel road up grade 1 in 44; covered with one-half inch wet sand; frozen beneath.	1.64	1.76	2.81
Dirt road frozen; thawing one-half inch; rather rough; mud sticky....................	1.34	1.59	1.85
Timothy and blue grass sod, dry, grass cut....	1 76	1.94	2.33
Timothy and blue grass sod, wet and spongy..	2.30	2.70	3.75
Cornfield, flat culture, with spring-tooth cultivator; across rows; dry on top	2 38	2.68	3.55
Plowed ground not harrowed, dry and cloddy.	3 37	4.23	4.98

From this table it appears that the hauling of two tons on a wagon, at the rate of 2.5 miles per hour, under the varying conditions of the farm, requires a team to develop energy at a rate ranging from 1.13 H. P. to as high as 4.98 H. P. of 550 foot-pounds per second.

611. Horse Power Required to Plow.—Taking the draft of the stubble plow as given in **(305)** and the mean rate of travel for the team 2.5 miles per hour, the mean H. P. required to do the work, for furrows of different widths and depth, is as given in the table which follows:

Table giving the H. P. required to draw the stubble plow when the soil is in medium condition.

Depth of furrow.	4 inches.	5 inches.	6 inches.	7 inches.	8 inches.
Width of furrow	10 inches.	10 inches.	10 inches.	10 inches.	10 inches.
Horse power...........	1.44	1.79	2.15	2.51	2.87
Width of furrow.......	12 inches.	12 inches.	12 inches	12 inches.	12 inches.
Horse power...........	1.72	2 14	2.58	3.02	3.45
Width of furrow.......	14 inches.	14 inches.	14 inches.	14 inches.	14 inches.
Horse power...........	2.01	2,51	3.02	3.52	4.02

From this table and the one of **(609)** it appears that two 1,600-pound horses find their full capacity for work taxed by the 14-inch plow cutting 4 to 5 inches deep; by the 12-inch plow running 5 to 6 inches deep and by the 10-inch plow running 6 to 7 inches deep. The team of 1,200-pound horses finds its full ability taxed by the plow cutting a 12-inch furrow 4 to 5 inches deep and a 10-inch furrow 5 to 6 inches deep.

612. Increased Speed Diminishes the Traction Power.—If the horse walks more rapidly than 2.5 miles per hour, or at a slower pace, the force which he can exert changes also and is less or greater than 100 pounds. Experience seems to indicate that at speeds between $\frac{3}{4}$ of a mile and 4 miles per hour, and continued 10 hours per day, the traction will be given by the following equation:

$$2.5 \text{ miles} \times 100 = n \text{ miles} \times \text{Traction.}$$

Thus at two miles per hour the traction would be:

$$2.5 \times 100 = 2 \times \text{Traction.}$$

whence $\text{Traction} = \frac{250}{2}$ or 125 lbs.

613. Diminishing the Number of Hours Per Day Increases the Power of Traction.—When the speed remains the same experience has shown that, between 5 and 10 hours per day, diminishing the time increases the possible traction in about the same ratio, or

$$10 \text{ hours} \times 100 = n \text{ hours} \times \text{Traction.}$$

Thus, if the horse is to be worked only 5 hours the traction he may exert will be

$$10 \times 100 = 5 \times \text{Traction.}$$

whence $\text{Traction} = \frac{1000}{5} = 200$ lbs.

PRINCIPLES UNDERLYING THE DRAFT OF THE HORSE.

The principles governing the draft of a wagon have been discussed in Chapter XX; there are others affecting the horse as a motor which need to be considered here.

614. Direction of the Line of Draft.—When a horse has a muscular development and a type of skeleton which permits him to utilize his full weight in hauling then the direction of the line of draft, or of the traces in pulling, exerts an important influence upon how much the horse can draw. In Fig. 235 is represented an apparatus for demonstrating this and other principles underlying the draft of the horse.

When the line of draft is horizontal, as represented in the figure, the spring balance will register a tension on the traces nearly equal to the weight of the model when the fore feet are raised from the ground, and this is the limit of its power to draw under these conditions. If the traces are moved downward it is clear that there will be less tendency to tip the horse up, and hence the greater the slope of the traces the more will be required to raise the horse from his front feet; and at an angle of 18 to 20 degrees the weight of the horse will permit him to draw double what he can with the traces horizontal. If the traces are carried above the horizontal then the horse is raised from his front feet more easily and his draft will be decreased.

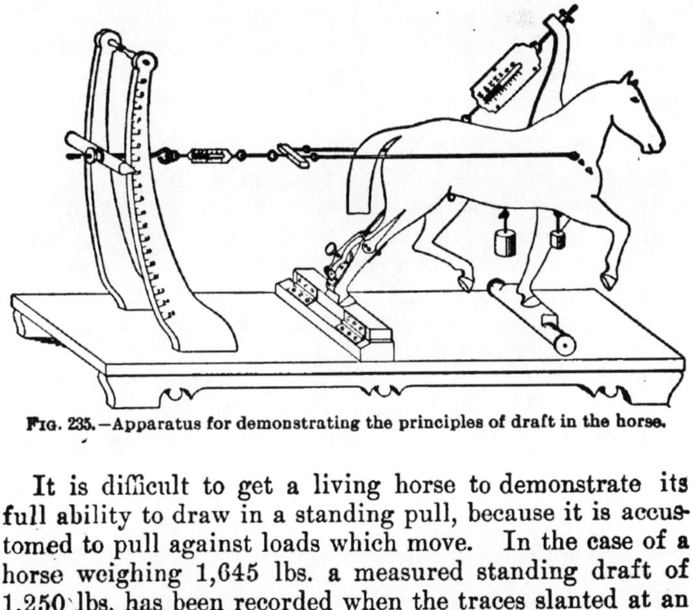

FIG. 235.—Apparatus for demonstrating the principles of draft in the horse.

It is difficult to get a living horse to demonstrate its full ability to draw in a standing pull, because it is accustomed to pull against loads which move. In the case of a horse weighing 1,645 lbs. a measured standing draft of 1,250 lbs. has been recorded when the traces slanted at an angle of 22°, and of 1,120 lbs. with them horizontal. It is doubtful if any of the heavy draft horses are able to utilize their full weight in hauling except when the line of draft is above the horizontal.

615. Influence of Weight on the Draft of the Horse.—Weight in a draft horse is as important a factor to his service as it is in a locomotive; there must be weight enough to make a secure footing. It can be demonstrated with the apparatus in Fig. 235 that two pounds at the place of the center weight in the figure increases the ability of the model to draw an equal amount when the traces are horizontal and, with the same added weight and the traces given a slant of 20 degrees, the ability to draw is increased 4 pounds.

These results mean that of two horses, each having a muscular power capable of utilizing its full weight, the heavier one will exert the stronger pull. The 1,200-pound horse may pull about 200 pounds more than the 1,000-pound horse of like build when the traces are horizontal and 400 pounds more when the traces slope at an angle of 20 degrees.

616. Influence of the Distribution of Weight on the Draft of a Horse.—It will be clear from an inspection of the model represented in Fig. 235 that to transfer the weight from the center ring to the forward one, giving what is in effect a horse with heavier shoulders, will make the weight count for more in preventing the horse being raised off his feet; so, too, will it be evident that if the weight is shifted to the hind quarters it must have much less influence on the draft.

Indeed, most horses in heavy draft, when given the freedom of the head, show that they understand this principle in practice, by both lowering the neck and extending the nose forward, thus giving this portion of their weight a longer leverage to hold the body down on the hind feet which are acting as a fulcrum.

617. Influence of the Strength of the Hock Muscle on the Draft of a Horse.—When a horse is drawing a heavy load a tremendous strain is brought upon the muscles which straighten the hock joint so as to force the body forward, and the load after it, in walking; and it is a deficiency of ability at this point oftener than at any other which limits the power of the horse as a draft animal.

With the spring balance represented above the back of the model in Fig. 235, which controls the hinged hock joint through a rod, it is possible to vary the tension which holds it rigid and thus demonstrate the ratio of muscular tension to the draft on the load and show that with too weak muscles only a portion of the weight of the body can be utilized in draft. In the model the tension of the hock muscle is about double the draft and while this is not intended to demonstrate the relation of strength of muscle to intensity of draft in any horse it illustrates the fundamental principle and shows how extremely powerful the muscles of the horse must be to permit him to make the draft he does.

618. Influence of the Width of the Hock on the Draft of the Horse.—Not less important than the strength of the hock muscle, in determining the qualities of a draft animal, is the "width of the hock joint" itself; or, stated in the language of mechanics, the ratio between the two arms of the lever upon which the hock muscle acts. If the projection of the heel bone backward, which forms the point of the hock, is long in comparison with the distance to the hoof, as represented in the diagram, Fig. 236, at the left, instead of short as shown at the right, then it is clear that

FIG. 236.—Diagram showing difference between wide and narrow hock.

with a given strength of hock muscle it will be possible to straighten the limb under a greater pull and the ability of the horse to draw is thereby increased. In the model represented in Fig. 235 the attachment at the hock joint is arranged so as to lengthen the power arm of the hock muscle lever different amounts and thus demonstrate an increasing draft when the strength of the muscle is maintained constant.

In fixing the attention upon the hock joint as influencing the draft of a horse it is not intended to convey the idea that other features are not important or that they do not vary in a marked degree in the different types; for it is true that the make-up of the whole body of the draft animal is notably different from that of the one built primarily for speed, but the type of variation shown at the hock joint runs through the whole framework.

619. Attachment of the Traces to the Hames at the Shoulder.—To enable a horse to utilize his full weight to the best advantage in draft it is important that the attachment of the traces at the collar should be as low as the comfort of the animal and other conditions will permit. When the traces are low at the shoulder there is less leverage for the draft to raise the horse off his front feet and hence his weight counts for more. For the same reason a horse low on his feet and with a relatively long body has greater leverage for his weight in draft.

It will not do to so lengthen the hame strap above and shorten that below as to bring the attachment of the traces down upon the point of the shoulder, for then the heavy pressure of the collar will irritate the shoulder and make it sore.

620. Two-Horse Evener.—There are three types of two-horse equalizers or eveners in use on the farm: (1) where the holes for the whiffletrees are in a line back of the hole for the draft pin; (2) where the holes for the whiffletrees are in a line in front of the draft pin; and (3) where all three holes are in the same straight line.

Each of these types of evener divide the work equally

between the two horses so long as the evener crosses the line of draft at a right angle, but as soon as one horse falls behind the other then only the third type remains a just equalizer. The truth of this statement can be readily demonstrated with the apparatus represented in Fig. 237, where the three types of eveners are combined in one piece.

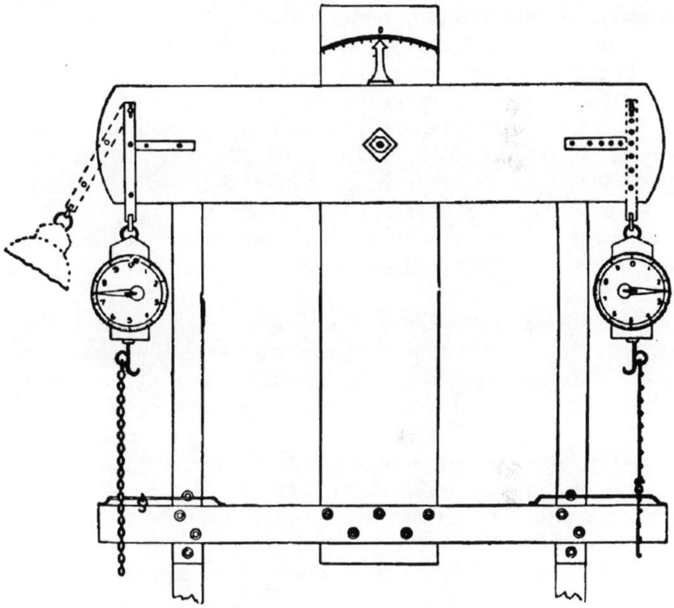

FIG. 237.—Apparatus for demonstrating the principle of eveners.

Referring to the figure it will be seen that as the clevises for the whiffletrees are there set the evener may be made to form various angles with the line of draft and the inequality of draft resulting may be measured with the pair of scales. With a 4-foot evener where the holes for the clevises are 4 inches behind the draft pin the horse which is ahead may have an advantage greater than 25 per cent., if the angle formed is as much as 20°.

Even in an equalizer where the three holes are only one inch out of line an angle of 20° for the evener with the line of draft may give the horse ahead nearly as much advantage as would result by setting the clevis of the other horse in toward the center one inch.

When the holes for the clevis pins are in front of the draft pin a similar inequality of division of labor occurs, but in this case the horse which is in front must pull the most, the differences measuring as great as with the other type. When the three pins are placed in a straight line there is nearly a true division of labor between the horses, even when the angle formed by the evener is large. This statement, however, is only true when the clevis pins and the draft pin fit the holes closely.

621. Giving One Horse the Advantage.—When it is desired that one horse shall do more work than the other this is accomplished by shortening the lever arm of the horse which it is intended shall do the larger share of the work. If it is desired that the off horse shall do 60 per cent and the near horse 40 per cent of the work then the clevis pin of the off horse must be set in until the two ends of the evener are in the reverse ratio, or as 40 to 60.

If the evener is 48 inches long the two arms would be each 24 inches. From the equation of the lever we have

$$P \times PA = W \times WA$$

and

$$60 \times PA = 100 \times 24$$

whence

$$60 \times PA = 2,400$$

and

$$PA = 40$$

From this is appears that the clevis must be set in

$$48 - 40 = 8 \text{ inches},$$

which leaves the off horse with a weight arm of 16 inches and the near horse with one of 24 inches.

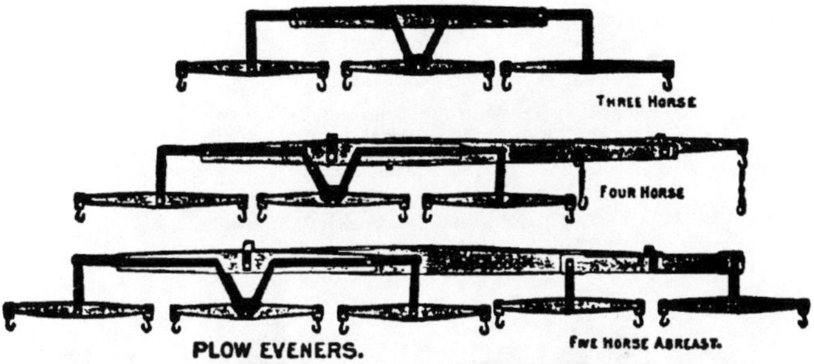

FIG. 238.—Equalizers for horses.

622. Three-Horse Equalizer.—There have been many forms of 3-horse equalizers devised, but the straight bar in which one horse pulls against two others is the simplest and generally the most effective. To make this evener the holes should all be as nearly in the same straight line as possible, and if the work is to be divided equally the hole for the draft pin should be placed at ⅓ of the length of the evener from one end. Fig. 238 represents a set of three, four and five-horse equalizers, and Fig. 239 represents a method of driving four horses abreast.

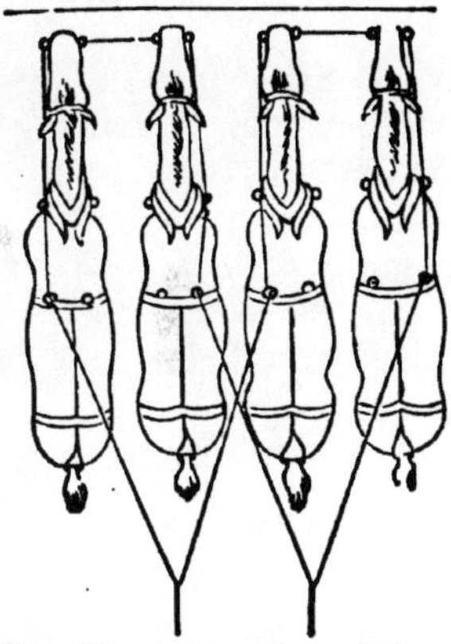

FIG. 239.—Arrangement of lines for driving four horses abreast.

623. The Tread Power. — The tread power is a rolling endless inclined plane so arranged that its motion is transferred to a shaft which is made to revolve and drive a belt. One form is represented in the upper portion of Fig. 240, and in the lower portion of the same figure are represented two forms of treads, the level at the right and the inclined at the left. The level tread has the advantage of permitting the horse to travel with its feet more nearly in the normal attitude to its limbs and on this account the fatigue is supposed to be and probably is less.

FIG. 240.—The tread power.

In order that a large per cent of the labor expended by the horse, when working on the tread power, shall become available it is very important to have all of the bearings clean, free from dust and grit and well oiled. There are so many of these bearings and they are of such a character that great care is required in running this power to avoid heavy loss of efficiency due to friction.

624. Working the Horse in the Tread Power.—When a horse is put into a tread power to work he accomplishes the result by lifting his own weight against the force of gravity and the more steeply the power is inclined and the faster the horses walk the more work they do. Inclining the power so that the bed rises 2 feet in 8 feet requires the horse to lift $\frac{1}{4}$ of his own weight, thus producing a pull equal to that on the belt when it travels at the same rate. This for a 1,600-pound horse represents a pull of 400 lbs.; for a 1,200-pound horse, 300 pounds; and for an 800-pound horse, 200 pounds. If, under these conditions, the horse walks at the rate of 2 miles per hour, the work done will be 2.13 H. P. for the 1,600-pound horse, 1.6

H. P. for the 1,200, and 1.07 H. P. for the 800-pound horse. These results are about double the horse power for corresponding weights where the draft is $\frac{1}{10}$ that of the weight of the horse, as given in the table of **(609)**.

It is a common practice to set the tread power as steep as 2 feet in 8 feet and when this is done it is clear that the horse is called upon to develop power faster than he is able to do and follow it day after day. It is clear also how a horse may be made to do more work in a tread power than when drawing on the sweep power, and why this form of power may appear more effective, when the chief difference is due to the fact that the horse is working harder, and horses are often overworked in a tread power without knowing it or intending to do so.

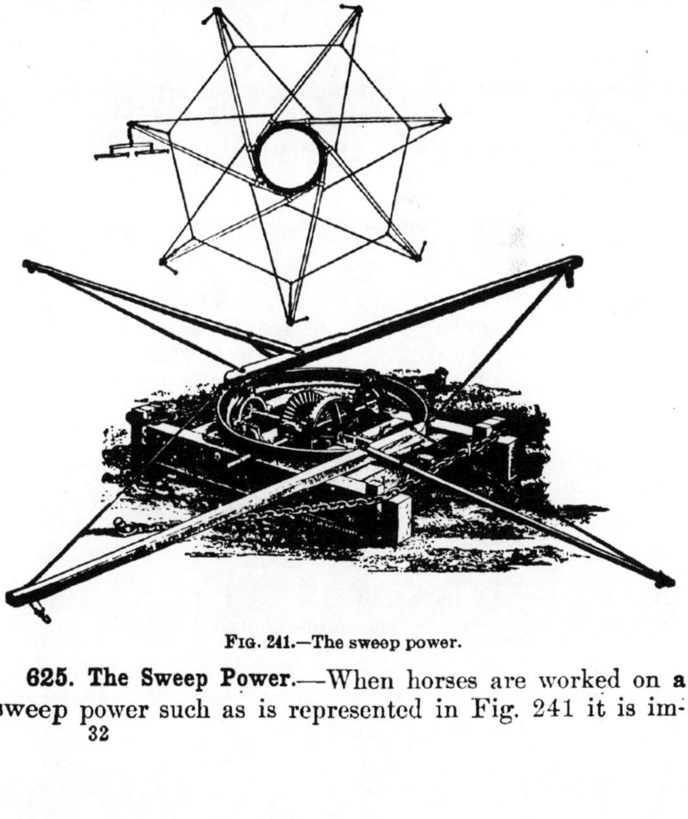

FIG. 241.—The sweep power.

625. The Sweep Power.—When horses are worked on a sweep power such as is represented in Fig. 241 it is important that the line of draft be as nearly as possible at right angles to the sweep, for it is this angle which renders the highest per cent. of the draft available. It will be clear from the upper portion of the figure, representing a plan of a 14-horse sweep, that the line of draft there cannot be at right angles to the sweeps and that it is impossible for it to be so in any sweep power. On this account, there is a considerable portion of the draft lost in producing pressure on the bull-wheel and this is greater the shorter the sweeps are and the longer the hitch is between the horses and the sweep. If the line of draft made an angle of 45 degrees with the sweep, one-half of the power would be lost in pressure on the bull-wheel and in increasing the friction.

STEAM ENGINES.

The steam engine is one of the earliest of man's inventions designed to utilize or transform molecular motion, in the form of heat, into useful work. The intense vibrations which are caused by the burning fuel in the combustion chamber are imparted to the water, converting it into steam capable of exerting greater or less pressure, according as its temperature is high or low.

626. Principle of Action in the Steam Engine.—It was shown in **(43)** that 966.6 heat units are required to convert one pound of water at 212° F. into steam at 212° under a pressure of one atmosphere; and in **(41)** it is shown that these heat units are equivalent to 752,305 foot-pounds of work.

The fuel value of one pound of coal is 14,000 heat units which, expressed in foot-pounds, is

$$14{,}000 \times 778.3 = 10{,}896{,}200 \text{ foot-pounds.}$$

The steam engine aims to utilize the power of coal or other fuel by transforming its enormous potential energy into that of confined steam, and if it were only possible to utilize 80 or 90 per cent. of this power the steam engine would be a very inexpensive motor.

627. Efficiency of the Steam Engine.—It is unfortunately true of the steam engine as a source of power, that in practical experience it is only able to render available from 2.5 to 20 per cent. of the full heat value of the fuel burned in the fire box, and it is still more unfortunate that there seems to be little hope that its efficiency can ever be made to much exceed 31.5 per cent. The reason this is so is because it has not been found practicable to use steam at very high temperatures nor to cool it much below that of the ordinary air conditions. To enable a water wheel to utilize the highest per cent. of the power of a falling stream it must be so arranged as to be able to take the water at the highest possible level and not to release it until it has reached the lowest possible level, and the principle is the same with the steam engine. If the steam could be taken into the cylinder at a temperature of 1,000° F. and released from it only after its temperature had fallen to 60° F. it is clear that much more work could be performed than when the temperature is only permitted to fall between 300° F. and 212° F.

Where heat is converted into work the efficiency is always equal to the quantity of heat taken into the engine minus the quantity given out divided by the quantity taken in; thus, if the steam entering the cylinder carries 100 heat units and it escapes from the cylinder with 90 heat units after moving the piston the efficiency of the engine has been only

$$\frac{100 - 90}{100} = 10 \text{ per cent.}$$

So, too, if steam enters a cylinder at a temperature of 300° F. and escapes at 212° F., the maximum efficiency would be only

$$\frac{(461 + 300) - (461 + 212)}{461 + 300} = 11.5 \text{ per cent.}$$

In this equation 461 is the number of degrees F. which the zero of the Fahrenheit scale is above absolute zero, and in such problems as these it is necessary to express the temperature in absolute degrees. When this is done 300° F. becomes 761° F and 212° F. becomes 673° F., and the above equation becomes

$$\frac{761 - 673}{761} = 11.5 \text{ per cent.}$$

From the results of this problem it is clear why it is not possible for the steam engine to utilize a very large per cent. of the total energy which the steam carries with it into the cylinder. Even if the steam could be carried into the cylinder at 1,000° F. and could do work on the piston until its temperature fell to 100°, the maximum efficiency would only be

$$\frac{(1000 + 461) - (100 + 461)}{1000 + 461} = 61.6 \text{ per cent.}$$

628. Pressure of Steam at Different Temperatures.—The temperature at which water is changed from a liquid into steam or invisible vapor varies with the pressure to which the water is subjected as stated in the table below:

Table showing the pressure of steam or water vapor at different temperatures.

Temperature of water.	Pressure of steam.
102° F.	1 lb. per sq. inch.
162	5 lb. per sq. inch.
194	10 lb. per sq. inch.
212	14.73 per sq. inch.
228	20 lb per sq. inch.
328	100 lb. per sq. inch.
432	350 lb. per sq. inch.
546	1000 lb. per sq. inch.

629. Dry and Wet Steam.—When steam contains no water held mechanically in suspension it is known as dry steam, but it is seldom possible to develop absolutely dry steam because as it escapes from the surface of the water in the boiler there is a tendency to carry away with it more or less water in the form of tiny drops, such as form the white cloud. Steam carrying much water in suspension is called *wet steam.*

It is important to keep this property of steam in mind when comparing the efficiency both of boilers and of engines. If, for example, the evaporating surface of the water in the boiler is small, and steam is forming rapidly, so that large quantities of water are carried over not evaporated, the boiler may be credited with evaporating a large amount of water with a comparatively small amount of fuel, when it is only carrying it away mechanically suspended in the steam.

Then, too, if an engine is being worked with *wet* instead of *dry* steam, and the fact is not known, it will appear that it is using much more steam for a given amount of work than it really is, because the water carried over in this way is not effective in developing power.

630. Causes of Water in the Cylinder of an Engine.—There are several causes for the presence of water in the cylinder of an engine, and these may be stated as—

1. Wetness of the steam coming from the boiler.
2. Wetness due to cooling of the steam when passing through pipes and steam chest on its way from the boiler to the cylinder.
3. Condensation of steam in the cylinder when the engine is first started, before the walls become heated to the temperature of the steam.
4. Condensation due to the work done by the piston after the cut-off has occurred.
5. Condensation due to cooling of the walls of the cylinder itself.

631. Wetness of Steam from the Boiler.—The wetness of the steam as it comes from the boiler is modified in several ways: (1) If the steam is generated rapidly the amount of water carried over is larger than when the generation is slow, because there is greater mechanical agitation. (2)

If the area of the water surface at the water level in the boiler is small in proportion to the pounds of steam delivered the water carried over will be large and for this reason the horizontal boilers of a given H. P. tend to supply dryer steam than the vertical boilers do because there is more surface from which the steam may escape and the agitation is less. (3) If the volume of steam space above the water is large there is more opportunity for the suspended water to fall back and leave the steam dryer, hence one means of preventing "priming," as carrying over water is called, is to work with the water level low in the gage glass. (4) If the size of the boiler compared with the amount of steam required for each stroke of the piston is small the tendency will be to cause the pressure in the boiler to vary and this variation will agitate the water and cause "priming." When this is the case priming may be lessened by throttling down the steam supply at the stop valve.

632. Wetness Due to Condensation in Steam Pipes and Valve Chest.—When the steam pipes are long, leading from the boiler to the steam chest, and when they are not jacketed and are exposed to the cold, priming is produced. Jacketing the steam pipes and the steam chest reduces the priming from this cause very materially because the loss of steam from uncovered iron steam pipes, per degree of difference of temperature between steam and outside air, is about 2.4 heat units per square foot of outside surface of the pipe per hour, while covering the pipe with wool felt one-half an inch thick reduces the loss to .7 heat units in the same time.

633. Initial Condensation.—The temperature of the walls of the cylinder is always, in practice, colder than the entering steam as it comes from the boiler and so there must be a greater or less condensation as it enters until both are brought to the same temperature. So great is the condensation of steam when the engine is first started that it is necessary to provide the cylinder with ***relief cocks*** at each end, shown at 8 in Fig. 250, which must be opened at the start to allow the water to escape. If these are not opened at the start enough water may collect in the cylinder to cause the piston to drive out the head of the cylinder or do some other injury.

As the temperature of the cylinder gradually increases less and less water is deposited and then the relief cocks may be closed, the water which is condensed afterward being so little that it is re-evaporated after the cut-off takes place and during the exhaust stroke because, as the piston travels, the space for the steam increases and this reduces the pressure so that at the lower pressure the heat in the walls of the cylinder is able to re-evaporate the water which had been condensed. In this way a well protected cylinder keeps itself empty after it has become heated.

634. Condensation Due to Work During Expansion.— When the steam expands and expends its energy in driv-

Fig. 243.—Horizontal boiler.

ing the piston forward its temperature is lowered in proportion to the amount of work which it does and on this account more or less of water tends to condense in the cylinder which, like the rest, must be removed by re-evaporation.

It should be clear from this and the preceding paragraphs that the cylinder should be well jacketed so as to reduce as far as possible all tendency to condense the steam in the cylinder before it escapes through the exhaust.

635. Engine Boilers.—The boilers of farm engines are commonly one or the other of two types, horizontal or upright, represented in Figs. 243 and 244. The horizontal boilers are best adapted to the engines of the larger sizes and are as a rule the most economical forms; but where a small engine is desired, and especially one which is compact and which occupies but little space, then the upright types may be used, such as represented in Fig. 244.

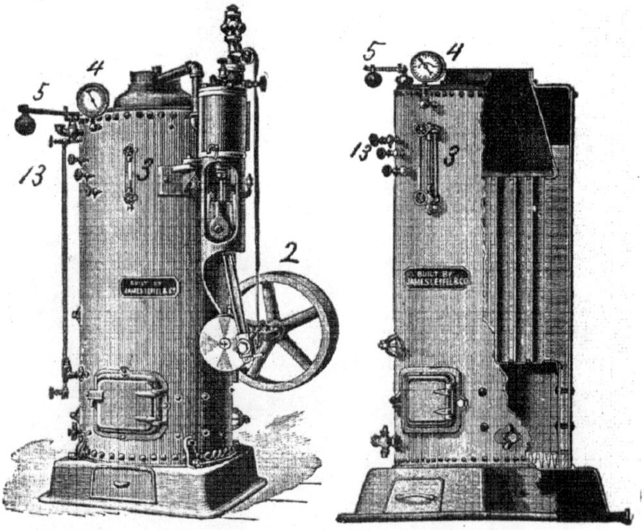

FIG. 244.—Vertical boiler and engine.

636. Construction of Steam Boilers.—Steam boilers are usually made of strong sheet steel $\frac{3}{8}$, $\frac{1}{2}$ or $\frac{5}{16}$ inches thick which are rolled into cylindrical forms, securely riveted and often braced as represented in Fig. 245. The fire-box is placed in one end and is entirely surrounded by water so as to lessen the loss of heat. The boiler represented in Fig. 245 is designed specially for burning straw as fuel, which is introduced into the fire-box A, from which the flame passes forward through the main large flue B into the combustion chamber C. From the combustion chamber the flame is sub-divided, returning to the smoke

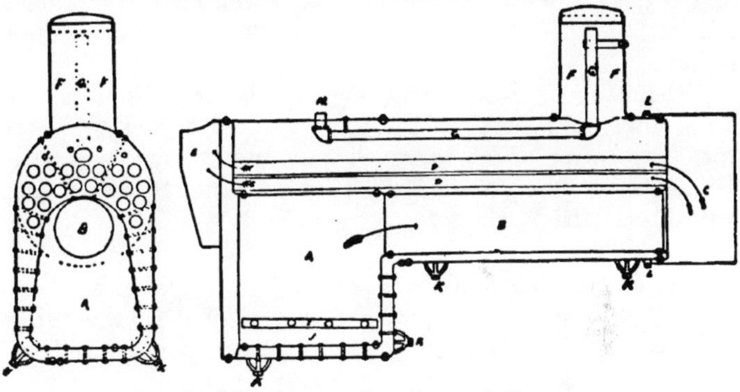

FIG. 245.—Construction of steam boiler.

stack E through the small flues D. In the same figure FF and FF represent the steam dome from which the dry steam is taken by the supply pipe G to the steam chest at H, not represented. At the bottom of the boiler at KK and KKKK are represented hand holes to be used in cleaning it out. The construction of the valve for closing the hand hole is shown at A in Fig. 243 and also the relation of the flues to the water being heated by them.

In the arrangement of the flues in the boilers, particularly in the horizontal forms, it is important to have them placed in vertical rows rather than one flue above the space between the two below, in order that there may be as free and rapid a circulation of water as possible. It is very important in the construction and management of a boiler to so arrange conditions as to have as little difference of temperature in all parts of the boiler as possible; because unequal temperature tends to develop strains in the metal and to tear or loosen rivets and cause leaks.

637. Gage Cocks.—Boilers are commonly provided with three gage cocks, represented at 13,13 in Fig. 244 and at 13 in Fig. 249. These are for the purpose of showing where the upper surface of the water is in the boiler at any time. The lower gage cock is placed about two inches above the upper surface of the upper flues in the horizontal boiler.

When the engine is running the water is held in the boiler near the level of the middle gage cock and is fed into the boiler so as to reach the upper gage cock only when the engine is to be shut down to stand for some time without allowing the fire to go out.

638. Gage Glass.—The object of the gage glass is to show at a glance just what the water level in the boiler is at any moment and its position is represented in Figs. 244 and 249 at 3.

It should always be kept in mind that it is not safe to rely entirely upon the indications of the gage glass because it is peculiarly liable to become clogged with sediment from the boiler; on this account the lower cock should be frequently opened to blow it off and clear out any sediment, and the water level in the boiler should frequently be tested by means of the gage cocks.

When the engine is to be stopped to stand with the fire on for any length of time the gage glass should be closed, shutting off the water first and then the steam; this is to lessen evaporation and to prevent escape of water from the boiler in case the gage glass should break. When opening the gage again the steam should be turned on first, the water last, and the pet cock opened to blow off any sediment and show that the gage is in proper working order

In case a gage glass should be broken when the pressure is on the water should be shut off first and then the steam, after which a new glass may be put in.

639. Pressure Gage.—The pressure gage is intended to show the number of pounds per square inch of steam pressure there is on the boiler at any moment and to serve as an indicator to the fireman of the condition of his fire. It is represented at 4 in Fig. 244 and the interior construction of this gage is shown in Fig. 246.

FIG. 246.—Construction of steam gage.

As the steam enters the hollow spring shown in the figure its pressure tends to straighten it because the pressure on the longer circumference is greater than that on the shorter one. The motion thus produced is communicated, through a segment lever and pinion, to the index which is made to revolve over a dial upon which the pressure may be read.

It should be remembered that a steam gage may get out of order and fail to show the true pressure. In such cases the operator must be guided by the safety valve. (**640**). Some pressure gages are provided with a siphon placed between the boiler and the gage, which prevents the dry steam entering the spring at too high a temperature and also automatically drains out the water, thus preventing injury from freezing.

640. Safety Valve.—The safety valve is connected with the steam chamber of the boiler where, when the pressure reaches a point as high as the boiler is intended to carry, it may be opened by the pressure and the steam be allowed to escape, thus relieving the pressure and at the same time warning the engineer by the sound of the escaping steam. The position of the safety valve is represented at 5, Fig. 244.

Care should be taken by the operator to see that this valve is in good working order by raising it gently at times to see that it has not become set in some way. The weight which has been provided by the manufacturers to hold the valve against the steam pressure should never be made heavier by loading or so set that it will oppose a greater pressure than the maximum intended for the boiler.

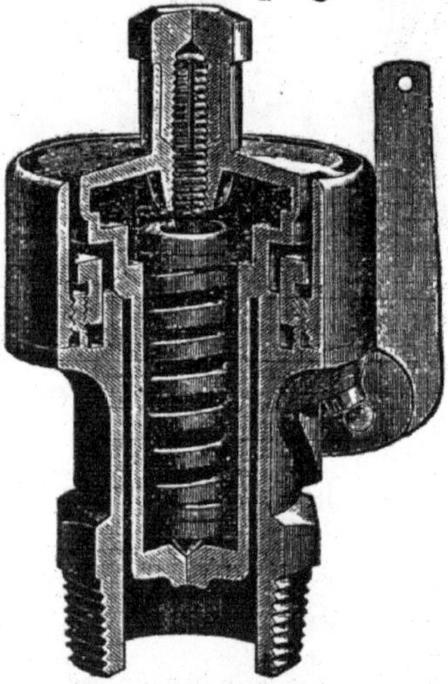

FIG. 247.—Kunkle lockup pop safety valve.

In Fig. 247 is represented the Kunkle lockup pop safety valve operated by a spring instead of a lever and weight.

641. Care of the Boiler.—In order to get the best results from a boiler it is necessary that the flues be often cleaned in order that there may be no soot or ashes to prevent the heat coming in direct contact with the metal. How often this should be done must depend entirely upon circumstances. Oftentimes it should be done daily, at any rate the flues should be kept clean and the draft perfect.

Periodically it is necessary to clean the interior of the boiler to remove the scale and sediment which accumulates from the water used in making the steam, (**649**). How often this must be done will depend entirely upon the character of the water. In some cases it must be done once a week but with clean soft water it may not be required oftener than once in six months.

When cleaning is to be done it is important to make sure that the fire is all out and the steam should be permitted to fall to as low as 10 pounds before the blow-off is opened. If the fire is not all out the flues may be made to leak and if the steam is too hot the mud will be caked on the flues so that it cannot be readily removed.

In replacing the plates for the hand holes it is important to see that they are clean and that no scale or dirt is on the seat. Sheet lead makes the best packing for these places. The nuts should be turned up tight at first and after steam is up and the metal expanded they may need tightening a little more.

642. Firing.—Care and skill are required to do good firing, whether with wood or coal. In firing with wood it is necessary to keep the fire-box nearly full all the time and it will occasionally require "knocking down" but it is a good plan not to use the poker more than necessary. The wood should be placed in the fire-box as closely as practicable.

In firing with coal the grates should be kept as evenly covered as possible with a thin fire, avoiding throwing on large lumps of coal or putting on large quantities at a time. If the coal forms clinkers these must be removed from the grate through the door but it is desirable not to use the poker when it can be avoided. The ashes must be kept removed form under the grate or the bars will be warped or melted.

It is well to allow the safety valve to blow off once a day to note how this and the pressure gage agree, but good firing will not permit this to occur unless the engine is stopped.

When the fire is too strong it may be controlled by opening the door to the fire-box an inch or less or leaving the damper open. It is not a good plan to open the fire door and close the damper at the same time when the engine is running.

643. Foaming.—Foaming in the boiler is a dangerous

symptom and should be avoided. The fact is indicated by the water in the gage glass becoming muddy and unsteady, rising sometimes very high and then falling again as quickly. It is often caused by dirty water, especially when it contains alkali or grease.

When foaming occurs it is difficult to tell just where the water stands in the boiler and here is where the danger lies. The tendency with foaming is to cause the heated surfaces of the boiler to become uncovered and become excessively hot so that when the water returns steam may be suddenly generated with explosive violence.

644. Low Water in the Boiler.—If by any chance the water should become too low in the boiler, cool judgment and quick action are called for, because if the crown sheet has become exposed it is liable to be weakened by overheating. In short, an explosion is imminent. The thing to do first is to cover at once the fire in the fire-box with three or four inches of wet ashes or earth so as to shut off the heat. Do not under any circumstances undertake to rake out the fire, as stirring it up fresh only makes the heat more intense for the moment.

At such a time the safety valve should not be opened as the sudden release of pressure which this would permit may cause an explosion by the agitation throwing water onto the overheated crown plate. The thing to do is to allow the engine to cool down and when cool enough to refill the boiler.

645. Soft Plug.—Boilers are provided with a "soft plug" which screws into the crown plate and is fitted with an alloy which melts at a low heat to allow the water to be forced upon the fire and extinguish it before the crown sheet could be injured. Such a plug, however, is not always reliable as the top of it may become coated with lime and thus rendered ineffective. On this account the plug should be removed and scraped occasionally and it is prudent to put in a new one each year or refill it.

If a soft plug blows out in the field it may be temporarily refilled with lead or Babbitt metal but the melting point of these is too high to prevent the plate from being injured. The soft metal is an alloy made by melting together equal weights of lead and tin, having a melting point of 420° F., that of lead being 610° and Babbitt metal 650° F.

646. Water Supply.—The water supply to the boiler must always be adequate and under complete control. The greatest care and vigilance should be exercised by the engineer and he should *know* that his pump and injector are in prime condition at all times. In the first place the cleanest water which can be had should always be used and if necessary the water should be strained when it is put into the supply tank. Be sure that the suction hose and connections are free from leaks. It sometimes happens that the nipples screwed into the boiler through which the injector and pump feed, lime up and these should be examined occasionally to see that they are free.

There are two methods of supplying the boiler with water (1) with a pump and (2) with an injector. Pumps are either driven by the engine when that is running or directly by steam pressure.

647. Cross-head Pump.—A common form of pump for supplying the boiler with water is known as the cross-head because it is driven from the cross-head of the engine. This being true it is of course only available when the engine is running and an engine with this sort of pump should also be provided with an injector.

The independent boiler feed pumps are some one of the steam types and are practically small steam engines which drive the pump cylinder.

648. The Injector.—The principle by which steam from the boiler is able to force water back into the same boiler against the same pressure and the action of the injector is as follows: When steam is issuing from the boiler under a pressure of eighty pounds and entering the injector at V, Fig. 248, it may have a velocity of nearly 1,800 feet per second; as this passes through R into S it produces a strong suction in through the supply pipe and when the steam strikes the cold water it is at once condensed. But when the steam is condensed into water it still has its high initial velocity and, striking the incoming water, drives a portion of it directly into the mouth Y through into the chamber O and from thence into the boiler.

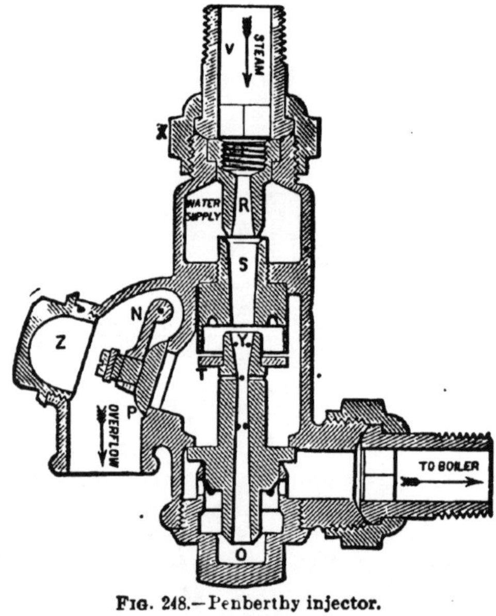

FIG. 248.—Penberthy injector.

When the injector is used at a steam pressure of 65 pounds the water supply valve is opened one turn, then the steam valve wide. If the injector does not start at once, and water runs from the overflow, throttle the water supply slowly until it picks up; but if hot steam and water issue from the overflow open the water supply valve farther.

649. Boiler Incrustation.—The use of hard water for making steam results in the precipitation of the carbonates of lime and magnesia, and their sulphates also, when these are present, on the flues and walls of the boiler in the form of a more or less resistant scale which may be harmful in several ways: (1) The incrustation on the boiler is not a good conductor of heat and both the capacity and efficiency of the boiler are decreased. (2) When a heavy crust forms on the boiler which prevents perfect contact of the water the boiler may become overheated and the scale thus weaken it by allowing it to "burn out." (3) It is thought that even boiler explosions may sometimes originate from the thick scale suddenly flaking off when the boiler underneath is overheated and thus letting the hot water come suddenly in contact with the hot surface, which results in the sudden evolution of a large volume of steam.

To prevent the formation of scale on boilers and to remove it when formed many methods have been proposed. A common one is to use the simple sodium carbonate or sal soda of commerce, dissolving a quantity in water and letting it be fed into the boiler with the water. Its action is to cause the carbonates to be precipitated in a more or less powdery form which does not adhere to the flues so firmly. It is possible that the influence of the sodium carbonate, besides taking up the excess of carbon dioxide from the bicarbonates of lime and magnesia, is to flocculate the lime and magnesium carbonates and sulphates, causing them to fall in larger granules which have not the power of adhering to the walls of the boiler and flues as the molecules do. Sometimes ammonium chloride is used and in this case the carbonates are converted into chlorides, which are very soluble in the water, while the ammonium carbonate is volatile and passes off with the steam. Where the steam is not to be used for any other purpose than driving the engine, kerosene is sometimes employed but its method of action is not clearly understood.

650. The Engine.—Most farm engines are mounted upon their boilers as represented in Fig. 244, at the left, and in Figs. 249 and 250. Its chief parts are the cylinder, 5; the steam chest with sliding valve, 4; the fly-wheel, 2; the eccentric; the governor, 6; and the throttle valve, 7.

The construction of the cylinder of the compound engine is shown in Fig. 252, where A is the high pressure cylinder, B the low pressure cylinder and 1, 1, 1, 1, 1, 1, 1, 1, the sliding valve which regulates the entrance and exit of

FIG. 249.—Portable steam engine.

the steam. As the steam from the boiler comes to the steam chest at E it first enters the compartment D of the slide valve by a port not shown in the section and from there is conveyed into the cylinder A along the passage O where it forces the piston toward G. While this is being done the steam on the other side of the piston at A, which has spent only a part of its energy, passes out through the passage 2,2, into the steam chest C from which it enters the large cylinder B on the side of the piston at 3 and

Fig. 250.—Portable steam engine.

in this position assists the high pressure steam in driving the piston toward G. On the opposite side of the large piston, in the low pressure cylinder B, is the steam which has spent its available energy in driving the large piston in the opposite direction and this is being forced out through the passage 4,4 into the exhaust 5. At the proper time, when the pistons are nearing the ends of the cylinders toward G, the eccentric reverses the action of the rod F and pushes the slide valve until 6 stands over 4 and 7 over 2, which permits the high pressure steam to enter A through 2,2 and the partly expanded steam to enter B by way of O to C and from thence through 4,4 to B, when the direction of motion of the piston is reversed.

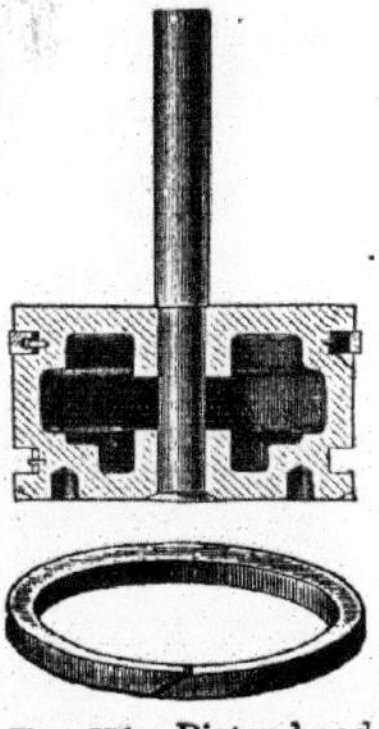

Fig. 251.—Piston head with metal packing

The construction of the piston head with its self adjusting metal packing rings are shown in Fig. 251.

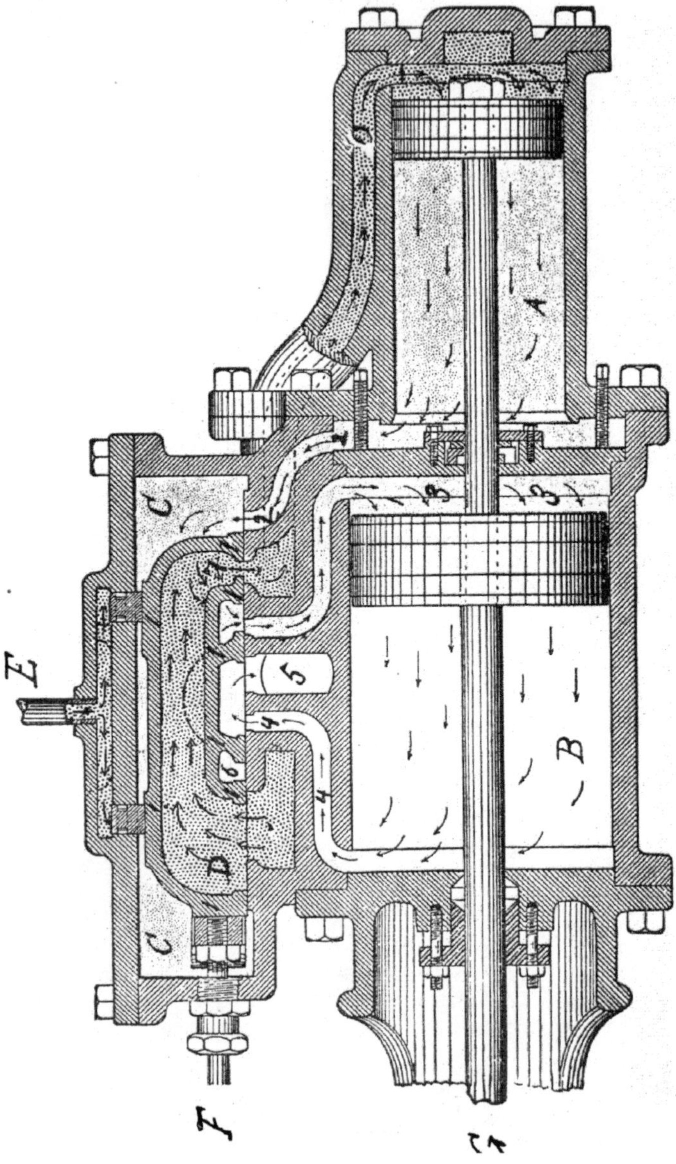

FIG. 232.—Construction of cylinder of compound engine and steam chest with slide valve.

651. Governor.—In order that the speed of the engine may be controlled it is necessary that the amount of steam admitted to the cylinder should vary with the work before the engine. To maintain a uniform speed there is provided a governor, one form of which is represented in Fig. 253, whose action is as follows:

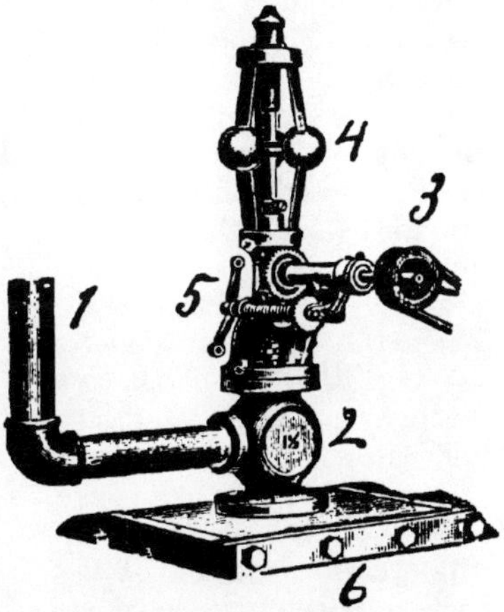

FIG. 253.—Governor of steam engine.

At the point 2 in the pipe 1 leading from the boiler there is a valve which can be opened and closed by the action of the balls 4, which are made to revolve by the belt working on the pulley 3. As the speed of the engine increases the balls of the governor are made to revolve more rapidly and by their centrifugal force bend the strips of elastic metal to which they are attached outward, and this draws the upper end of the spindle downward, partly closing the valve at 2. By means of the spring at 5 the resistance the governor must overcome to close the valve may be varied and in this way the governor may be set so as to cause the engine to run steadily at different speeds.

652. Lubricator.—To keep the valves in the steam chest and the cylinder well oiled, a special form of lubricator is required, and one of these is represented in Fig. 254 and is seen in place on the engine at 10, Fig. 249. This is screwed into the steam pipe leading to the steam chest at the threaded end H. The oil receptical is the cylinder above I which must be filled by removing F, but first closing E and G and removing I so as to drain out the water. After returning I the oil cup is filled entirely full above the level of the sight-feed D, when F is again closed and E and G opened.

The action of the lubricator is caused by steam rising into the bend B and condensing in the left leg. The water being heavier than oil flows through G across the glass face D and falls to the bottom of the oil reservoir, thus forcing a like amount of oil up and out through valve E and on into the steam pipe where the steam carries it into the steam chest and cylinders. When the oil is all out of the cup the water shows through the face D, and the lubricator must be refilled.

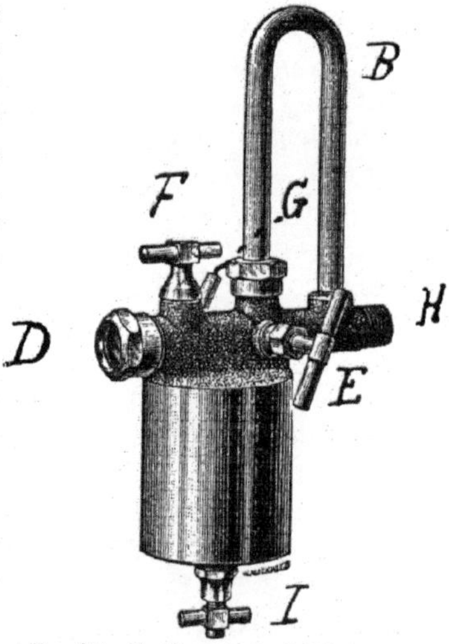

FIG. 254.—Swift sight feed lubricator.

653. Fly Wheel.—In all single crank engines it is very important to have a well designed and ample fly wheel in order to ensure steady running of the engine. It will be clear that as the piston rod passes through the course of its stroke its efficiency must rise and fall as it approaches and recedes from the dead centers. The fly wheel, represented at 2 in Figs. 244, 249 and 250, enables energy to be stored in its heavy fast-moving rim when the crank shaft has the greatest efficiency and this may be given out again to maintain the speed when the dead centers are being approached and passed.

GASOLINE ENGINES.

Within the past ten years there has been a strong movement to place upon the market for farm use motors of the internal combustion type and many kinds of gasoline engines, ranging from 1 to 15 and 20 horse-power, are now offered for sale by manufacturers. While it cannot be said that these motors have in general earned for themselves the reputation for reliability that steam engines possess, it is now acknowledged that there are upon the market gasoline engines which are efficient and quite satisfactory for farm purposes.

654. Gasoline and Steam Engines Contrasted.—Gasoline engines are widely different from the steam types described in the last section. In those the power is derived from a steadily burning fire converting water into steam, which transmits the power to the working parts of the engine; in these the fire is an intermittent one which is almost instantaneous in duration and which begins and ends like an explosion. Indeed, the gasoline engine may be likened to a cannon which loads and fires itself at determined intervals and where the ball is a piston whose motion is arrested by a crank shaft and transformed into rotary motion in the fly wheels of the engine, to be used as a source of power. After the first charge has been fired a portion of its energy is used to reload the piece again, making it ready for a second explosion, to be repeated as often as needed.

655. Principal Parts of a Gasoline Engine.—The gasoline engine, like the steam engine, has its cylinder and piston, and its fly wheel and governor, but it has no boiler or firebox and is much more simple in its construction and management. There are provisions for supplying the engine with gasoline and air as needed for the explosions, for igniting the charge when ready and for disposing of the waste products after the explosion has taken place.

656. The Working Cycle.—The working cycle of most gasoline engines consists of five operations:

1. Charging the cylinder with the explosive mixture of air and gasoline vapor.

2. Compressing the charge preparatory to explosion.
3. Igniting the compressed charge.
4. Expansion of the charge after its explosion.
5. Expulsion of the waste products of the explosion.

657. Arrangements to Prevent Over-heating.—The continual repetition of the explosions in the cylinder of the engine results in so much heating of the parts, where any considerable work is done, that it is found necessary to provide means for absorbing the heat not changed into mechanical motion. This is usually done by providing the working parts which come in contact with the heat with water jackets in which water or oil is kept circulating to absorb the heat imparted to them.

Where water is used to cool with it is necessary in freezing weather to draw it off when the engine is shut down to avoid injury, but where a lubricating oil is used as the circulating medium there is no danger of this sort.

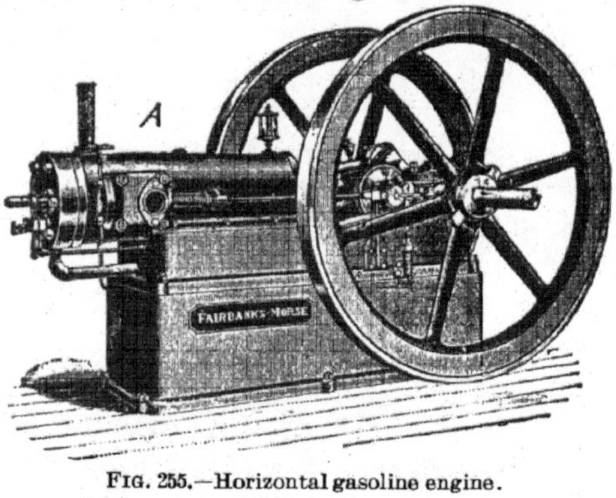

FIG. 255.—Horizontal gasoline engine.

658. Types of Gasoline Engines.—Gasoline engines, like the boilers of steam engines, are spoken of as vertical or horizontal according as the cylinder is upright or horizontal. It is possible to make the floor space occupied by the upright engines less than with the horizontal forms, but with few exceptions all the larger engines belong to the horizontal type. These two types of engines are represented in Figs. 255 and 256.

CONSTRUCTION OF THE GASOLINE ENGINE.

659. Cylinder.—The cylinder of the ordinary gasoline engine with its piston is not widely different from that of the steam engine, except that here there is nothing which corresponds to the steam chest and the slide valve, and the cylinder has a double jacket through which water is kept circulating to prevent over-heating. In Fig. 255 A represents the cylinder and the opening on the side is the exhaust port.

The piston has essentially the same construction as that of the steam engine represented in Fig. 251, using similar elastic metallic packing rings. There being no head in one end of the cylinder the piston can usually be seen.

660. Pumping Mechanism.—Formally it was the practice to arrange the gasoline supply tank so that the oil would flow by gravity to the engine, but this practical experience has proved to be unsafe on account of the tendency for leaks to develop and flood the engine room with the explosive oil. The plan now generally followed is to use an automatic pump, represented in connection with the engine in Fig. 257, where D is the plunger and A, B, C parts for working it when it is desired to throw a charge into the reservoir H. The gasoline comes from a tank outside the building through the valve F, and is discharged from the pump through the pipe E into H.

The disk with the hand wheel J is used to regulate the amount of oil going to the engine and when the pointer I is over the letter O the valve is wide open, but the proper amount of oil is supplied when the pointer is at R in this engine. The air is drawn in through the same chamber H by means of a pipe not shown in the cut, which ends under the base of the engine where as little dust as possible will be sucked in.

661. Governor.—The governing mechanism for gasoline engines varies in detail, but is usually a device by

Fig. 256.—Vertical gasoline engine showing governing mechanism.

which the pump is made to supply a charge of gasoline whenever an explosion is desired and the essential parts of the mechanism are represented in Fig. 256, where E E are a pair of governing balls which revolve with the fly-wheel and operate a finger in such a way as to prevent a charge being given to the engine whenever its speed is running too high. As the speed runs up the balls fly apart and this brings the finger C down upon the catch B

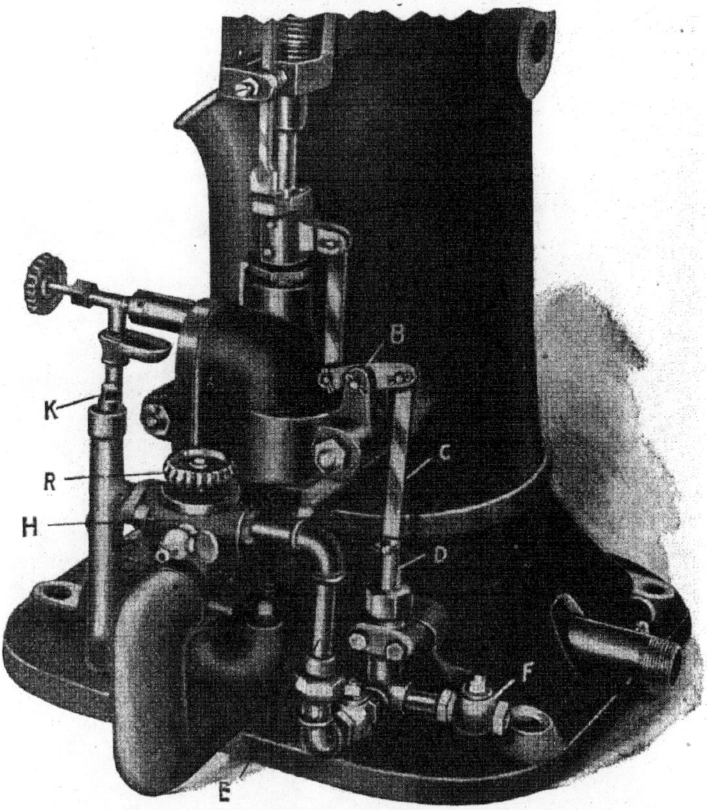

FIG. 257.—Pumping mechanism for supplying gasoline to gasoline engine.

which holds the exhaust valve open and prevents the pump being worked. The catch and finger are more clearly seen at M in Fig. 257 where the upper K is the valve stem which also works the gasoline pump.

662. Valve Mechanism.—The supply and exhaust valves for the engine of Fig. 256 are represented in Fig. 258 and

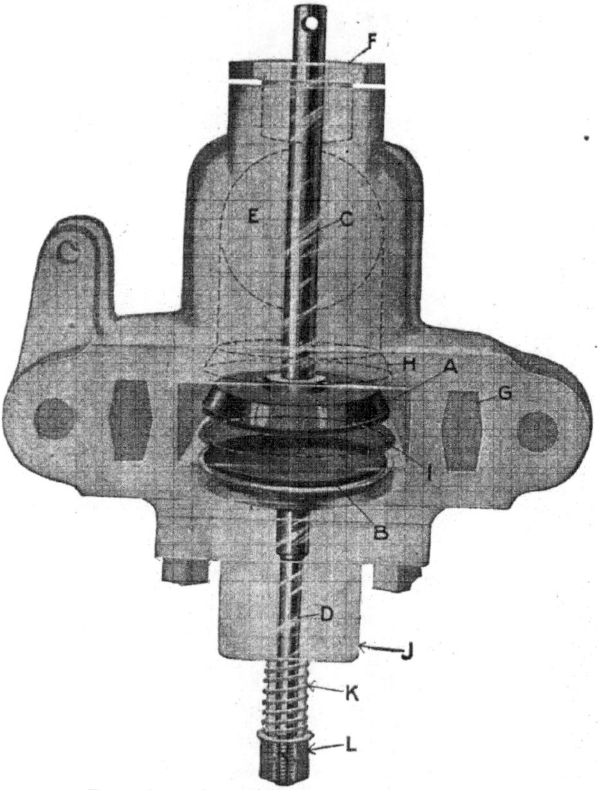

FIG. 258.—Valve mechanism of gasoline engine.

are located in the chamber A of Fig. 256. The upper valve A is the exhaust and is represented forced down so as to open the port, allowing the burnt charge to escape upward to reach the opening E in Fig. 258, which is the same as F in Fig. 256. When this valve is closed it is at H and is always controlled by the stem C worked by the revolutions of the fly-wheels.

The supply valve B is represented closed and is held down by the spring K, which can be regulated by the tension given through the jamb-nut L. This valve is lifted by the suction produced by the up-stroke of the engine piston. The opening G is a water jacket around the valves to keep them cool.

663. Igniting the Charge.—There are two methods of igniting the charge at the proper time, in these engines: one is by means of an electric spark which is produced at just the right time by means of a devise worked by the engine; the other is by means of a hot tube which rises out of the chamber A, of Fig. 256, into the curved chimney standing just to the left of C B. This tube is kept at the proper temperature by means of a Bunsen burner fed through the cock shown above F and at L, Fig. 257. After the charge has been drawn in and the piston is coming down in the cylinder so as to compress the gas, this compression forces a part of the explosive mixture up into this hot tube and when this is done the gas ignites and an explosion follows. If this tube becomes too hot the tendency will be for it to explode the charge too soon and either lessen the power of the engine or reverse its motion. If it is too cold the explosion will be too late. After the tube has been used for some time a scale may form over it which prevents the inner wall from taking the proper temperature and it is then necessary to replace it with a new one. In replacing this tube it is necessary to use one which is adapted specially to the engine because if it is too large or too small, or too long or too short, its capacity will affect the time of the explosion and it will not be correct.

664. Lubrication.—Cleanliness of all working parts of the engine and proper oiling are matters of prime importance and should receive the most careful attention. It requires a special lubricating oil for gasoline engines and only this oil should be used. It is known on the market as gas or gasoline engine oil. All parts should be carefully wiped clean at frequent intervals to free them from grit or gummy products and the operator should always have an ear to the sounds of his engine and should know what are normal and what are not in order that he may the quicker discover when anything is getting out of order and remedy it in time.

665. Gasoline.—Only the best quality of gasoline should be used with these engines, that known as the "74° test gasoline."

666. Size of Engine.—In purchasing a motor of any kind it should be remembered that it is much better to get one which has a little greater capacity than will be needed than one which is a little too small; and this caution applies with special force to the gasoline engines, for the reason that their capacity cannot be increased above the normal. With the steam engine it is possible to increase the steam pressure and the rate of firing, and the horse may for a short time develop two, three or even four horse-power, but if you overload a gasoline engine it must stop. If, therefore, it is desired to use steadily two full horse-power from a gasoline engine it should be not less than a three horse actual to do at all times perfectly satisfactory work.

It should be said in this connection, however, that it is never economical of fuel to use a large engine to develop a small horse-power. A 10 H. P. engine could not be economically used when it is desired to simply pump water from an ordinary well or to run a small separator which a man can turn. There should be a rational relation between the engine and the amount of work it is expected to perform.

WINDMILL.

If we except horse-power and that of cattle there is no form of motor which has been so generally or so widely used on the farm as the windmill and its use is daily increasing, especially now since all parts are made of steel well galvanized to protect them from rust, and their relative efficiency has been increased.

667. Work to Which the Windmill Is Adapted.—It must not be understood that a windmill is well suited to furnish power for any and all kinds of farm work if only it is made large enough. On the contrary it is only adapted to certain lines where the work done can be accumulated at times when the wind is favorable.

The windmill is peculiarly well adapted to pumping water for stock and for the supply of the house if only a suitably placed reservoir of sufficient capacity is provided. It must be remembered, however, that in many localities there may be periods of calm of three or even occasionally of seven days' duration when there will not be wind enough to permit the mill to do any work.

For grinding grain for farm stock the windmill is peculiarly well suited, provided arrangements are made so that the grinder is automatically fed and the meal allowed to drop into a bin where it may accumulate without personal attention. Arrangements of this sort may easily be made but it requires a special form of grinder which is not only automatic in its feed, but in the rate at which it feeds as well, supplying the mill heavily when the wind is strong and leaving the burrs empty whenever the wind falls so that no work can be done.

Where an abundance of water is available, with a lift of only 10 to 20 feet, the windmill may be used to advantage in irrigating small areas of two to five acres, but in such cases it will usually be necessary to provide a reservoir of suitable size into which the water may be pumped and stored.

For wood sawing also the windmill may often be used to advantage, by getting everything in readiness to do the work on those days when the wind shall be strong, but for this kind of work mills as large as 12 to 16 feet in diameter are required.

668. Wind Pressure.—The pressure which the wind may exert upon a surface depends primarily upon (1) its weight per cubic foot, (2) its velocity, and (3) the angle at which it strikes the surface. The weight of the wind per cubic foot is greater when the air temperature is low and when the barometric pressure is high; this being true, the capacity of a windmill in a given place varies with the season, being greatest in winter and least in summer, for like wind velocities.

As the weight of a cubic foot of air decreases with altitude windmills at sea level can do more work than those at hights of 1,000, 2,000 or 3,000 feet, when the air temperatures and wind velocities are equal.

669. Relation of Wind Pressure to Wind Velocity.—When conditions are similar wind pressures increase as the squares of the wind velocity. Thus, if the wind pressure at 5 miles per hour is taken as 1, then at 10, 15, 20, 25, 30, 35 and 40 miles per hour the wind pressure will increase in the ratio of the squares of the numbers 2, 3, 4, 5, 6, 7, 8; that is to say, a 10 mile wind may exert 4 times the pressure that a 5 mile wind does, and a 40 mile wind a pressure 64 times as great.

Taking the air at a pressure of 2,116.5 lbs. per sq. ft. the wind pressures at different velocities and temperatures will be as stated in the table below:

Table giving the pressure of the wind per sq. ft. at different velocities and temperatures when the barometric pressure remains the same. (*Wolff.*)

Wind velocity, miles per hour..	5	10	15	20	25	30	35	40
Pressure at temperature of 30° F	.126	.505	1.135	2.018	3.156	4.548	6.195	8.099
Pressure at temperature of 60° F	.1187	.475	1.069	1.902	2.973	4.284	5.836	7.628

670. Ability of Wind to Do Work.—The work which wind can do depends upon the amount which passes through a given windmill per minute and the pressure which it exerts. But as the pressure varies with the square of the velocity, and the quantity passing the mill varies directly as the velocity, the theoretic working capacity of the wind must increase as the cubes of the wind velocity.

Thus with miles per hour of....	5	10	15	20	25	30	35	40
Or, taking 5 = to 1 they are as .	1	2	3	4	5	6	7	8
The relative horse-powers are as	1	8	27	64	125	216	343	512
Theoretical horse-power is	.025	.2	.675	1.6	3.125	5.4	8.575	12.8

Perry regards it approximately correct to state that a 12 ft. windmill in a 5 mile wind may develop $\frac{1}{40}$ of a horse-power and the figures in the last line in the table above are his.

671. Relation of Diameter of Wheel to Its Efficiency.—In increasing the horse-power of an engine it is not usually necessary to increase its weight and strength much more than in proportion to the increase of power which is to be developed, but in the case of two wind wheels, having the same type of construction, the one which is to develop double the horse-power must have a strength of resistance practically 8 times as great in order to withstand the highest wind pressures to which it is liable to be subjected. This is so because doubling the diameter of the wheel not only makes the surface of wind pressure four-fold, but at the same time carries the center of pressure farther from the axis of the wheel, causing it to act upon a longer lever arm. But to increase the strength of resistance of the wheel 8-fold makes it necessary to build it much heavier and this detracts from its relative efficiency.

Besides this, with wheels of large diameter there are much greater differences in the wind pressure on the different parts of the wind sails because the actual velocity of the sails increases with the distance of their points from the center of the wheel. But the angular velocity must be the same in all parts of the sail, and this causes the wind sail to be forced around away from the wind passing through the wheel with very different velocities, and this difference reduces the relative efficiency so that large windmills of like pattern do not increase the available horsepower as much as the size is increased.

672. Unsteadiness of Wind Velocity.—It should be understood that the wind rarely blows with anything like uniform velocity for even a single minute, and an anemometer which gives the total number of miles of wind in an hour furnishes no sufficiently reliable data from which to calculate the work which the windmill should be expected to do. It very often happens that a wind which is registered as 10 miles per hour may have been blowing during a considerable portion of the time at the rate of 20 miles per hour and these high velocities are very much more effective than the mean 10-mile wind, and this would cause the wheel to show a relatively high efficiency in such a case.

673. Hight of Towers.—The wind velocity near the earth's surface is not only less than at higher elevations at the same time, but near the ground it is very much less uniform, so that for both of these reasons mills should be placed upon as high towers as practicable when the greatest efficiency is desired. If there are obstructions to the wind movement even within 1,000 feet of the windmill the tower should carry it several feet higher than these. Observations indicate that, taking the velocity of the wind at a hight of 50 feet as 1, at 25 feet its velocity would be nearly .8; at 75 feet it would be 1.2 and at 100 feet it would be nearly 1.4. These are deduced from Stevenson's formula,* which is

$$V = v\frac{H + 72}{122}$$

* Journal of the Scottish Meteorological Society, 1881.

where V is the velocity at the hight of the tower, **v** the velocity at 50 feet, and H the hight of the windmill.

Taking the efficiency of the wind as increasing with the cube of the velocity, the relative efficiency of the same mill at the four hights would be at 25 feet .51, at 50 feet 1, at 75 feet 1.73 and at 100 feet 2.74, from which it appears that a mill placed on a 100-foot tower may have more than 5 times the efficiency of one placed at 25 feet, and a mill on a 75 foot tower is likely to do three-fourths more work than one on a 50-foot tower.

674. Observed Amount of Work Done by a Windmill in Pumping Water.—We have measured the amount of water which was pumped during one entire year by the 16-foot geared windmill represented on the cover of this book.* This mill was provided with three pumps arranged so as to lift water 12.85 feet whenever there was wind enough to enable it to do any work. When the wind was lightest it was given the pump of smallest capacity, when stronger the one of next size, when still stronger both pumps together, the third pump being used only in the very highest winds.

The water was pumped into a large tank holding 141.2 cu. ft., so arranged that when full it emptied itself automatically in $\frac{3}{4}$ of a second, and at the same time recorded the time of emptying. In connection with this an automatic U. S. Weather Bureau anemometer made a continuous record of the miles of wind passing through the mill each hour of the day for a whole year and the amount of water pumped during the same intervals.

The amount of work done by this windmill during 10-day periods for the whole year is computed in acre-inches of water lifted to a hight of 10 feet and expressed in the table below:

*Bulletin 68, Wisconsin Agricultural Experiment Station.

Table showing computed amount of water lifted 10 feet high during consecutive 10-day periods for one full year, expressed in acre-inches.

Date.	Water pumped.	Date.	Water pumped.	Date.	Water pumped.
	Acre-in.		*Acre-in.*		*Acre-in.*
Feb. 28-Mch. 10	33.54	July 8-18.......	21.53	Nov. 15-25......	52.77
Mch. 10-20......	36.62	July 18-28......	29.73	Nov. 25-Dec. 5.	47.46
Mch. 20-30......	52.77	July 28-Aug. 7..	9.87	Dec. 5-15........	39.52
Mch. 30-Apr. 9..	47.01	Aug. 7-17.......	36.26	Dec. 15-25......	31.18
Apr. 9-19........	54.11	Aug. 17-27......	20.20	Dec. 25-Jan. 4..	51.22
Apr. 19-29.......	63.05	Aug. 27-Sept. 6.	21.27	Jan. 4-14.......	33.92
Apr. 29-May 9..	59.97	Sept. 6-16	18.00	Jan. 14-24......	29.16
May 9-19.	28.69	Sept. 16-26	40.42	Jan. 24-Feb. 3..	59.36
May 19-29.......	51.38	Sept. 26-Oct. 6.	23.79	Feb. 3-13.......	33.45
May 29-June 8..	40.54	Oct. 6-16	55.07	Feb. 13-23......	75.73
June 8-18.......	27.59	Oct. 16-26.......	18.45	Feb. 23-28......	16.20
June 18-28......	13.82	Oct. 26-Nov. 5..	36.71		
June 28-July 8..	26.68	Nov. 5-15.......	49.49		

It will be seen from this table that the smallest amount of water lifted ten feet high, in 10 days, was enough to cover 9.87 acres one inch deep and this occurred from July 28 to August 7, at the time when water for irrigation is most needed. The largest amount pumped occurred during the 10 days, from Feb 13 to 23, and was enough to cover 75.73 acres one inch deep.

675. Observed Amount of Work Done by a Windmill in Grinding Feed.—Another set of trials, aiming to measure the amount of feed which may be ground with a 12-foot geared windmill, was made at the Wisconsin Experiment Station,* and using the observed amounts of corn ground under a wide range of wind velocities and the observed hourly wind velocities, as recorded for the pumping experiment, the amount of feed which could have been ground, had it been fed automatically and kept running continuously, has been computed and given in the table which follows:

* Bulletin 82, Wisconsin Agricultural Experiment Station.

Table showing the amount of corn which could have been ground by the 12-foot aermotor windmill during the year, from March 6, 1897, to March 6, 1898, with all winds from 9 miles to 30 miles per hour.

Wind, miles per hour.	No. of hours of wind.	Amount ground per hour.	Total meal ground.	Wind, miles per hour.	No. of hours of wind.	Amount ground per hour.	Total meal ground.
		lbs.	lbs.			lbs.	lbs.
9	480	20.61	9,891	20	195	515.10	100,400
10	559	38.31	21,410	21	144	592.8	85,360
11	495	61.46	30,430	22	114	675.9	77,050
12	425	90.07	38,280	23	112	764.4	85,610
13	406	124.10	50,400	24	92	858.4	78,970
14	401	164.00	65,770	25	71	957.8	68,010
15	341	208.60	71,130	26	70	1,063.0	74,390
16	328	259.00	84,950	27	57	1,173.0	66,870
17	264	314.90	83,120	28	44	1,289.0	56,710
18	223	376.10	83,880	29	40	1,410.0	56,400
19	193	442.90	85,480	30	33	1,537.0	50,710

The total footing of this table shows that the mill might have ground an average of about 75 bushels of corn per day for the entire year, but this figure would represent the maximum amount of work possible. The minimum could hardly have been less than ⅓ of this amount.

CHAPTER XXIII.

FARM MACHINERY.

FRICTION.

It has never been practicable to devise a machine which could transmit the energy imparted to it without sustaining a certain amount of loss through friction and in some forms of machines the loss of power through friction is necessarily very great under the best management. In other cases ignorance of the laws of friction or carelessness leads to much larger losses than are necessary.

On the other hand friction may be a very essential condition to the accomplishing of important results. How essential it is in walking we appreciate when we attempt to move over very smooth ice and it is the friction of the drivers of the locomotive upon the rails which enables it to haul the enormous loads it does. In transmitting power by means of belts it is friction which enables it to be done. The service of nails and screws in holding parts together depends upon the amount of friction developed in forcing them home.

676. Friction Between Solids.—When one surface rests upon another there is a tendency for the inequalities of one to fit into those of the other, producing an interlocking not very unlike, except in degree, what would be produced by putting the cutting edges of two saws together. When such an interlocking has occurred it is only possible to move one surface over the other by either separating the two surfaces slightly or else by tearing off the interlocking portions, and it is the separating of the two bodies or the abraiding of these inequalities which causes the chief part of friction between solids. Because of the molecular structure of bodies, no matter how smoothly the two friction surfaces are polished, there must still remain elevations and depressions which permit interlocking. There is also a slight adhesion between the two surfaces which adds a small amount to the friction; from this it follows that no two surfaces can slide over one another without developing the resistance known as friction.

677. Friction of Rest or Static Friction Between Solids.—Whenever one body is brought to rest upon another over which it is sliding, the jarring which takes place at the time causes the upper body to fall from the higher inequalities into the depressions of the lower one, thus developing the maximum interlocking due to this cause. Then again, after long standing, if there has been considerable pressure, the plasticity of the substances will cause them to interlock still more completely, and so it happens that the amount of force required to overcome friction between two solid surfaces at rest is always greater than the force required to keep the body moving after the static friction has been overcome, and Thurston states that it is commonly 40 per cent. greater. The case is not profoundly unlike the difficulty in starting a loaded wagon after it has stood over night and formed a depression under each wheel.

678. Friction of Motion Between Solids.—After one solid surface is once started sliding over another time enough does not intervene in passing from inequality to inequality to permit maximum interlocking to take place, and the result is the smaller amount of friction of motion, compared with that of rest, stated in (**677**).

The amount of sliding friction can be simply illustrated by using inclined planes with different kinds of surfaces, by first placing the plane and then putting upon this the object whose friction is to be determined. By gradually elevating the plane and jarring it a little an inclination will be reached at which the body will slide down the surface with a uniform velocity. Suppose the hight to which the end of the plane has been raised is just $\frac{1}{10}$ of its base, then the friction of motion is 10 per cent.,—that is, it will require a force equal to $\frac{1}{10}$ of the weight of the load to maintain motion.

It has been found that usually the sliding friction between non-lubricated surfaces increases directly as the weight but is independent of the area of surface, provided the weight is not great enough to cut or tear them. Gen. Morin found, in his experiments, that wood without lubricant showed a friction of 25 to 50 per cent., but when soap was used between the surfaces the friction was reduced to from 4 to 20 per cent. In the case of metal on wood, without lubricant, he found it to be from 50 to 60 per cent. With the smoothest and best lubricated surfaces he found it as small as 3 to 3.6 per cent.

679. Rolling Friction.—When one solid surface rolls over another, no matter how smooth they may be, there is always friction, but the amount is much less than that of sliding, and the fundamental reason is the same as that which permits a load to be carried over bare ground on wheels with less friction than when carried upon a sled. The roller bearings in common use on the grindstone illustrate the smaller friction due to the use of rollers, but we get the most perfect example of reduced friction in the ball bearings of bicycles.

680. Friction Between Liquids.—The laws governing the friction of liquids are very different from those of solids. In the first place it increases with the square of the velocity, instead of directly as the velocity in the case of solids, and it decreases with increase of temperature of the liquid, often to a very important extent.

The liquids used as lubricants have generally comparatively high viscosity, as fluid friction is technically called, and they vary between wide limits among themselves, especially when they experience wide changes of temperature. The oils and fats suitable for lubricants belong to both the animal and vegetable and the mineral non-drying types. Any oil or fat which becomes permanently gummy and stiff after the exposure of service makes a very unsatisfactory machine lubricant, and although such oils can be found upon the market, selling at a lower price than the best lubricants, it is seldom economy to use them because they soon gum up the bearings, greatly increase the friction and increase the labor necessary to put the machine into condition again. In special cases finely divided solids, like graphite and soapstone, are used as lubricants.

681. The Action of Lubricants.—When liquids are brought into contact with solid surfaces to which they adhere, as in the case of water flowing through pipes, there is a thin layer held to the walls of the pipe so rigidly that it hardly takes part in the flow, so that instead of having the friction of a liquid upon a solid, the slipping takes place between layers of the liquid itself. So, too, when an oil is poured into the bearings of a machine there come to be two comparatively stationary layers against the two metal surfaces and the sliding or friction takes place between layers of oil rather than between layers of metal, and it is because the friction of lubricants upon themselves is so much less than that between solids that they are so serviceable.

The lubricant may act in two different ways in lessening the friction: (1) by causing the chief part of the resistance to be that due to the slipping of oil over oil, which is usually less than the friction between solids, and (2) by the lubricant acting to fill up the smaller inequalities of the two friction surfaces and in this way preventing so much interlocking.

682. Adaptation of Lubricant to Place of Service.—Where the speed of the sliding surfaces is relatively high, and especially if the pressure between the bearings is not heavy, one of the thin oils will render the best service, securing the least friction. On the other hand, if the bearings are carrying heavy pressure at a slow rate of sliding, as in the case of the axles of wagons, then the heavy thick oil or grease will last longer, maintain less wear of the bearings and ensure a smaller friction. But in any case the rate of revolution of parts and the amount of friction must be so related that sufficient heat is not developed to burn the oil. Further than this it may sometimes happen that the oil does not feed rapidly or completely enough to all parts of the bearing to ensure perfect lubrication and constant watchfulness is required on the part of every operator of a machine to keep things in proper condition.

683. Scrupulous Cleanliness of Bearings.—Next in importance to good lubrication of all slipping portions of a machine stands the maintenance of scrupulously clean bearings, where the friction surfaces are free from both grit or any gummy substance. Where sand or grit of any sort has found its way to the bearings of machinery its grains cut through the film of oil on both friction surfaces and tend to lock the two together in the same manner that sanding the rail under the driver of a locomotive does and thus prevent slipping.

Because of this tendency of dirt to get into bearings, even under the best management, it is necessary to occasionally overhaul the bearings of important machines and carefully clean them, and very special attention should be given to all those parts where the speed is high, because not only is here where power is absorbed most rapidly by needless friction but the wearing and injury to the machine is most rapid.

684. Hot Boxes.—The heating of boxes in machinery may result from one or more of several causes: (1) insufficient lubrication, (2) dirt in the journal, (3) the box may be screwed down too tight, (4) the belt may be too tight, producing unnecessary friction, (5) the box may have gotten out of line with the shaft, (6) the collar or pulley may bear too hard against the end of the box.

BELTING.

The use of belting, ropes and cables, in transmitting power from a motor to the machine being driven, is a practical utilization of friction.

685. Action of Belting.—When machinery is being driven by a belt its two sides are not under equal tension and the efficiency of the belt depends upon the difference in tension between the two sides and the rate at which the belt is traveling. Suppose the effective tension of the belt is 66 lbs. and that the belt is traveling with a velocity of 1,000 feet per minute, then the energy it is transmitting, or its activity, is equal to

$$66 \times 1000 = 66{,}000 \text{ foot pounds per minute.}$$

or

$$\frac{66{,}000}{33{,}000} = 2 \text{ H. P.}$$

It is clear from this that the more rapidly the belt is driven the larger is the horse-power transmitted when the effective tension of the belt is the same.

686. Efficiency of Belting.—The highest efficiency is attained from belting when there is least stretching and least slipping of the belt and when there is the least unnecessary pressure developed by it on the shafts of the driving pulleys.

Good leather belts usually give a higher efficiency than rubber or other types and when they are used where they can always be kept dry are most satisfactory. To get the highest service from a leather belt it should be run with the hair side next to the pulley and over a pulley faced with leather with the hair side out. Under these conditions there is the least slipping of the belt and the strain on the belt in bending around the pulley is least, so that it wears more slowly when being bent and straightened.

687. Size of Belt for Transmission of Given Horse-Power.—In order not to over-strain a good two-ply leather belt it ought not to be subjected to an effective tension of more than 40 pounds per inch of width. On this basis the width of belt for a given number of horse-power will depend upon the speed of the belt. Suppose the driving pulley of an engine is 9 inches and that it makes 350 revolutions per minute, developing 3 horse-power. What width of belt would be required? This may be calculated from the following equation.

$$\frac{3 \times 33,000}{3.1416 \times .75 \times 350 \times 40} = 3.001 \text{ inches, width of belt.}$$

In this case 3 is the number of H. P., 33,000 is the number of foot-pounds in one H. P., $3.1416 \times .75 \times 350$ gives the velocity of the belt in feet per minute and 40 is the effective tension to which the belt may be safely subjected. From this solution a general equation for calculating the width of belt for any H. P. may be stated as follows,

$$\text{Width of belt} = \frac{\text{No. H. P.} \times 33,000}{\pi D \times \text{No. rev.} \times 40}$$

D = circumference of driving pulley in feet.
No. rev. = number of revolutions of driving pulley per minute.

Some belt manufacturers allow a strain of 60 pounds per inch of width for a two-ply leather belt as safe but it is in the direction of economy to have the belt stronger than is really necessary, as it will wear enough longer to pay.

688. Condition of Belt.—It is very important to keep the belt in a good, soft, pliable condition, as a flexible belt will not only transmit power with less loss but it will wear much longer. If for any reason belts have become hard and stiff, they should be softened with neatsfoot oil. New belts always stretch more at first than after they have been used.

689. Pulley and Shaft.—In order that a belt may run well on the pulley it is essential that the shafts of the two pulleys connected by the belt shall be rigidly parallel with each other and that the pulleys shall turn true on the shafts. When the pulleys are properly placed the belt will run to the center of the pulley and stay there.

Belts tend to run to the largest part of the pulley and for this reason pulleys are commonly made a little crowning in the center so as to cause the belt to run to the center. But where tight and loose pulleys run side by side, so as to throw off the belt without stopping the power, then the faces of the pulleys are flat.

In running a belt onto a pulley, especially when it is wide, heavy and tight, care needs to be taken not to overstrain it as there is danger of stretching the edge so that it will never run true afterwards, or even of cutting or tearing it, especially if the edge of the pulley happens to be sharp.

690. Lacing a Belt.—In lacing a belt care should be taken to make the lacing plenty strong enough, but to make it unnecessarily so is worse than to have it a little light, especially if it has been done in a bungling manner so as to form an enlarged place in the belt which brings undue strain when the lacing is passing the pulleys. The lacing should be so nicely done that this portion of the belt passes the pulley without a jar or extra strain. To secure this the ends of the belt should be cut true and square by using a try-square. Holes should be punched just large enough to allow the lacing to fill them well without danger of tearing them out by wedging. Space the holes equally, leaving the outside ones just far enough from the edge to be safe against tearing out, and they should be not more than $\frac{1}{4}$ to $\frac{1}{2}$ an inch from the ends.

Sew with the smooth side of the lacing out, beginning

at the center of the belt, and never cross it or have more than two thicknesses of lacing on the side next to the pulley. Fasten the lacing by running the ends through small holes punched in line with the lace holes where they will be in the right place to serve as lacing holes when the belt needs to be shortened.

691. Calculating the Length of Belts.—To ascertain approximate length of a belt to connect two pulleys measure exact distance between the centers of the two pulley shafts. Then add the circumferences of the two pulleys together, dividing the sum by two; add this sum to twice the distance between the centers of the two shafts and the total is the length of belt required.

FARM PUMPS.

There are several forms of devices used in lifting water on the farm, chiefly for the use of the stock and as a water supply for the house; these are known as suction and force pumps and hydraulic rams.

692. Suction Pump.—The common suction pump consists of a cylinder and piston connected below with a suction pipe, and above with a discharge pipe. At the upper end of the suction pipe, usually in the lower end of the cylinder, below the piston, there is the suction valve which opens upward by the force of the water but closes with the down stroke of the piston, preventing the return of water to the well. In the piston is a second valve, also opening upward which permits the piston to be forced downward through the water in the cylinder, held there by the suction valve, but which closes the moment the piston begins to rise and thus lifts whatever water is above the valve, at the same time tending to produce a vacuum below the piston into which the pressure of the air on the water of the well lifts the water through the suction pipe and past the suction valve already described.

The piston is usually worked by a simple lever in the form of the pump handle and the water is discharged through the spout in the pump-head near the level of the ground.

693. Size of Piston.—The size of piston which should be used in a well depends upon the hight to which the water must be lifted and the power which is available to work the pump. In working a common pump a man can comfortably exert a pressure of only 15 to 20 pounds upon the pump handle and, as the power arm of this lever is only 5 to 7 times that of the weight arm, the pressure exerted by the water to be lifted at one stroke cannot much exceed 75 to 100 pounds. Water at ordinary well temperature exerts a pressure of .43 pounds per square inch for each foot of depth. This being true, whenever the piston is called upon to lift water through a hight of 40 feet the pressure on the piston would be at the rate of

$$40 \times .43 = 17.2 \text{ lbs. per sq. in.}$$

A 2-inch piston has an area of 3.14 square inches, a 2.5 inch piston 4.9 square inches, a 3-inch piston 7.07 and a 3.5-inch piston 9.62 square inches, so that in lifting water through 40 feet with each one of these pistons the force required to be applied to the piston rod would have to be

For the 2 inch piston....................................	54.01 lbs.
For the 2.5 inch piston....................................	81 28 lbs.
For the 3 inch piston....................................	121.60 lbs.
For the 3.5 inch piston....................................	165.46 lbs.

It will be clear from these figures that for ordinary hand pumping a 2 to 2.5-inch piston is as large as can be comfortably worked in a well where the lift must equal 40 feet. If the well has such a depth that the water must be lifted 100 feet the 2-inch piston would sustain a pressure of 135 pounds and hence would be larger than could be comfortably worked in such a well.

694. Rate of Pumping.—The rate of discharge from a common single acting suction pump is determined by the area of effective cross-section of the cylinder, the length of the stroke and the number of strokes per minute. Taking a 2.5-inch cylinder, which would have an effective cross-section of about 4.7 square inches and supposing it to make a 5-inch stroke at the rate of 30 per minute, the amount of water pumped per hour would be

$$\frac{4.7 \times 5 \times 30 \times 60}{231} = 183.1 \text{ gallons.}$$

or enough for about 21 cows allowing 72 pounds per head.

695. Relation of Size of Suction and Discharge Pipe and Piston to Power Required to Work the Pump.—When a large piston is worked on a small suction and discharge pipe it is necessary for the water to travel much faster through these than when they have an effective diameter equal to that of the piston; but to increase the velocity of flow through a pipe requires an increase of pressure so that more power is required to pump the same quantity of water through a 1-inch pipe, using a 3-inch cylinder, than to pump the same amount in the same time through a 3-inch pipe. In the apparatus represented in Fig. 259, when the pump with the 3-inch cylinder C is worked, discharging water through the 3-inch discharge pipe 3, lifting the water to a hight of about 18 feet and working the pump at the ordinary rate, the pressure gage E shows that the pumping is developing a pressure of about 9 pounds to the square inch. If now the pump is kept working at the same rate and the gate valve in 3 closed, while that in the 2-inch pipe is opened, the pressure is seen to rise to nearly 11 pounds per square inch. Then on opening the gate valve in the 1-inch pipe and closing that in the 2-inch the pressure rises to between 13 and 14 pounds, but when closing the 1-inch gate and opening that in the ¾-inch pipe the gage registers a pressure of between 18 and 19 pounds per square inch of the piston, when the same

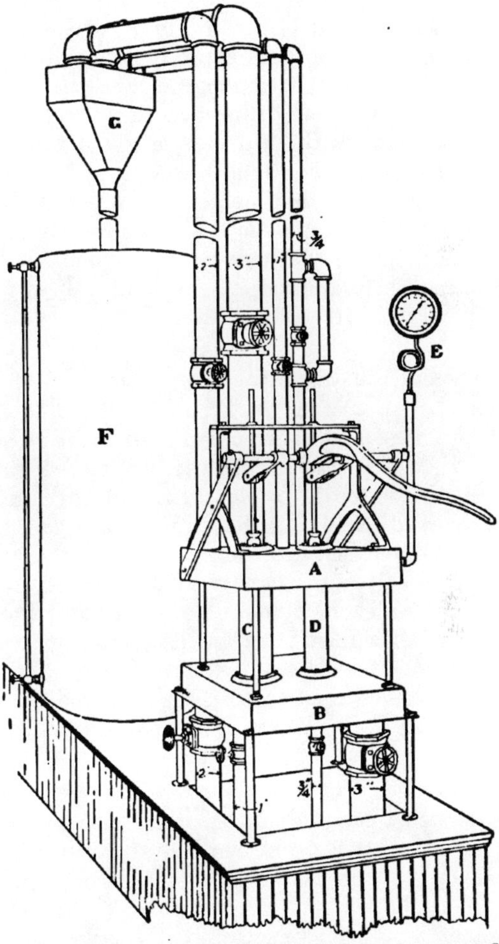

'IG. 259.—Apparatus for demonstrating the resistance of pipes to flow of water in pumping.

amount of water is being discharged per minute as when the pump was being worked on the 3-inch discharge pipe.

From these observations it is clear that the size of the suction and discharge pipe, compared with the piston, may make a very material difference in the amount of power necessary to work the pump at a given rate. The area of the 3-inch piston contains the area of the ¾-inch pipe nearly 16 times and this means that when the piston is driving the water through the ¾ inch pipe its velocity must be 16 times that of the piston, while when forcing it through the 3-inch pipe the water travels at only the same rate and therefore requires less power.

696. Influence of Elbows on the Power Required to Work a Pump.—In the apparatus of Fig. 259 there is represented, leading out of and back into the ¾-inch pipe, a side tube, so that when the ¾-inch gate valve is closed and the pump worked the water is forced to travel through four right angles instead of taking the straight course possible when the gate is open. Under these conditions the gage E shows an increase of pressure amounting to ¾ of a pound per square inch for each right angle, or a total increase of three pounds per square inch on the piston and, as the piston has an area of over seven square inches, the extra power which had to be applied to the piston rod, in order to pump around the four elbows, exceeded 28 pounds.

697. Double-Acting Suction Pumps.—In the ordinary suction pump, practically all of the work has to be done with the up stroke of the piston and this requires a heavier pressure than would be necessary if the work could be divided between the up and down strokes. An effort is sometimes made in the construction of the pump to divide the labor between the two strokes and one of the methods employed is represented in the double acting pump of Fig. 260. In this pump there are two cylinders, the upper one without a valve and having one-half the cross-section of the lower one. With this arrangement, when the piston is raised one-half of the water passes into the discharge pipe, while the other one-half rises into the smaller cylinder which, with the down stroke, must be forced out through the discharge pipe, in this way dividing the labor between the two strokes of the pump. In other forms of pumps the same result is accomplished by arranging an air chamber in connection with the cylinder in such a way that, when the up stroke is made, a part of the water rises into the air chamber, compressing the air to such an extent that while the down stroke of the piston is being made the air expands, forcing the water out, thus securing double action.

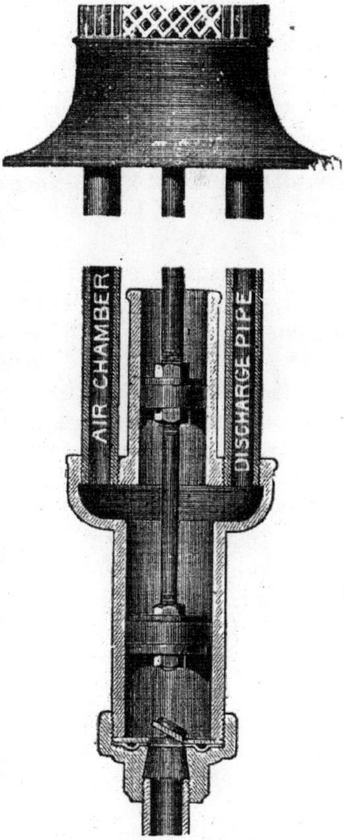

FIG. 260.—Double acting suction pump

698. Proper Place for the Cylinder in the Well.—The maximum hight to which the air pressure can sustain a column of water at sea level is only about 34 feet. But the imperfect action of the best suction pumps, together with the pressure exerted by the water vapor and the air escaping from the water when a high vacuum is produced over it, makes it impracticable to have the piston placed more than 16 to 20 feet above water in the well.

The best place for the cylinder in any well, if there is a sufficient depth of water to permit of it, is several feet below the level of the water. When the cylinder is placed beneath the water it is where it will always be "primed" and where there is little danger of lowering the water by pumping to a level at which the pump works imperfectly. The general rule to follow then is to place the cylinder as low in the well as practicable, or so far below the surface of the water that it will always be covered.

699. Hydraulic Ram.—Where the conditions are favorable for the use of the hydraulic ram for domestic water supply it is one of the cheapest, most satisfactory and efficient means yet devised for lifting water.

The hydraulic ram consists of (1) a drive pipe, (2) an air chamber, (3) an impetus valve, (4) a discharge pipe and (5) a discharge valve. The principle of the hydraulic ram is that of using the inertia or momentum of a large volume of water to raise a fraction of the same water to the desired hight. The water is allowed to flow through the drive pipe until it acquires velocity enough to close the impetus valve and this immediately stops the column of water in the drive pipe, causing it to act like a water hammer to force open the discharge valve leading into the air chamber, compressing the air by means of a portion of the water which is driven in. As soon as the column of water in the drive pipe is brought to rest the impetus valve falls down of its own weight and this allows the water in the drive pipe to flow at full velocity again until it is finally moving fast enough to close it once more, when the sudden stopping forces another charge into the air chamber. In this way the steps are continually repeated, thus maintaining a steady supply of water, the compressed air forcing the water into the discharge pipe leading to where the water is desired.

The hydraulic ram can be used where there is only a comparatively small fall, of even two or three feet, but where it is used to supply drinking water from a spring the chief difficulty is in arranging to convey the water in such a way that it will not become too warm, even if the pipe is carried 3 or 4 feet under ground where it would be safe against frost in winter. The ground becomes warm enough in summer time to leave the temperature unsatisfactory if the water must be carried any considerable distance. The water may be carried in this way to any distance and any hight but the per cent. of the stream conveyed decreases with both the hight and the distance.

PRINCIPLES OF WEATHER FORECASTING.

CHAPTER XXIV.

THE ATMOSPHERE.

As the life processes of all plants and animals are dependent upon the air, and are greatly influenced by changes in it, it is eminently proper that the atmosphere and its changes should be considered in their relations to agriculture. From the standpoint of food supply the clover crop, for example, containing at maturity 70 per cent. of water, has—directly or indirectly—obtained all but its ash ingredients from the atmosphere. The water is brought to the soil as rain, the carbon comes from the carbon dioxide and the nitrogen is obtained from the soil air by the free-nitrogen-fixing bacteria. The relations stand

Water from the atmosphere as rain	70.00 per cent.
Nitrogen from the soil-air	.70 per cent.
Carbon and oxygen from the atmosphere as rain and carbon dioxide	26.57 per cent.
Ash ingredients from the soil	2.73 per cent.
Total	100.00 per cent.

Thus 97.27 per cent. of the plant food is derived from the constituents of the atmosphere, either directly or indirectly.

700. Relation of the Atmosphere to the Earth.—The earth consists of three concentric spheres, (1) at the center, the solid, or earth-sphere; (2) surrounding this is the liquid or water-sphere, (3) and outside of all is the gas or air sphere. These have been named—

1. Geosphere.
2. Hydrosphere.
3. Atmosphere.

701. Interpenetration of the Three Spheres.—The materials of the three spheres are neither entirely separated from one another nor stationary. Beneath the oceans and beneath the surface of the continents the solid earth is permeated by water. Even under desert skies there may be wells and the soil contains moisture. With the water, too, goes more or less of air from the atmosphere; the fishes of the oceans and lakes find air to breathe wherever they go and the spaces in rock and soil not occupied by water are filled with air. Floating in the water and drifting in the atmosphere even at great hights are solid particles of silt and dust broken from the earth-sphere, and nowhere is air so dry that it contains no moisture.

Drifted by the currents of air and water on land and at sea solid particles are continually being moved from place to place. The water of the ocean, of the lakes or of the atmosphere is never at rest, neither is that which has penetrated tho solid crust of the earth. So, too, the air of the atmosphere, of the water and of the soil is continually changing and upon the rate of these changes depends the well being of plant and animal life.

702. Relation of the Life Zone to the Three Spheres.—The living forms of the earth make their homes in the bottom of the atmosphere and in the top of the water sphere or of the earth sphere. This relation is necessitated by the fact that all living forms derive their food from the air, from the water, and either from the earth or from other forms which take their ash ingredients from the earth. This relation is further necessitated by the fact that all living forms must dwell where they can have a certain amount of direct sunshine or else where they can live upon other forms which depend upon it, for this is the moving power of the world and all life implies motion. Deep in the solid earth no life exists. In the greatest depths of the ocean, where the air changes are slow and where little or no light can come, life is nearly absent; and high in the atmosphere only latent forms of life, like the spores and germs of microscopic forms are drifted by the winds.

In brief the life zone is that portion of the three spheres where the largest amount of sunshine is transformed into heat motion and therefore where there is the largest amount of energy available for the use of plants and animals.

703. Depth of the Atmosphere.—We are living at the bottom of an ocean of air whose depth is at present unknown. Judging from the rate of decrease of pressure, as measured by the barometer, its depth would be placed at something less than 50 miles, for at 30 miles, could an instrument be placed at that level, it is calculated that its reading would be only .005 of an inch of mercury. Observations which have been made upon the hight at which shooting stars or meteors become visible shows that this is even more than 100 miles and it is believed that these bodies become visible only after they have traversed enough of our atmosphere to develop sufficient heat by friction and compression to make them white-hot; and although the velocity of these bodies is very great yet the upper air is so rarified they must pass through great depths before sufficient heat can be developed to make them white-hot. From these considerations it appears likely that air may be found at hights even exceeding 500 miles.

704. Composition of the Atmosphere.—The air at different times and in different places contains a great variety of gases and volatile products but there are certain constituents which are found everywhere in the explored regions and in pretty constant ratios. These are, for dry air:

1. Nitrogen, forming about 77.18 per cent. by volume.

2. Oxygen, forming about 20.61 per cent. by volume.
3. Water vapor, forming about 1.40 per cent. by volume.
4. Argon, forming about .94 per cent. by volume.
5. Carbon dioxide, forming about .03 per cent. by volume.

Besides these ingredients there are usually present in the air small amounts of ammonia and of nitric acid, which are brought down with the rains to the extent of 3.37 pounds per acre per annum at Rothamstead, England; 1.74 pounds at Lincoln, New Zealand; and 3.77 pounds in the Barbadoes Islands.

Oxygen often occurs in the allotropic form of ozone, which is much more active as an oxidizing agent than the ordinary condition.

705. Materials Mechanically Suspended in the Atmosphere.—In the gaseous body of the atmosphere there are always mechanically suspended varying amounts of solid and liquid particles and bodies. These are:

1. Inorganic dust grains or soil particles.
2. Organic dust fragments.
3. Microscopic germs and spores.
4. Pollen grains from various plants.
5. Snow or water crystals.
6. Water particles in cloud forms.

PARTS PLAYED BY THE DIFFERENT INGREDIENTS.

The atmosphere as a whole, in its relation to living forms, plays the important function of an equalizer of temperature, preventing the occurrence of such excessively high and extremely low degrees as would otherwise be produced when the sun is above or below the horizon.

706. Oxygen.—Oxygen is essential to both plants and animals, it being indispensable to the activities of the protoplasm of living cells, whether this be in the root, stem or leaf of plants or in the tissues of animals. In the development of muscular and nervous energy large quantities are used by the animal kingdom, and other large volumes are used by man with fuel as a source of power and heat.

707. Nitrogen.—The nitrogen of the atmosphere is primarily the source of all nitrogen compounds of living forms; and by its dilution of all the other ingredients it modifies their physiological effects.

708. Water.—Moisture in the atmosphere greatly influences the temperature of the earth's surface, as it is very opaque to dark heat waves radiated back into space. The frosts forming under clear skies and the absence of them when the air is damp are evidence of this influence. But the chief function of water is found in its large movement to the land in the form of rain and snow and its return from the fields through springs and rivers to the seas. As it falls it is food for plants and drink for animals, as it returns it carries away soluble salts which, if left, would develop sterile "alkali" lands.

709. Dust.—The dust particles give to the sky its blue color and by their radiation of heat into space become cold centers upon which moisture condenses and snow flakes form. In this way they greatly influence the precipitation, making it less violent than it might otherwise be.

710. Carbon Dioxide.—Carbon dioxide is the source of all the carbon entering into the constitution of the tissues of both plants and animals, and it is a constituent of the great majority of feeding stuffs and of most organic compounds.

From recent investigations it is held that carbon dioxide plays an important part, with water, in lessening the transparency of the atmosphere to dark heat rays radiating from the earth into space, and in this way holds our temperature much higher than it could be with this gas absent; and Chamberlin has proposed the working hypothesis that long period changes in the amount of carbon dioxide in the atmosphere may be the cause of the recurrent glacial periods to which the earth has been subjected.

711. Pressure of the Atmosphere.—The air, like all other substances, has weight, and this weight causes it to exert pressure proportional to the amount above a place. Its mean pressure at sea level is equal to 14.73 pounds per square inch. A cubic foot of air at this pressure and at a temperature of 62° F. weighs about .08 pounds, 100 cubic feet would weigh 8 pounds, and 10,000 cubic feet 807.28 pounds. The air of a stable 50x50 feet, 10 feet high, weighs a ton.

As the hight increases above sea level the amount of air to exert pressure is less, the weight of a cubic foot becomes less and it is necessary to breathe a larger volume to supply the system with the same amount of oxygen. In the next table are given in round numbers the hights above the sea at which the pressure would fall from 30 to 16 inches and the hight to which these pressures would sustain a column of water, could a perfect vacuum be maintained.

Hight above sea level.	Barometric pressure.	Hight of water column.
0	30 inches.	34.0 feet.
1,800 feet.	28 "	31.7 "
3,800 "	26 "	29.4 "
5,900 "	24 "	27.2 "
8,200 "	22 "	24.9 "
10,600 "	20 "	22.6 "
13,200 "	18 "	20.4 "
16,000 "	16 "	18.1 "

712. Applications of Atmospheric Pressure.—The most general application of atmospheric pressure by the animal world is in bringing air into their respiratory organs. Where animals are constituted so as to take advantage of this, a reduction of pressure is made about the lungs, as in raising the ribs and lowering the diaphragm, and then the greater pressure of the air outside expands them and causes a fresh supply to enter and fill the space.

In drinking and in sucking animals take advantage of the air pressure to perform these operations, which would be impossible without the pressure, and difficult where the pressure is small.

Even in eating, animals with lips and cheeks take advantage of air pressure to force the food from between the teeth after it has been masticated, and a man would make awkward work eating for the first time in a vacuum.

In the common suction pump and the siphon air pressure is an essential factor, as it is in the low pressure steam engine.

All of the machines invented for milking cows develop a vacuum and depend upon atmospheric pressure to force the milk from the udder.

713. Temperature of the Atmosphere.—The air is warmed in three ways: first, and chiefly, by contact with the earth's surface and with solid objects upon it, this heating giving rise to ascending currents of warm and descending ones of cold air. Second, by dark heat radiations outward, which are absorbed by the atmosphere as water absorbs light. Third, by absorption of the direct rays from the sun on their way to the earth's surface.

When air descends from a higher to a lower level the pressure upon it becomes greater and its volume is reduced. This reduction of volume causes it to have a higher temperature, and so if the air rises it expands, and this expansion results in lowering the temperature. A rise or fall of 100 feet causes a change of temperature of .55° F. in dry air.

If dry air crosses a mountain range and falls 2,000 feet its temperature is raised 11° F.

When moisture is condensed or frozen in the atmosphere the air temperature is raised by the heat generated during condensation. So, too, if water is evaporated in the air, or snow melts, the temperature falls. This is why the weather is always warmer in winter when it snows, and cooler after showers.

CHAPTER XXV.

MOVEMENTS OF THE ATMOSPHERE.

714. Primary Cause of Winds.—Winds usually begin in one of two ways, represented in Fig. 261. In the lower

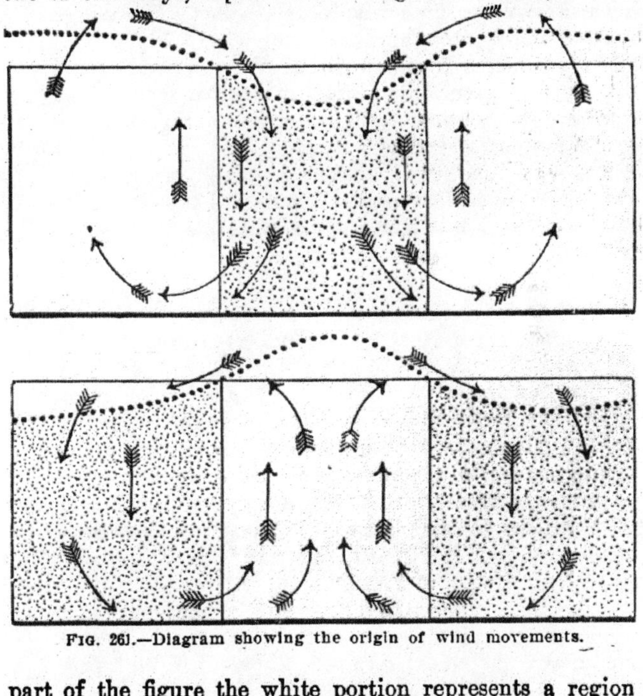

FIG. 261.—Diagram showing the origin of wind movements.

part of the figure the white portion represents a region where the air is expanding. When this occurs the lower and heavier air is carried upward and brought alongside that which is lighter; then because of the resulting unbalanced pressure the air above flows over outward, as represented by the upper arrows. But as soon as some air has left the expanding area the whole column is made lighter, while the shaded areas become heavier from the added amount, and there is an unbalanced condition through the whole hight. At the center there is an area of low pressure and around it one of high, hence the winds set inward from all sides at the surface and outward above, as shown by the arrows in the diagram, and we have what is called a cyclonic system of winds, where the currents are moving inward toward a low pressure area below and outward above toward one that is higher.

If the central area is one where the air is contracting and becoming denser then air will flow in upon it from above, as shown in the upper part of the diagram of Fig. 261. But as soon as air moves from the surrounding area upon the central one the inner region becomes a high area, where the greater pressure forces the air outward below and inward above. Such a wind system as this has been named an anticyclone.

GENERAL CIRCULATION OF THE ATMOSPHERE.

715. The World System of Winds.—In the region of the equator, where the heat is greatest, the air is continually expanding, and flowing toward the poles above; this makes the pressure greater on either side, resulting in surface winds setting toward the equator, as represented in vertical section in Fig. 262, which it will be seen is essentially the cyclonic system of Fig. 261. Farther toward the poles on either side, where the overflowing air from the equator accumulates, a high pressure belt is developed, from under which part of the air flows toward the equator below and another toward the poles; these are the tropical high pressure belts.

At the poles, where the air is continually cooling, it is

steadily descending and flowing outward below, maintaining an anti-cyclonic system of winds like that of the upper part of Fig. 261. Between the high area at the poles and the tropical high pressure belts, where the two systems of

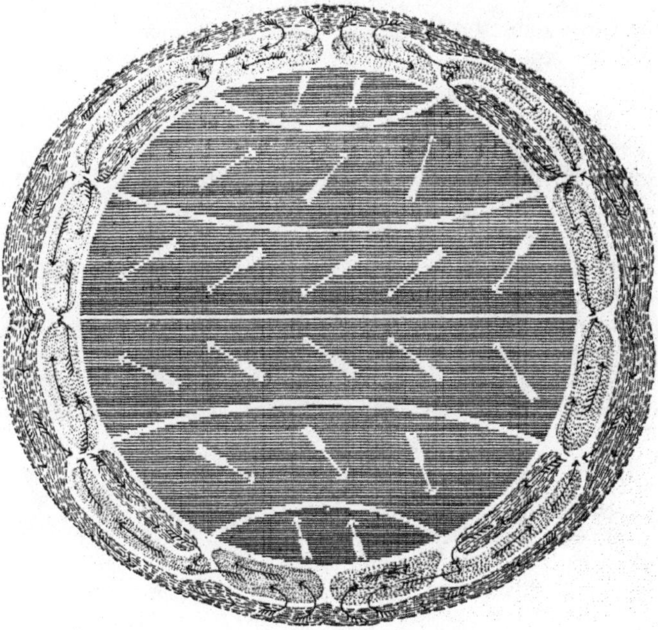

FIG. 262.—Diagram of the World system of winds. (Adapted from Ferrel.)

surface winds meet, there is, in the judgment of Ferrel, a tendency to develop a third or polar calm belt, over which the air rises to return as an upper current to the tropical calm belts, or else back again to the poles.

716. Wind Zones.—There is thus a tendency for the surface winds of the globe to divide into six zones, separated by five calm belts,—two tropical or trade wind zones, two temperate or anti-trade wind zones and two polar zones, as represented in Fig. 262. In the two tropical and two polar zones the winds move toward the equator, while in the two temperate zones they move away from the equator.

Above the earth's surface the directions of the wind are the reverse of those found below, that is, over the tropics and in the polar regions the upper winds move toward the poles, while over the temperate zones the upper winds are toward the equator.

717. Direction of Wind Modified by Form and Rotation of the Earth.—The shape of the earth and its rotation upon its axis greatly modify the direction of winds. The rotational velocity of the earth's surface at the equator is about 1,000 miles per hour toward the east. As the distance toward the poles increases the eastward velocity decreases. When therefore air moves toward the poles it travels eastward faster than the land it approaches and hence blows from a westerly toward an easterly direction.

On the other hand air moving toward the equator passes over land traveling eastward more rapidly than it does, and hence these winds fall behind and appear to blow from some easterly toward some westerly direction.

The surface winds in the tropics and polar regions are northeast or southeast, according to which hemisphere they are in, while the upper winds of the same zones have the reverse direction. In the temperate zones the winds are southwest or northwest at the surface and northeast or southeast above, according as they are north or south of the equator.

718. Character of the Winds.—Winds blowing toward the equator or descending from the upper regions have a tendency to be dry and to maintain a clear sky. On the other hand winds moving toward the poles, or rising to greater altitudes, tend to become more and more nearly saturated with moisture and hence to produce cloudy skies and precipitation.

The reasons for these relations are found in the fact that

air rising or moving toward the poles is passing toward a colder region. Lowering the temperature of the air, without changing the amount of moisture in it makes it more nearly saturated, while raising the temperature without changing the amount of moisture makes the air dryer.

Besides this, air is cooled by expansion and warmed by compression, and on these accounts ascending currents tend to become damp and descending air more dry.

719. Weather of the Wind Zones.—It will be evident from **718** that, so far as the world system of winds are not interfered with by local conditions, they must give to the countries over which they blow characteristic types of weather. Under the tropical high pressure calm belts, where the air is descending, and for a long distance to the south and a shorter one to the north, there must be a region of clear skies and dry weather, and it is under these two zones that the deserts of the world are found.

In the polar regions also the cloudiness and precipitation are relatively small for the same reason.

But at the equator, where large volumes of air are rising into the upper regions and after doing so pass toward the poles, the air having become very moist before rising, quickly becomes saturated and throws back to the earth large amounts of rain. The heaviest rainfalls of the world are under the equatorial calm belt of ascending currents.

In the two temperate zones also, where the winds cool as they move northward, frequent rains and showers and much cloudy weather are the rule.

There is thus a tendency for the systems of world winds to develop three rainy or cloudy zones and four clear weather or dry zones. The dry zones are under the tropics and about the poles; the wet and cloudy zones are under the equator and between the tropical and polar circles of both hemispheres.

720. Shifting of the Zones.—Because the vertical rays

of the sun fall alternately 23½ degrees north and south of the equator, the regions of greatest heating must also move north and south with the apparent shifting of the sun, and this causes the equatorial and tropical calm belts to move north and south. As a result of this shifting there is a tendency to develop two rainy and two dry seasons each year in the regions over which the calm belts travel twice.

CONTINENTAL WINDS.

721. Continents Disturb the World System of Winds and Weather.—The small specific heat of the land, its opaque nature and the absence of currents of all kinds in it cause the land surface to warm rapidly in the day and during summer, and to cool rapidly at night and during the winter. On the other hand the transparency of the oceans, which allows the sunshine to be distributed through a great depth of water; their high specific heat and the horizontal and vertical currents to which they are subject, all conspire to make the oceans, relative to the lands in the same latitude, warm in winter and cool in summer.

During the long days of summer and short nights, in high latitudes, the land becomes much warmer than the water and tends to develop ascending currents and a low air pressure, causing the winds to tend to blow toward the land at the surface and away from the land above in summer; but in winter, when the nights are long and the days short, the ground becomes very cold and the air contracts, causing the upper air to blow in over the continents above, thus developing high pressure, which forces the surface winds to move from the land toward the ocean in winter.

There is therefore a tendency for the weather of continents to be rainy and cloudy in summer and dry and sunny in winter, and for the oceans to be dry and sunny in summer and wet and cloudy in winter. This is a very fortunate relation, because it diminishes the evaporation on the land and increases that on the ocean and thus makes the rainfall heaviest at just the season when crops need most water.

722. The World Winds of January.—The prevailing winds of the world, as they are observed during the month of January, are represented in Fig. 263, the lines of black circles showing where the modified tropical high pressure calm belts are situated, and the light circles showing where the equatorial calm belt and other low pressure areas are.

In the southern hemisphere, where it is summer, and where the amount of land is small compared with the water, the tropical high pressure calm belt is crowded toward the pole on the land and the air is heaped up on the water, and the arrows show that the wind blows toward the land; but in the northern hemisphere, where it is winter, and where the amount of land is much larger, it is also drawn toward the poles by the extreme cold of the land, while a low area is formed over each of the northern oceans. The wind blows off both continents onto the two oceans and there are upper currents tending toward the land from the low areas.

The equatorial calm belt is farther south everywhere, but especially so over South America and over Africa and Australia, where the land becomes warmest.

723. World Winds in July.—At this time of the year, when the northern hemisphere has the vertical rays of the sun and the longest days, the large masses of land have become over-heated, the equatorial calm belt has been drawn northward and expanded into wide continental low areas, crowding the high pressure belt of the Tropic of Cancer upon the Atlantic and Pacific oceans, as represented in Fig. 264. The warm air rising over the continents and flowing over upon the oceans makes high pressure there and low pressure over the land, and this brings surface winds and moisture from the sea, giving rains to the land in the summer season.

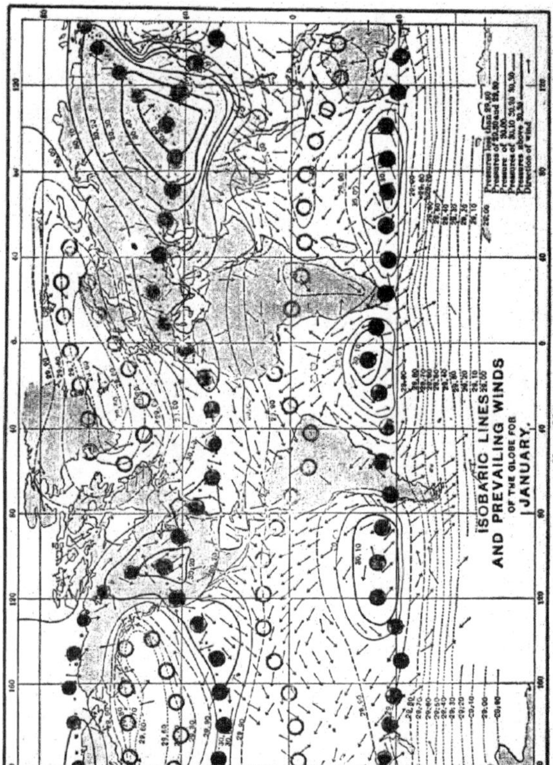
ISOBARIC LINES
AND PREVAILING WINDS
OF THE GLOBE FOR
JANUARY.

FIG. 263.

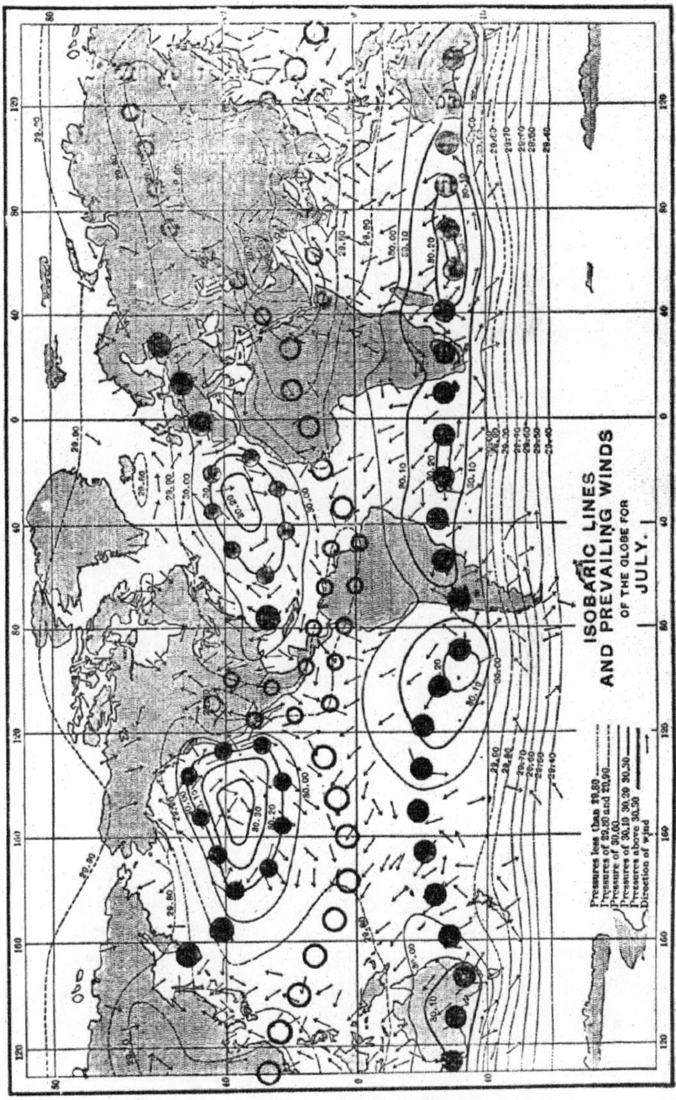
ISOBARIC LINES
AND PREVAILING WINDS
OF THE GLOBE FOR
JULY.
Pressures less than 29.80
Pressures of 29.80 and 29.90
Pressure of 30.00
Pressures of 30.10 30.20 30.30
Pressures above 30.30
Direction of wind

FIG. 264.

South of the equator, where it is winter, the high pressure calm belt has moved nearer the equator so that the air is blowing off the three continents and they are èxperiencing their dry season.

724. Monsoon Winds.—Where the world system of winds is so strongly influenced by the land areas as is the case notably in the region of the Indian Ocean they have been given the special name of monsoons, and these give to India its rainy season, when they blow from the ocean, and its dry season, when they blow from the land.

ORDINARY STORMS.

Besides the world system of winds, which have been described, and the continental winds with their intensified forms called monsoons, which change with the seasons, there are others of smaller magnitude and shorter duration which give rise to our ordinary storms and the still more local tornadoes and thunder storms which are associated with them. These are technically called cyclones or cyclonic storms.

725. Cyclones.—Most of the rainfall of temperate climates and much of that which falls between the tropics and the equatorial calm belt, occurs during the passage of these cyclonic systems of wind movement, represented in Figs. 265 and 266.

In these winds the surface air moves spirally about a center, going to the east as it passes toward the poles and to the west of the center when it comes toward the equator. Air coming from the eastward of a cyclonic center always passes to the polar side, while that coming from the west always passes to the equatorial side.

726. Cause of Wind Directions in Ordinary Storms.—The cause of the wind directions in ordinary storms is the same as that of the direction of the general earth currents, that is,—the form and rotation of the earth. As the air leaves the equator it passes over land moving eastward slower than it and hence outruns, appearing to blow from the

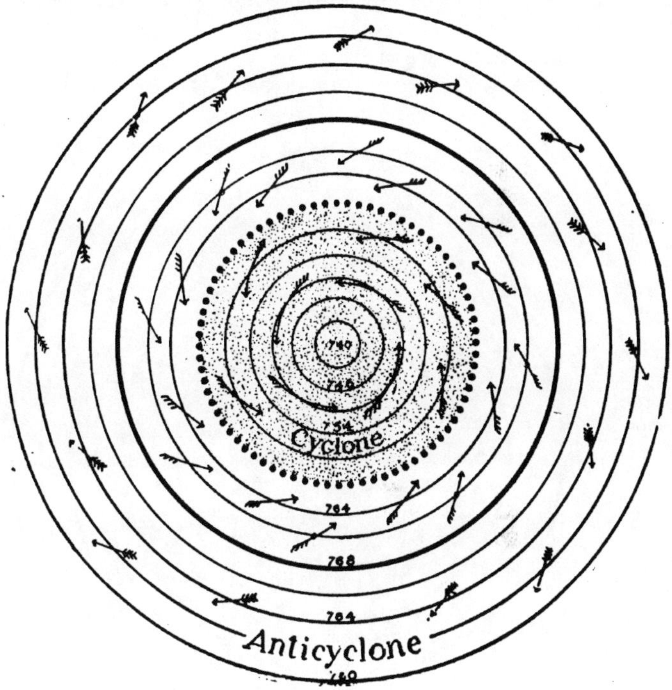

FIG. 265.—Diagram of surface winds in a typical cyclone. (After Ferrel.)

S. W. toward the N. E. in the northern hemisphere, and from the N. W. toward the S. E. in the southern hemisphere. If it approaches the equator it travels over land moving eastward faster than it does and hence appears to come from the N. E. in the northern hemisphere and from the S. E. in the southern.

Where the wind approaches the center from the east it can only do so by having its eastward motion with the earth made slower than the earth's surface in the same latitude:

while if it approaches the center from the west it can only do so by traveling eastward faster than the earth itself and these changes in velocity cause winds from the west to move toward the equator side of the storm center, while those from the east always go to the polar side. The effect is the same as would result from checking or increasing

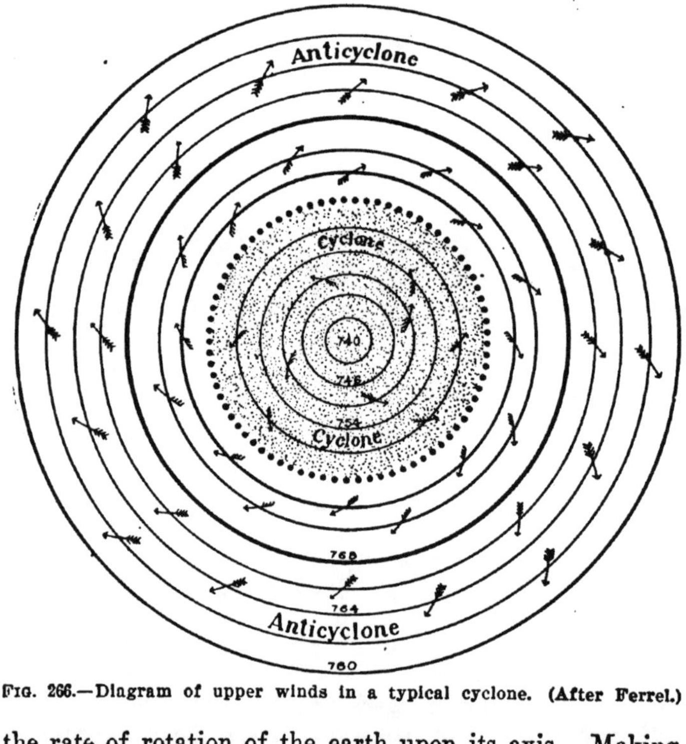

FIG. 266.—Diagram of upper winds in a typical cyclone. (After Ferrel.)

the rate of rotation of the earth upon its axis. Making it rotate faster would throw the air and water also toward the equator, while slackening its speed would permit both air and water to move toward the poles.

727. Progressive Movements of Storms.—Cyclonic storms in all parts of the world have a progressive movement

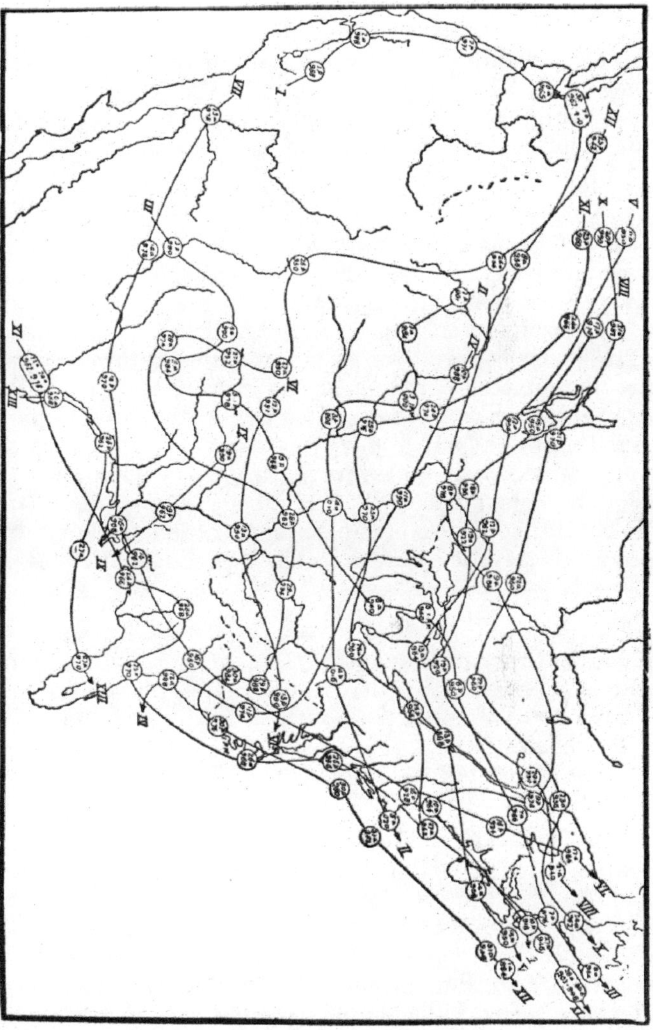

FIG. 267.—Chart of paths of Low areas across the United States, March, 1900.

across the earth's surface and the general direction is that of the prevailing winds of the part of the earth in which they are. That is, in the temperate zones they tend to move away from the equator and toward the east, while in the tropical zones they tend to move toward the equator and toward the west.

728. Direction of Storms in the United States.—In the great majority of cases the storms of the United States travel from some westerly toward some easterly point and the mean direction is a little north of east. Very many of these storms travel for a time from the northwest toward the southeast until they near the longitude of the Mississippi river, when they very often turn their course strongly to the northeast, and Fig. 267 represents the course of the storm centers as they traversed the country during March, 1900, there being 13 of them in all. Wherever the storms of the United States originate or enter the territory they nearly all leave it by crossing the New England states.

729. Rate of Travel of Storms in the United States.—There is a very wide range in the rate at which the storm centers progress across the United States, but the average is from 26 to 30 miles per hour. The circles in the paths of the several storm tracks in Fig. 267 mark the positions of the storm centers at intervals of 12 hours.

730. Diameters of Storms.—The diameter of these cyclonic wind systems in the United States is generally from 1,500 to 2,000 miles, the longest diameter being usually from the southwest to the northeast. A typical one of these storms is represented in Fig. 268, where the heavy lines are drawn through places having the same weight of air above them, while the dotted lines are lines of equal temperature. It will be seen that this wind system reaches from north of the Great Lakes to well into Texas and from North Dakota to Tennessee.

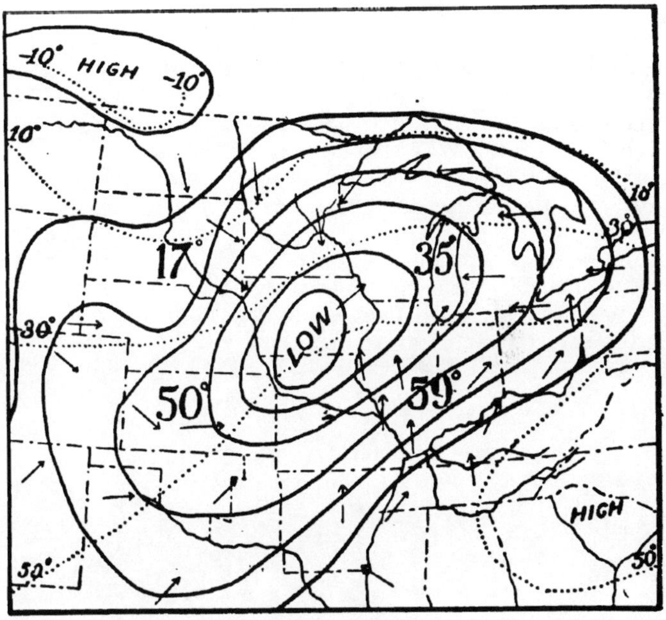

FIG. 268.—Chart of strongly developed low area in the vicinity of the Great Lakes.

731. Duration of Ordinary Storms.—The length of time one of the ordinary cyclonic storms of the atmosphere lasts is very variable. In some cases they are of but a few days duration; at other times they last for weeks together and in that time travel long distances.

It is common for them to cross the United States, the North Atlantic and the whole of Europe; and one, unusual at least in the completeness of its known history, has been followed from the vicinity of the Philippine Islands, across the Pacific, across North America and the North Atlantic; across Europe and well on toward the central portion of Siberia, where lack of sufficient observations prevented following it farther.

732. Relation of the Region of Precipitation to the Storm Center.—The region over which rain or snow falls during the passage of cyclones across the United States lies usually in advance of the central LOW, much as represented by the heavily shaded area in the diagram Fig. 269, and at a distance of 200 to 700 miles from the center.

In this area the precipitation is most continuous and steady over the eastern and northern portion, where the surface winds range from S. E. to N. E. in direction. To the southeast and south of the low center, where the winds are S. and S. W., there is a general tendency for the precipitation to occur in the form of showers, to be more violent in character, and local rather than wide spread.

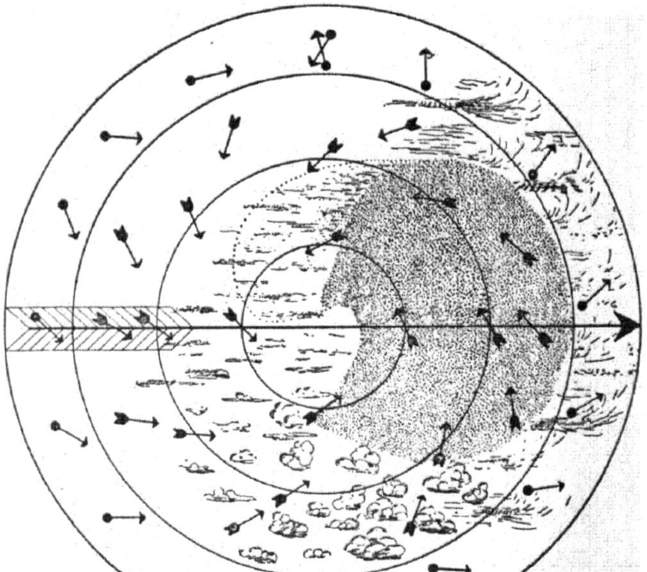

FIG. 269.—Diagram of storm area.

733. The Origin of Ordinary Storms.—There is as yet no general agreement among students of meteorology regarding the origin of cyclonic winds and storms, some thinking that the low areas are primary and that the areas of high pressure result from the overflow of air from one or more of these which overlap; while others maintain that the high areas are primary and that the low areas are secondary. At the present time the former view is able to bring much the stronger evidence to its support, so far as the operation of well established physical principles are concerned, and, with some modifications, seems likely in the end to prevail.

CHAPTER XXVI.

WEATHER CHANGES.

The forecasting of weather changes from 24 to 36 hours in advance is based upon several well-established facts: (1) Rainy or cloudy weather is usually associated with areas of low pressure, about which the winds move as represented in Fig. 269. (2) Fair or clear weather is usually associated with regions of high pressure. (3) Both low and high areas have prevailing dimensions and move in the United States from the west toward the east.

If areas of low pressure always had the same diameter, and if they traveled at the same rate and in the same direction, it would be possible for anyone to forecast the weather changes with much certainty 12 to 36 hours in advance. But with all the irregularity of form, dimension, intensity, rate and direction of motion, it is possible for even a local observer to form a rational judgment of the approach, time of arrival and passage of an ordinary storm. Indeed, it will seldom happen that a strongly developed storm can approach a locality without giving sure signs of its coming 12 to 24 hours in advance.

734. Prevailing Winds.—In the forecasting of weather changes it is important to have clearly in mind the direction of the prevailing winds of the locality, or those which are not due to the storm whose approach is to be forecast.

In most parts of the United States east of the Rocky Mountains the prevailing fair weather winds are from some westerly quarter and they should be the southwest winds of the general world and continental system unless modified by local conditions, such as give rise to "land and sea breezes" or "mountain and valley winds."

735. Locating the Storm Center.—When the weather has been for some time fair and the prevailing winds are blowing, the first indication of an approaching storm is usually to be found in the long thread-like or hair-like curved cirrus clouds represented in the outer front side of Fig. 269. If these are seen strongly developed in any quarter of the sky it is usually true that a more or less strongly developed low area exists in that direction.

If these appearances first develop to the east of a north and south line the first probability is that this storm will not reach the observer because it is already past and traveling away from rather than toward the place.

On the other hand if the cirrus clouds show themselves well developed to the west of a north and south line, and especially if between the southwest and northwest, then a storm center is located where its future course may bring it over the locality.

736. Change in Wind Direction.—If a storm is approaching from the westward in the direction of the cirrus clouds these will advance and in a few hours will overspread the sky, the wind will decrease and finally shift to a direction which will indicate the approach of the storm, and more definitely the direction of the low area from the observer.

737. Direction of the Storm Center Indicated by the Wind.—When a storm has advanced far enough to give definite direction to the wind it is then possible to judge from this the location of the storm center.

Standing with the back to the wind and extending the right arm directly in front, and the left arm at right angles to this, the storm center is usually in a direction somewhere between the two hands; this will be clearly seen from a study of Fig. 269 and also of Fig. 268.

It will sometimes happen that winds blowing outward from a HIGH, or region of heavy pressure which has passed to the eastward, may be mistaken for those due to an approaching storm, because they are easterly, but the character of the sky and the weather, with experience, will usually serve to distinguish these anticyclonic winds from those belonging to the cyclone or storm proper.

738. Discovering the Course the Storm Is Traveling.—After having observed the existence and direction of a storm center it is important to know whether it will pass to the north or south of the locality or whether it will move directly across it. This can be foretold by the changes in the direction of the wind. Referring again to Fig. 269 it will be observed that if the storm center comes directly toward the observer the direction of the wind will hold steady in the S. E. until after the storm has passed, when it will shift abruptly to the N. W., as indicated by the arrows laid on the axis of the storm track. If, however, the storm center is passing considerably to the north of the observer the winds will shift toward the south, finally becoming S. W. But if the low area is passing to the south of the observer then the winds will shift around by the north, becoming finally N. W. and then W.

If the winds hold steady, or if they shift to the north, a general rain or snow may be expected, unless the storm center is too distant, but if it is shifting toward the south, showers, rather than widespread precipitation, may be anticipated. After watching the progress of storms during two or three months, comparing them with the daily weather maps, one becomes able to recognize with much certainty the approach of all well marked storms and to forecast their course and the character of the weather 12 to 24 hours in advance. Mistakes will occur, just as they do with the Weather Bureau expert having a much wider knowledge before him, but with a little experience the judgment becomes much more reliable than would at first be expected.

739. Temperature Changes Connected with Storms.—During the colder portions of the year the temperature changes, which are associated with the progress of a storm across the country, are often very marked. The general rule is that with the approach of a storm the temperature rises above the normal for the place and season, if it is the cold part of the year, but after the storm passes the temperature falls below the average.

The rise in temperature is due to three causes: (1) The warming of the air by the heat due to the condensation of moisture; (2) the checking of radiation by the moisture in the air; (3) the importation of warmer air from farther south under the influence of the storm center.

It was shown in (41) and (42) that the formation of a pound of water at 212° from a pound of steam at 212° is associated with the development of 966 heat units, and the freezing of a pound of water is also associated with the appearance of 142 heat units. When, therefore, a pound of snow forms in the air from a pound of water vapor there is imparted to the air in which this occurs

$$966 + 142 = 1108 \text{ heat units}$$

and if snow enough falls to represent an inch of rain the heat produced in the air is at the rate of about

$$1{,}108 \times \frac{62.4}{12} = 5761.6 \text{ heat units}$$

per square foot of the surface upon which the snow falls. The warming of the atmosphere when it snows heavily must be very considerable and this is why it is seldom more than a few degrees below freezing when a heavy snow is in progress.

The low temperature following a storm is due to three chief causes: (1) The rapid loss of heat by radiation from the ground under the clear sky; (2) the descent of cold air from high altitudes; and (3) the importation of colder air from farther north under the influence of the storm center.

If reference is made to Fig. 268, it will be seen that the southeast quadrant has a mean temperature of 59° F., while the northwest quadrant has a mean temperature of only 17° F., 42° colder. In the northwest HIGH there is a temperature of —10° F., while to the east of the LOW, above 60°, or a difference of 70° F., and while a part of this difference is due to difference of latitude, most of it is due to the effect of the storm.

740. Barometric Changes Connected with Storms.—During the progress of a storm across a given station the barometer falls more or less gradually until the center has reached the place and then it begins to rise, and may continue to do so until a pressure greater than is normal has been attained. The changes of the barometer, therefore, become indices of the approach, progress and passage of a storm, and so, too, in a less degree, may temperature changes also, during the winter. If the barometer falls faster than usual, if the wind velocity increases rapidly and rapid changes in the wind direction occur, the indications are either that the storm center is approaching at a high rate of speed or that its diameter is small and hence that it is likely to arrive sooner after indications have developed.

741. Cold Waves.—Cold waves in the United States are usually the result of a strongly developed storm which has traversed somewhat slowly the southern and eastern states. When these conditions prevail a HIGH area with clear sky and descending cold air from above forms over Manitoba, or the northern boundary of the United States, and the strongly developed LOW area, traveling slowly, sets this body of cold air in motion toward it, which often attains a velocity of 25 to 40 miles per hour. Under these conditions intense cold is rapidly transported southward and eastward with the speed of an express train, and occasionally temperatures even below zero are transported as far south as northern Alabama.

Besides the extremely cold waves just referred to there are others more common, which are due principally to the first two causes named, and are usually coincident with the

HIGH areas, following them in their course across the country.

742. Forecasting Warm and Cold Weather.—Since strongly developed storms tend to draw the air into themselves across long distances, it is clear that when they pass to the south during the cold months of the year cold waves are likely to follow their passage. On the other hand, if the low area has passed to the north it can only bring air from the south northward, importing but little cold with it. To be able to forecast the path of a storm then is also to be able to forecast the temperature changes which are likely to follow.

743. Long Warm and Dry Periods.—It frequently happens that a series of storms follow along a single track, one after another for several weeks together, and Fig. 270 rep-

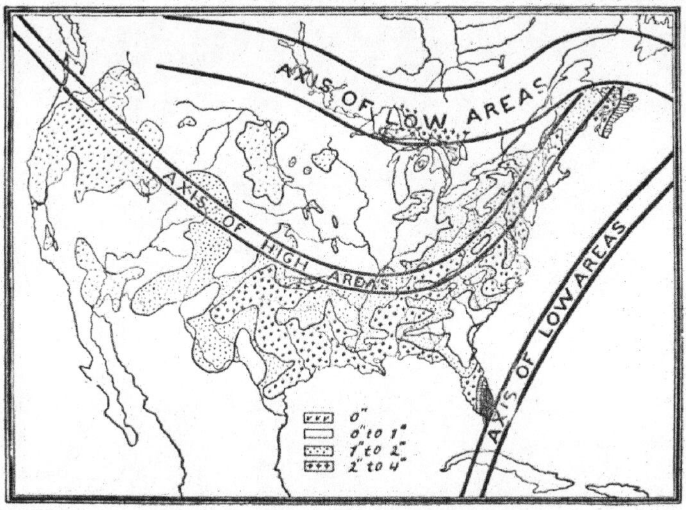

Fig. 270.—Chart showing conditions which determine dry weather in the eastern United States.

resents one of these sets of conditions. During the month of October, 1895, all but four of the fifteen low areas recorded by the Weather Bureau, moved along axes within the northern belt marked "axis of low areas."

It is clear that so long as such conditions as these prevail but little rain could fall in the United States, and all the northern portion must have unusually warm weather. The weather must be clear and dry because along the axis of high pressure the air is descending from the higher altitudes where it is already dry, and in descending must become still dryer because of increasing temperature due to compression. As this is the air which must be drawn toward the low areas on either side of the axis it could contribute but little moisture for rainfall in either system of lows, and the map shows that but little fell.

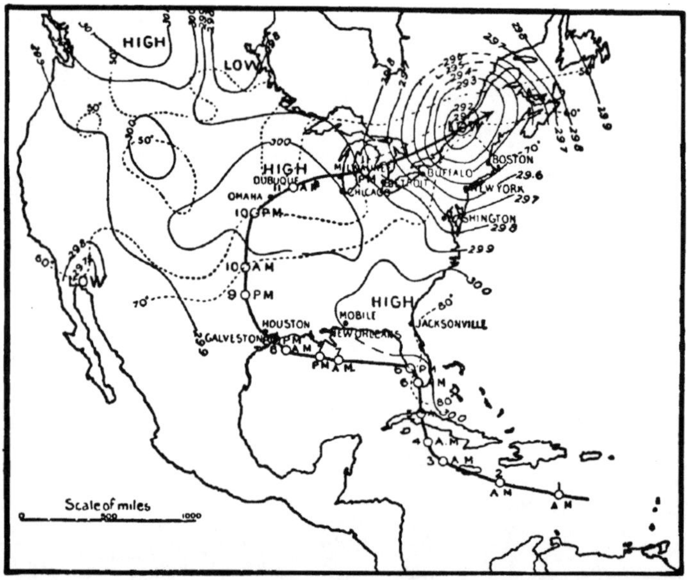

FIG. 271.—Path of the West Indian Hurricane of Sept. 1–11, 1900.

So long as a high pressure occupies the Gulf and Atlantic states, this effectually shuts off the moist gulf and ocean air and forces the storm centers to maintain a high northerly course. Then, too, as long as storms pursue a course off the Atlantic border they also must shut off the moisture from the northern states and tend to maintain warm, dry weather there.

Whether in this case the two systems of low areas were the cause of the belt of high pressure which prevailed, or whether the high pressure belt simply marks the place where, for some reason, the upper air from the general wind system was falling to the earth, the outcome, so far as the weather is concerned, must be essentially the same.

744. Tropical Cyclones.—During the latter part of August, September and the fore part of October it frequently happens that storms of unusual magnitude, intensity and destructiveness originate in the north tropical zone of trade winds, somewhere in or to the east of the Carribean Islands and, after traveling westward with the prevailing

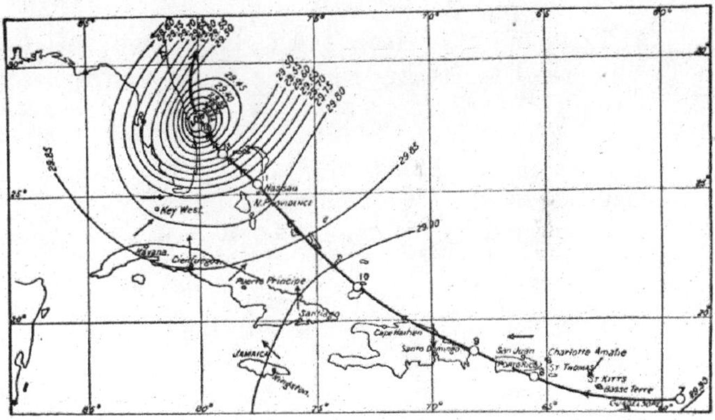

FIG. 272.—Path of West Indian Hurricane of Aug. 7–14, 1899.

winds of that zone, they finally make their way northward across the tropical calm belt and break into the zone of southwest winds, making their way northward and eastward, as represented by the two storm tracks in Figs. 271 and 272, the former being the storm which produced the terrible destruction of life and property at Galveston on

September 8, 1900, when more than 5,000 human lives and $20,000,000 of property were lost.

The severe cold winds which are designated as the "Northers" of Texas owe their origin to storm centers of unusual intensity off the Gulf coast, which set large bodies of air in motion from the northward, drawing it into themselves as they pass along to the southward and eastward.

THUNDER STORMS, HAIL STORMS AND TORNADOES.

Associated with the ordinary storms which have been described in a preceding section there are others much more local in their character, shorter in duration, but often more violent in wind movement and precipitation. These are thunder storms, hail storms and tornadoes.

745. Relation of Tornadoes and Thunder Showers to Ordinary Storms.—Careful study of the time of occurrence and distribution of these storms has shown that they are almost always associated in a definite way with some cyclonic wind movement, and that they usually originate to the southeast, south or west of south of a storm center, in the region designated by the cumulus clouds in the diagram, Fig. 269.

746. Tornadoes.—Tornadoes are whirling winds of extreme violence which last but a short time, progressing almost always from the southwest toward the northeast, often at the rate of a mile per minute, sweeping a belt 40 to 80 rods wide and several miles long. Sometimes the width of the zone of destructive winds may reach a full mile. At the center of the tornado the moisture is swept together by the revolving winds into a dark funnel-shaped cloud, where the velocity of the whirling air may be so great that few structures can withstand the enormous pressure they develop.

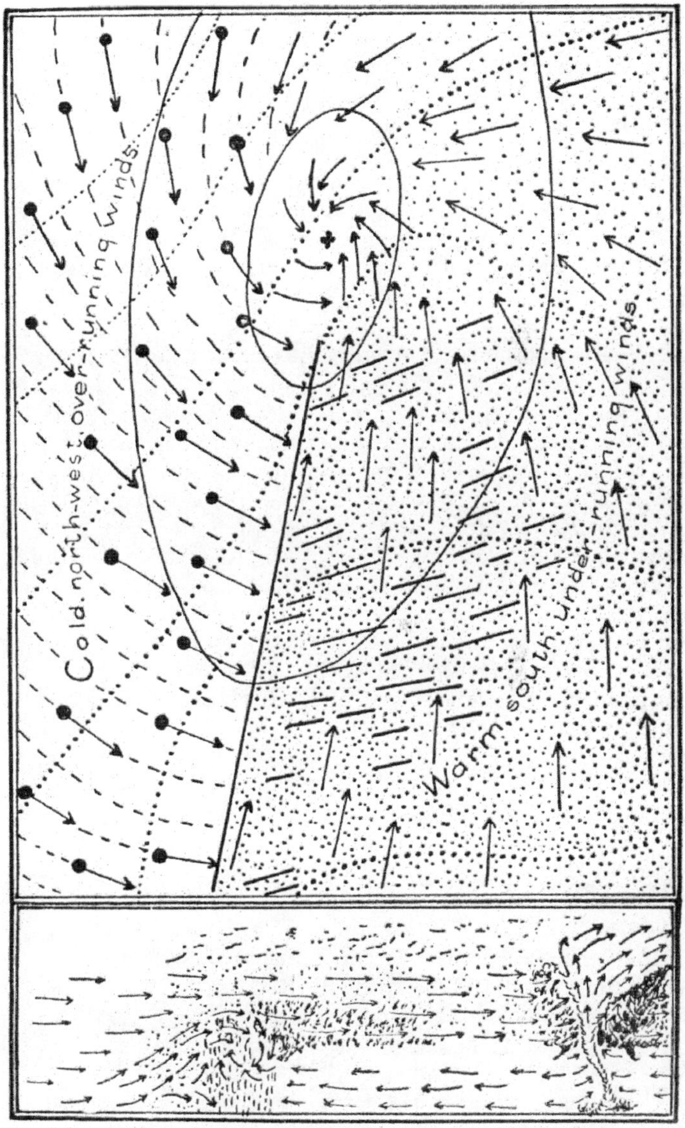

FIG. 273.—Diagram showing the origin of tornadoes and thunder storms

747. Schools of Tornadoes.—When the conditions are extremely favorable for the formation of tornadoes they often appear in schools, originating one after another or simultaneously, as the main storm center progresses across the country, and Fig. 273 shows how these local but violent storms are related to a storm center and how many may develop in the southeast quadrant as it travels along. In this figure the short, heavy straight lines to the southeast of the center represent the paths of tornadoes which developed during its course.

748. Distribution of Thunder Showers.—Thunder showers, like tornadoes, originate in the great majority of cases to the southeast and south of a well developed storm center and often large numbers of them, scattered over considerable areas, form as the storm progresses, much as is the case with tornadoes, and Fig. 274 is a diagram showing the advance of the front along which thunder showers originated in a storm of early May, 1892, as recorded in the Monthly Weather Review of that month, p. 138.

On May 3 a long low area had advanced from the south and west and at 8 P. M. its lowest portion was central north of Lake Huron. The front of the thunder shower line had reached the east end of Lake Erie at 2 P. M. of the same date and showers were in progress along the line marked 2 P. M. in Fig. 274. As the storm center advanced the thunder-shower-front also moved forward and swept across the state, as shown by the curves on the diagram, reaching Long Island at 2 A. M. on the morning of May 4th, the front thus progressing from 20 to 30 miles per hour.

749. Conditions Under Which Thunder Showers and Tornadoes Originate.—In the diagram of Fig. 273 are represented the wind directions and temperature relations which exist when conditions are favorable for the formation of both of these classes of storms. There is a region of warm moist southerly winds to the south and east of the low area and another region of decidedly colder winds blowing from the west and north of west; and it is along the meeting of these two systems of winds that thunder showers tend specially to form, and in advance of it that the tornadoes have their birth.

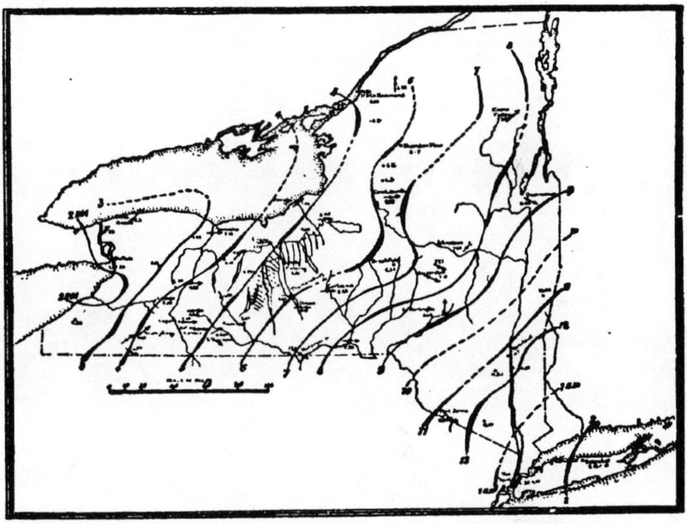

FIG. 274.—Diagram showing the progressive development of thunder storms.

750. Formation of Tornadoes.—The most satisfactory explanation of the formation of tornadoes is represented in the lower portion of Fig. 273, which is a cross-section of the lower portion of the atmosphere at right angles to the line dividing the two systems of winds shown in the upper portion of the same diagram.

It is supposed that, under these conditions, the cold west and northwest winds at times over-run the moist warm and lighter southerly stratum, thus producing a condition of unstable equilibrium. When such conditions have been developed the warm air, at some point, is supposed to break up through the over-running colder layer, as shown in the lower right-hand corner of the diagram, and in doing so is thrown into a rapidly whirling movement in the same manner that water runs into whirls in discharging through the bottom of a wash-bowl. When the volumes of air which must change places are large and the stratum of cold air deep, there comes ultimately to be developed an enormous rotary velocity which gives to the air an extremely destructive power.

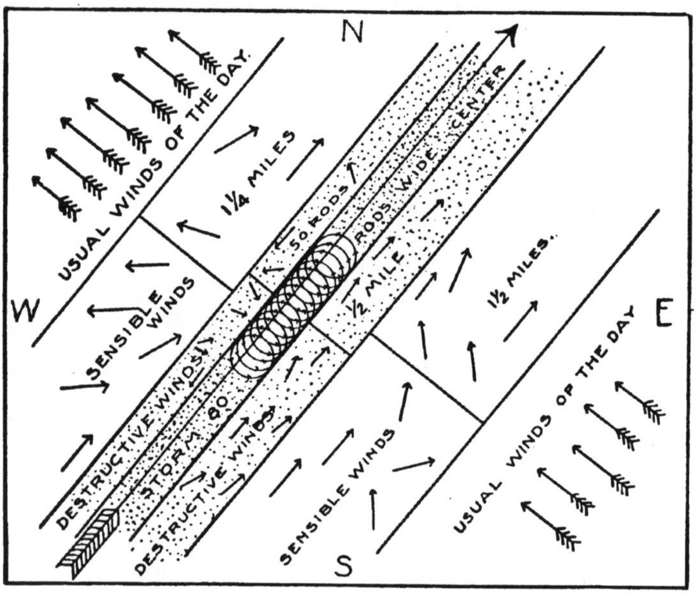

FIG. 275.—Diagram of the path of a tornado.

751. Explosive Violence of Tornadoes.—At the center of a tornado cloud the rapidly whirling motion reduces the air pressure at the center of the funnel so much as to produce a high vacuum, and when a building lies in the path of the funnel the vacuum surrounds it so suddenly that often the great pressure of air within the building will throw the walls outward or lift the roof off before the air has time to escape into the vacuum formed by the tornado.

752. Unsteady Action of Tornadoes.—A tornado seldom displays a uniformly destructive power and oftentimes the point of the funnel fails to reach the ground and considerable gaps are passed in the path where little damage is done. This unsteady action is often due to the slowing up of the rotary motion in the cloud due to the great friction developed at the ground. After withdrawing to the upper air the speed increases sufficiently to allow the funnel to grow to the surface again and resume the destructive work.

When the funnel reaches the surface it does not always describe a straight path along the ground, but tends to cross and recross the main axis of movement.

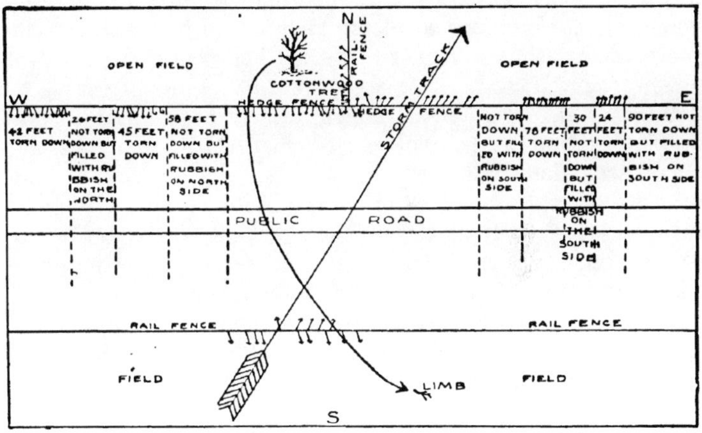

FIG. 276.—Diagram showing the rotary movement of winds in a tornado.

753. Character of the Tornado Path.—It is usually true that the path of a destructive tornado is not symmetrical, one side being wider than the other, as represented in Fig. 275, where it will be seen that the northwest side is narrower than the southeast side. Not only is the zone of destructive winds wider on the south side but that of the sensible winds is also. On account of this character of the tornado track it is clear that if one has an occasion to escape from an ordinary tornado, the shortest path would lie to the northwest, at right angles to the line of progress. The evidences of a rotary motion of the air in a tornado are abundant and conclusive, and in Fig. 276 are represented some of these.

754. Formation of Thunder Showers.—Thunder showers appear to have an origin similar to that of tornadoes, but evidently occur where there is less air to change places, and probably also where the depth of the overlying stratum is less. Indeed, it appears very often, if not generally, true that a volume of cold heavy air has dropped directly to the ground and is moving bodily against the warmer moist air, which it is forcing upward, as represented in the lower left-hand corner of Fig. 273. The rapidly ascending warm moist air is cooled by expansion and by mixing with the cold air, thus giving rise to the heavy precipitation so often observed.

The rolling movement shown in the diagram is often violent enough and involves so great a hight in the atmosphere, that often raindrops are carried round and round until they become very large before they are able to fall. If the vertical circulation reaches above the zone of freezing temperature the raindrops freeze, forming hail. These hailstones, in the most violent storms, are often carried around with such force and so many times that they become very large before they are able to overcome, by their weight, the velocity of the air, and fall to the ground.

INDEX.

E.

F

T

(Issued December, 1908)

Ventilation for Dwellings, Rural Schools and Stables

By F. H. KING

Professor of Agricultural Physics in the University of Wisconsin, 1888-1901; Chief of the Division of Soil Management, U. S. Department of Agriculture, 1901-1904.

Author of "The Soil," 1895; "Irrigation and Drainage," 1899; "Physics of Agriculture," 1901; "Tillage, Its Philosophy and Practice," "The Necessity and Practice of Drainage" in Cyclopedia of American Agriculture, 1907; "Drainage" and "Irrigation," in The Standard Cyclopedia of Modern Agriculture, (British) 1908.

128 pages, 7½x5½ inches, 63 illustrations.

Published by the author, Madison, Wis. Price 75 cents, by mail on receipt of price.

CONTENTS

"The subject is so widely and strongly, and so profoundly treated and so profusely illustrated by new drawings never before used, that we are sure that it will be regarded by all classes of society as one of the most important contributions to human knowledge that has yet appeared. All the knowledge Prof. King uses exists in the great domain of scientific investigation. But it is here brought down to the average comprehension in a way that makes it of the greatest importance in every household in the land."

"Every man that builds or keeps a home, a place of business, a factory, or work shop, or a stable, where living, breathing life is assembled, should procure one of these books and study it long and well."—Hoard's Dairyman.

THE SOIL

By F. H. KING

Professor of Agricultural Physics in the University of Wisconsin. 1888-1901; Chief of the Division of Soil Management, U. S. Department of Agriculture, 1901-1904.

Author of "Irrigation and Drainage," 1899; "Physics of Agriculture," 1901; "Tillage, Its Philosophy and Practice," "The Necessity and Practice of Drainage", in Cyclopedia of American Agriculture, 1907; "Drainage" and "Irrigation," in The Standard Cyclopedia of Modern Agriculture, (British) 1908.

303 pages, 7x5 inches, 45 illustrations.—$1.68 prepaid.

CONTENTS

"I consider it a most desirable addition to our agricultural literature, and a distinct advance over previous treatises on the same subject, not only for popular use, but also for students and specialists." * * *
Dr. E. W. Hilgard, Director Calif. State Agr. Exp. Station.

"For practicability and entertaining power combined this work is at the head of its class."—The Boston Traveller.

"The manual is brief, accurate, comprehensive and hits the practical point every time."—Independent.

"It is a book which progressive farmers will come to regard as one of the essential implements of farm life."—Boston Daily Advertiser.

"The great point about the book, in our opinion, is its thorough practical nature. Personally the writer is acquainted with probably all modern works on this vitally important question * * * ; but we certainly never derived so real benefit from the perusal of any two, nay, even three or four works of this character, as from the one now under consideration."—E. Kemp Toogood, F. R. H. S. in Royal Cornwall Gazette.

IRRIGATION AND DRAINAGE

By F. H. KING

Professor of Agricultural Physics in the University of Wisconsin, 1888-1901; Chief of the Division of Soil Management, U. S. Department of Agriculture, 1901-1904.

Author of "The Soil," 1895; "Physics of Agriculture," 1901; "Tillage, Its Philosophy and Practice", "The Necessity and Practice of Drainage", in Cyclopedia of American Agriculture, 1907; "Drainage" and "Irrigation in The Standard Cyclopedia of Modern Agriculture, (British), 1908.

502 pages, 7x5 inches, 163 illustrations.—$1.66 prepaid.

CONTENTS

"To the ordinary farmer the title of this book is somewhat misleading. If he is not in an irrigating district, and has no wet lands, he will at once conclude, on seeing the title, that the subjects treated in the book do not concern him to the extent of $1.50. If, however, by chance he has the opportunity of reading the book he will change his opinion. The proper amount of water available at the right time is essential to successful or profitable farming in any country, and, therefore, Professor King opens his books with some general remarks on the importance of water; on the texture of the soil necessary to conserve the moisture; and follows it up with the report of some experiments showing the amount of water used by plants, which will be a surprise to the farmers who have not investigated the subject. The method by which the water is obtained by plants and exhaled; the remarkable way in which the plants themselves control the demand, economize water, so to speak; the mechanism by which the roots get hold of the moisture; the extent of the root surface; all these are treated in a wonderfully interesting way in this book, and are of all-absorbing interest to the man who is farming for dear life, and, if he is properly awake to the importance of the subject, will prove as interesting as a novel . . . We regard it as one of the most valuable contributions made to the science of agriculture in recent years."—Wallaces' Farmer.

"But although the author travels far and wide in search of examples and results in illustration of the principles which he advances, and so far introduces matter which is of great importance in the discussion of irrigation proper, yet the bulk of what he has written is full of instruction of the most practical character for the rent-paying farmer."—Manchester (England) Guardian.

Spotted something that isn't quite right?

Let us know and we'll fix it for future reprints. And as a thank you for your diligence, we'll send you a free Sothis book of your choice.

publisher@sothis.press

This page intentionally left blank

Except of course for that line above. And this one.
And whilst we have your attention, perhaps you'd
like to follow us on Instagram @sothis.press. Go on.

Made in United States
Orlando, FL
07 November 2026

90337369R00363